Safety, Societal Problems and Citizens' Perceptions

New Empirical Data, Theories and Analyses

GofS Research Paper Series

Safety, Societal Problems and Citizens' Perceptions

New Empirical Data, Theories and Analyses

Editors

Marc Cools
Brice De Ruyver
Marleen Easton
Lieven Pauwels
Paul Ponsaers
Gudrun Vande Walle
Tom Vander Beken
Freya Vander Laenen
Gert Vermeulen
Gerwinde Vynckier

Maklu
Antwerpen/Apeldoorn/Portland
2010

Editors: Marc Cools, Brice De Ruyver, Marleen Easton, Lieven Pauwels, Paul Ponsaers, Gudrun Vande Walle, Tom Vander Beken, Freya Vander Laenen, Gert Vermeulen & Gerwinde Vynckier

Safety, Societal Problems and Citizens' Perceptions. New Empirical Data, Theories and Analyses
Governance of Security Research Paper Series (GofS), Volume 3.
Antwerpen – Apeldoorn – Portland
Maklu
2010

312 pag. – 24 x 16 cm
ISBN 978-90-466-0327-7
D/2010/1997/7
NUR 824

Maklu Publishers
Somersstraat 13/15, 2018 Antwerpen, Belgium, info@maklu.be
Koninginnelaan 96, 7315 EB Apeldoorn, The Netherlands, info@maklu.nl
www.maklu.eu

USA & Canada
International Specialized Book Services
920 NE 58th Ave., Suite 300, Portland, OR 97213-3786, orders@isbs.com, www.isbs.com

Table of contents

The view of the police on community policing in Belgian multicultural neighbourhoods   **161**
Marleen Easton and Paul Ponsaers

Population density, disadvantage, disorder and crime.
Testing competing neighbourhood level theories in two urban settings.   **183**
Caroline Mellgren, Lieven Pauwels and Marie Torstensson Levander

The continuum of conflicts of interest: from corruption to clubbing and the underlying risks at victimisation   **203**
      Gudrun Vande Walle

Corruption as a judgement label   **221**
      Arne Dormaels

"Governance of Security" –
Monitoring contemporary security issues

The governance of security has undergone profound changes over the past thirty years. Security has become a commodity in our mind-set and daily life and the governance aspect of security is more complex and expanded than ever before. Since 2007, the Research Unit Governance of Security of the Ghent University Association is a critical observer of the new developments in governance of security and addresses the complex question regarding efficient management and control (governance) of contemporary security issues. The touchstones for these observations are human rights and the implications of insecurity (policy) on human rights.

The GofS association research unit is interdisciplinary composed, consisting of three research units: (1) the Research Unit Social Analysis of Security of the Department of Criminal Law & Criminology, Faculty of Law of Ghent University (SVA) (co-directors Marc Cools, Lieven Pauwels and Paul Ponsaers), (2) the Institute for International Research on Criminal Policy of the Department of Criminal Law & Criminology, Faculty of Law of Ghent University (IRCP) (co-directors Brice De Ruyver, Tom Vander Beken and Gert Vermeulen), and (3) the Research Unit Governing and Policing Security (GaPS) of the Department of Business Administration and Public Administration of Ghent University College (director Marleen Easton). All members of these research-units participate in the GofS Research Unit since 2007 (Consult: http://www.gofs.ugent.be).

Within this interdisciplinary composed association research unit GofS develops three key research lines. The first research line focuses on the change of the concept of insecurity/security (the interaction between objective and subjective feelings of insecurity and the notions of nuisance, of victimisation, of mistrust in governments, ... alongside the changing areas of social relevance which the idea of insecurity has become associated with in late modern society (from traditional forms of criminality to food safety, environmental threats, corruption, terrorism).

The second research line sheds light on the administrative and judicial policies which relate to the changing paradigm of insecurity/security, paying particular attention to the problems of integration between: public versus private conduct of policy; centralised versus decentralised conduct of policy; international, rural and local conduct of policy and the development of sector specific policies in contrast to integral or integrated policy making.

The third research line concentrates on the implications for law and order and crime prevention policy execution of this evolving paradigm of insecurity/security, with a specific focus on the decreasing rigidity of organisational borders and competence displacement in its implementation with reference to police forces; administrative supervision holders; new security professions; the army; information services; inspection and detection services; private security, guarding and detection, self-regulating agencies and active citizen participation.

The GofS Research Unit wishes to disseminate the results of the research it conducts in this research area as widely as possible. It is with this intention that the group started the GofS publication project. On the one hand, GofS will encourage the writing of English contributions by GofS-members in the "**GofS Research Paper**

Series". On the other hand, it is the intention to occasionally publish English or Dutch GofS research reports in the **"GofS Research Report Series"**. With this initiative the Governance of Security Research Unit contributes to a better understanding of contemporary governance of security by presenting recent research results and scientific reflections, by devising new approaches and by re-evaluating criminology's heritage. It implies a new openness with regard to other disciplines and to the normative questions resulting from the commission of crime and the reaction to it by actors in the criminal justice system and beyond.

After release in early 2009 of an initial set of two volumes in the GofS Research Paper Series, the editorial board is proud to issue a set of two more volumes, comprising papers (again all reviewed by international peers, the list of which is set out in the appendix) clustered around two well-profiled research axes. Volume 4 focuses on topical issues in EU and International Crime Control. Its table of contents is provided below the brief description of the papers comprised in the current book, which constitutes Volume 3, providing new empirical data, theories and analyses on Safety, Societal Problems and Citizens' Perceptions. Some articles in Volume 3 focus especially on issues of conceptualisation and measurement of key constructs in the study of security in its broadest meaning (from fear of crime to corruption) some articles present tests of theoretical models derived from theoretical criminology, and finally some articles focus on different institutional reactions towards crime and drug-related problems (e.g. policing, the conflict of interests between private companies and authorities and restorative justice).

January 2010, The editorial board:

Marc Cools, Brice De Ruyver, Marleen Easton, Lieven Pauwels,
Paul Ponsaers, Tom Vander Beken, Freya Vander Laenen,
Gudrun Vande Walle, Gert Vermeulen, Gerwinde Vynckier.

Volume 3: New Empirical Data, Theories and Analyses on *Security, Societal Problems and Citizens' Perceptions*

Different measures of fear of crime and survey measurement error

Wim Hardyns and Lieven Pauwels

The measurement of fear of crime is somewhat problematic. Many studies have focused on the cognitive and behavioural components of fear. The emotional affective component of fear of crime has been studied rather less. Traditional measures of fear of crime fail to address the complexity of this concept. Following an alternative question structure, previous research has shown that 'old'-style questions overestimate the everyday experience of fear (see Farrall, 2004; Farrall and Gadd, 2004; Gray, Jackson and Farrall, 2008). Furthermore, gender differences in fear of crime seem to be influenced by socially desirable answers by men (Sutton and Farrall, 2005). In this article, we study differences in outcomes when measuring fear of crime using 'old'-style questions ('avoidance behaviour') and an alternative question structure introduced by Stephen Farrall (three-part questions treating prevalence, frequency and intensity of fear). Measuring the emotional affective component of fear with an alternative question structure presents a totally different picture than can be found by measuring the behavioural component of fear of crime with a traditional scale such as avoidance behaviour. Second, different measures of fear of crime are especially differentially related to previous victimisation. Third, we found rather surprising effects of social desirability on gender differences in fear of crime.

Mobility and distance decay at the aggregated and individual level

Stijn Van Daele

Most crimes are committed near to where the offender lives; this has been observed both at the aggregate and at the offender level. At the aggregate level, as the distance increases there is a decline in the number of offences committed, and initially this decline is quite slow. This pattern has been described by a number of scholars, and results in a distance decay curve. Near-home offending has also been observed at the level of the individual offender, although it has been debated whether distance decay actually exists at the level of the individual offender. We therefore believe it is important to distinguish near-home offending from decay, i.e. the gradual decline in offences as distances increase. This paper studies mobility patterns and decay curves on serious property crimes in Belgium. First, aggregated patterns are discussed and categorised. Second, individual offenders are analysed. It becomes clear through studying offender patterns that offender mobility and decay are not intertwined at the individual level to the same extent as they are at the aggregate level. This suggests that it is important, particularly when studying individual offenders, to clarify whether (average) distances or decay are being considered.

Exploring the role of exposure to offending and deviant lifestyles in explaining offending, victimisation and the strength of the association between offending and victimisation

Gerwinde Vynckier and Lieven Pauwels

While victimisation surveys have been fostering since the 1960s, young adolescents (below the age of 15) have often been neglected or simply forgotten in such surveys. As a consequence, less is known about victimisation prevalence rates of young adolescents. This study consists of a descriptive and an explanatory part. The descriptive part focuses on levels of victimisation of young adolescents in Sint-Niklaas, a Belgian city of approximately 70.000 inhabitants. The explanatory part assesses the role of characteristics derived from an integrated self-control/lifestyle-exposure theory and evaluates the well-known association between offending and victimisation. In this study, we apply the integrated self-control/lifestyle-exposure model to the school context of exposure, since adolescents spend a lot of time at school. The results show that exposure to offending can explain individual differences in both victimisation and offending, but overall the results suggest that this classic framework originally designed to explain victimisation, is far more suited to be used as a theory of offending.

"Safety: everybody's concern, everybody's duty"? Questioning the significance of 'active citizenship' and 'social cohesion' for people's perception of safety

Evelien Van den Herrewegen

The catchphrase "Safety: everybody's concern, everybody's duty" implies that in order to safe-guard the social order and safety we, the professionals as well as the public, need to unite and work together. In this sense, social connectedness and civic engagement are perceived as the prime sources to counter crime and people's perception of safety. In this paper, we clarify that the references to 'active citizenship' and 'social cohesion' in criminal policy discourse are the result of the development of 'perception of safety' as an autonomous subject for research and policy. Policymakers have come to see (in)security as a phenomenon that needs to be explained by taking into account crime and non-crime related factors. We describe the emergence of 'social cohesion' and 'active citizenship' as natural barriers against crime and other deviant behaviour and as prerequisites for people's perception of safety. We point out that both concepts are not necessarily positively interlinked with people's 'perception of safety'. Moreover we will indicate that activating civic engagement and stimulating social cohesion can even be detrimental to people's perception of safety. Finally we suggest that in order to understand people's perception of safety, we need to consider the process of identity formation and social categorization.

Institutional distrust in Flanders. What is the role of social capital and dimensions of discontent?

Maarten Van de Velde and Lieven Pauwels

Trust towards governmental institutions such as the local government, the police and the criminal justice system are very important issues in a political democracy. It has been argued before that without trust political democracy is at stake, mainly because of the doubted legitimacy of the governmental system. This study aims at both describing and explaining individual differences in institutional trust in Flanders. The unique contribution to the literature is that it integrates social capital theory and social identity theory in the study of trust. We test the hypothesis that background characteristics such as the educational level are especially related to trust through their effect on social capital, anomia and ethnocentrism. Social capital has been repeatedly argued to be one key sociological mechanism in understanding trust, while anomia and ethnocentrism are important psychological mechanisms that refer to discontent and have been cited as important mediators of the effect of background characteristics.

Conceptualising the role of police culture in change strategies

Marleen Easton and Dominique Van Ryckeghem

This paper aims to conceptualise the role of police culture in change strategies, a topic for which policymakers show a lack of concern these days. The first part aims to clarify the concepts culture and police culture in particular by addressing the theories or beliefs that underpin the study of both. This results in an overview of concepts on (police) culture developed since Westley's first sociological analysis of the American police organisation. This overview shows that studying and investigating culture is a very complicated business which involves a variety of theoretical and ideological premises. Insight into these premises is necessary to fully understand research on police culture and to make an adequate interpretation of the findings of these types of research. This is the necessary theoretical background to move away from an unidisciplinary approach to police culture. In the second part of this article, we argue the importance of an interdisciplinary approach which enables culture to be factored in, against the background of change strategies alongside other variables including policy-making, the environment and actual policing. This approach is more suited for the ambitions of the 21st century.

The view of the police on community policing in Belgian multicultural neighbourhoods

Marleen Easton and Paul Ponsaers

As part of Belgium's police reforms, the government opted to introduce Community (Oriented) Policing (COP) as the official policing model. Despite the fact that the COP model is elsewhere historically rooted in a number of ethnically-coloured conflicts, the most pressing question was that of how this model might be applied in multicultural neighbourhoods with extremely complex and diverse social contexts in our country. The relations between police officers and ethnic minorities have become strained as the result of problematic mutual perceptions, ascribed meanings, visions and expectations. Factors such as these, combined with structural neighbourhood factors (discrimination and a heterogeneous social and ethnic mix), are a severe hindrance to the implementation of COP, but further obstacles are encountered in the intrinsic ambiguities wrapped up in the COP model itself as regards the meaning to be attached to "the community". This contribution aims to come to grips with these factors and processes by examining the extent to which the Belgian COP model takes shape (or not) in Belgian multicultural neighbourhoods.

Population density, disadvantage, disorder and crime. Testing competing neighbourhood level theories in two urban settings

Caroline Mellgren, Lieven Pauwels and Marie Torstensson Levander

Geographical patterns of official and survey based crime data show that crime levels are often higher in urbanised areas. In this study, we examine the population density – crime relationship empirically in two large cities (Malmö in Sweden and Antwerp in Belgium). We examine the extent to which neighbourhood disadvantage, social trust and disorder have equal effects on neighbourhood crime levels. Both studies are cross-sectional and combine official data with an independent community survey. Block-wise regression analyses suggest that neighbourhood disadvantage, social trust and disorder are key factors in understanding why population density is quite similarly related to neighbourhood crime levels in two different settings. Differences nonetheless exist as to the extent to which the effect of disadvantage on crime is mediated by social cohesion and disorder.

The continuum of conflicts of interest: from corruption to clubbing and the underlying risks at victimisation

Gudrun Vande Walle

This paper explores the continuum of conflicts of interest between private companies and public authorities. Recently, many initiatives have been taken in the domain of one variety of conflicts of interest, being the illegal act of corruption. We investigated the changing political-economic power relations and more specifically the growing impact of the market on political decision making. This led us to the hypothesis that corruption is no longer the most straightforward mechanism for private companies to have an impact on political decision making and that the changed political-economic relation promotes other legal mechanisms of conflicts of interest. This hypothesis is illustrated with three mechanisms that are not illegal but run the risk of becoming as harmful as corruption can be: lobbying and networking in the interest of one's own business and revolving doors. These legal mechanisms become risk factors because they pass unnoticed and without disagreement. If the hypothesis that legal mechanisms are more attractive today for western companies accurately reflects reality, it is maybe a task of criminologists to extend their analysis to include these legal mechanisms of conflict of interest and their potential risks associated with victimisation.

Corruption as a judgement label

Arne Dormaels

The study of corruption always meant that the researcher adopted a specific viewpoint on corruption, which influenced the definition debate and the debate on methodologies used to research the concept. The aim of this contribution is to use the lessons learned from previous research on corruption in developing a methodology aimed at assessing the judgement of a situation as corruption or not. Using Peters and Welch's methodology a quantitative survey is constructed using a questionnaire composed of different scenarios. An experimental design is used to examine individual judgement to a series of hypothetical scenarios as corrupt or not. The compilation and selection of those scenarios is illustrated as first step in a research design.

Towards an integral and integrated drug policy: pearls and pitfalls

Liesbeth Vandam, Charlotte Colman, Freya Vander Laenen and Brice De Ruyver

At the European Union level, the focus is placed upon an integral and integrated approach of the drug problem. Cross-cutting interventions – interventions between several domains involved in drug policy – are often locally implemented and evaluations are seldom reported on in international databases. An international literature review was executed to identify effect- and process evaluations of these cross-cutting interventions. Best practices and pearls and pitfalls were identified for the development of a comprehensive and integrated drug policy. 27 evaluation studies were selected (twelve effect studies (classified by the Maryland Scientific Method Scale), eight process studies and seven studies evaluating the effect and process of an intervention).

Explaining violence and aggression on public transport from the perpetrator's perspective – Literature on typology and etiology applied

Neil Paterson, Patrick Moreau, Gert Vermeulen and Marc Cools

Questions concerning crime, safety and security have become and continue to be a hot topic in many western European countries with Belgium being no exception. A number of high profile incidents, although atypical in their severity, have focused attention on problems of violence and aggression on public transport in Belgium. As part of a wider research project aiming to improve knowledge of violent incidents in this area from the offender's perspective and thus contribute to their prevention, this article explores a number of related questions. The authors conclude by questioning whether situationally based crime prevention policy in this area is sufficient and whether such initiatives should be complemented with more socially focused interventions.

Myths and reality in the history of restorative justice

Nikolaos Stamatakis and Tom Vander Beken

The second half of the 20th century saw a 'blooming' in restorative theories and victim-oriented practices of criminal justice. However, the notion of Restorative Justice and the egalitarian conceptions of decision-making processes are not as contemporary as they may appear to certain policy makers and supporters of just deserts principles. Nevertheless, various historical and anthropological facts could help us determine the origins of restorative justice and draw a clear borderline between myths and reality. The (re)discovery of the historical roots of restorative justice (or what then was conceived as such) might be able to provide credible answers to contemporary conundrums concerning both its origin and future evolution. This article considers restorative justice principles as they apply to prisons and measures the capacity of a prison to progress in restorative justice matters. The methodological part is consisted of both qualitative and quantitative tools which evaluate the prisoners' perceptions on the impact of their crimes on their victims and their possible openness or willingness to engaging in restorative activities.

―――

Volume 4: Topical Issues on *EU and International Crime Control*

Appreciating approximation. Using common offence concepts to facilitate police and judicial cooperation in the EU – *Wendy De Bondt and Gert Vermeulen*

Approximation and mutual recognition of procedural safeguards of suspects and defendants in criminal proceedings throughout the European Union – *Gert Vermeulen and Laurens van Puyenbroeck*

Shaping the competence of Europol. An FBI perspective – *Alexandra De Moor and Gert Vermeulen*

Towards a coherent EU policy on outgoing data transfers for use in criminal matters? The adequacy requirement and the framework decision on data protection in criminal matters. A transatlantic exercise in adequacy – *Els De Busser and Gert Vermeulen*

The international private security industry as part of the European Union security framework: a critical assessment of the French EU presidency White Paper – *Marc Cools, Dusan Davidovic, Hilde De Clerck and Eddy De Raedt*

The anti money laundering complex on a crime control continuum: perceptions of risk, power and efficacy – *Antoinette Verhage*

Police officers' views and fears about some criminals' threatening reactions to police investigations – *Fien Gilleir*

Police torture in China and its causes – *Wei Wu and Tom Vander Beken*

Different measures of fear of crime and survey measurement error

Wim Hardyns
Lieven Pauwels

1   Introduction

Large-scale surveys of a general population are very popular in the social sciences. In criminology, a growing interest in survey methodology has been observable since the 1950s. One major reason for the growing body of research in this tradition can be found in the discovery of bias in official measurement instruments, more specifically the bias in police statistics. Official statistics tend to underestimate true rates of victimisation in the population, and are said to be seriously biased with respect to race, gender and social class. As a consequence, the validity of earlier research concerned with the causes of crime has been called into question. Another important reason for the widespread interest in surveys lies in their potential to serve as a means to empirically test causal theories of offending, victimisation and fear of crime. Studies on the causes of crime and victimisation have disappeared from the agenda of criminologists in Belgium to make room for the study of the criminal justice system (Goethals, Ponsaers, Beyens, Pauwels and Devroe, 2002). In Anglo-Saxon countries, victim surveys and to a lesser extent self-report studies were periodically repeated on a large scale and were used to describe the epidemiology of crime and to address theoretical issues. The British Crime Survey and the (US) National Crime Survey are well-known examples. In one of the first sweeps of the National Crime Survey widespread anxiety about crime was discovered, and the 'fear of crime' was born (Ditton and Farral, 2000; Hale, 1996). 'Fear of crime' as a distinct field of criminological research can be traced back to Lyndon Johnson's 1967 Crime Surveys. Originally the level of public concern about crime was interpreted as an indicator of the importance politicians should attach to crime rates. High levels of concern were taken to imply the need to reduce crime levels. They were not read as a diagnosis of a public malaise to be treated in its own right. Concern for crime seemed to be confused with fear of crime. Although a large body of research concentrates on the subject internationally, in Belgium this is not the case and therefore we follow Pleysier, Vervaeke and Goethals (2002) and prefer to talk about a research tradition under development (Pauwels and Pleysier, 2005a).[1]

---

1   Apart from some local studies in the 1980s and 1990s the first general population victim survey at the federal level was not conducted prior to 1997. This Federal Victim Survey is generally referred to as the 'Security Monitor' (Board of the Operational Police Information (Dutch: CGO), Police Policy Support, Department of Policy Data). The Security Monitor is a two-year large-scale Belgian population survey and is mainly conducted for policy reasons: the Security Monitor is an instrument designed to evoke new policy directions towards crime and victimisation and even to evaluate past policy strategies. Besides fear of crime and victimisation questions, the Security Monitor also includes questions concerning perceived disorder, the reporting of crimes to the police and public attitudes towards the police.

Notwithstanding the massive body of research since the 'fear of crime' concept was 'empirically discovered', two issues in this research tradition stand out in explaining the overall pessimism in most reviews of the literature (Ditton and Farrall, 2000; Hale, 1996; Pleysier, Pauwels, Vervaeke and Goethals, 2005; Pleysier, Vervaeke and Goethals, 2004; Vanderveen, 2006). A first issue, which we will deal with in this paper, concerns the weak theoretical and conceptual framework surrounding studies of 'fear of crime'. Obviously, both the policy-driven character of the early – and later – large-scale victim surveys and the positivistic approach of the era in which 'fear of crime' research originated are largely indebted to this. A second issue – which follows from the first issue – is related to problems of measurement: studies of fear of crime, especially the studies that are conducted on demand of the Ministry of the Interior, use somewhat conservative outdated methodologies and measures of fear of crime. Because the 'fear of crime' concept in these large-scale surveys, as well as in many smaller initiatives, for reasons of 'comparability', was predominantly measured with one single indicator (i.e. 'How safe do you, or would you, feel walking alone in this area after dark?'), all claims concerning the reliability and validity were considered questionable. In recent years, numerous authors have indeed objected to this conservative approach, and has resulted in a tendency to use scaling techniques, instead of the traditional standard items. Nevertheless, the complexity of a concept such as fear of crime demands further studies on the different components of fear, before one moves on to an explanation of fear based on survey data.

2   Measuring fear as the emotional affective component of fear of crime

In this paragraph we reconstruct some of the important problems that remain in the measurement of fear. It has been said many times before that the widespread public anxiety about crime, later known as 'fear of crime', became a research subject of its own, and led to a new tradition in criminology. A representative historical overview of this tradition can be found in Hale (1996), Ditton and Farrall (2000) and Vanderveen (2006). For an overview of this research tradition, including its 'growing pains' in Belgium we refer the reader to Pleysier (2009). Despite the weight of publications that have appeared since the early years, it is striking that the content and theoretical 'body' are less imposing. Ferraro and LaGrange (1987, p. 76) argue that *'conceptual cloudiness and inappropriate operationalization taints the majority of this literature, thereby distorting the meaning and the utility of the fear of crime concept'*; although perhaps somewhat pessimistic, the bottom line of this critique is not outdated, and has been repeated ever since. In fact, Ditton and Farrall (2000, p. xxi) state in their review of the literature that, despite more than thirty years of research on the subject, *'surprisingly little can be said conclusively about the fear of crime'*. Nowadays, this statement remains true and keeps challenging scholars to study measurement issues related to fear of crime.

This pessimistic view has largely to do with the use of fear of crime as an umbrella concept. The distinction between *'feelings of insecurity'* and *'fear of crime'* particularly leads to some essential conceptual confusion (Pleysier, 2009). Too often these concepts are mixed up. The key difference is that 'feelings of insecurity' are not necessary related to crime. They can refer to health, globalisation, lack of informal control, antisocial behaviour, etc. The measurement of 'feelings of insecurity' and 'fear of crime' in the

Belgian Security Monitor is in many ways indicative of this general malaise in 'fear of crime' literature and research (Pauwels and Pleysier, 2005a; Pleysier, 2009; Pleysier, Vervaeke and Goethals, 2004). The 'feelings of insecurity' item used in the Security Monitor contains a single, general or 'formless' item on 'how safe the respondent feels', without reference to crime, a specific offence or situation, and tapping more into a general feeling of unease (see Appendix 1, v63).

Another well-known example of a recurrent single-item question is: 'How safe do you, or would you, feel walking alone in this area after dark?' A classic criticism in the international literature is that the measurement of a complex concept such as fear of crime by a single-item instrument causes some existential problems in terms of reliability and validity. An empirical measure of fear of crime through a single-item measurement instrument lacks theoretical depth and conceptual clarity (Ditton and Farrall, 2000; Farrall, 2004; Hale, 1996; Pleysier, 2009). Numerous authors have questioned this conservative approach with profound persistence, resulting in recent years in a growing tendency to use scaling techniques as a far better way of measuring a complex and multidimensional concept such as 'fear of crime'. A well-known example in the Belgian Security Monitor is the 'avoidance behaviour' scale, which measures the behavioural component of fear of crime and consists of four questions; the respondent is asked whether he or she avoids certain areas in the neighbourhood, does not open the door to strangers after dark, hides valuable things at home, or avoids leaving home after dark because he or she does not consider it safe otherwise (see Appendix 1, v57-v60).

In contrast with 'feelings of insecurity', the 'fear of crime' concept has a more situational and concrete character (Pleysier, 2009). Ferraro's definition (1995, p. 4) describes the content very well: *'an emotional response of dread or anxiety to crime or symbols that a person associates with crime'*. In other words, fear of crime is an emotion, fear of crime is a reaction, and in particular fear of crime refers to crime or symbols that can be associated with crime. A narrow interpretation of this definition points to the *emotional affective component of fear of crime*. Following some well-known classifications (Ferraro and LaGrange, 1987; Gabriel and Greve, 2003) a broad interpretation of the fear of crime concept also demonstrates the importance of a cognitive and a behavioural component. The *cognitive component* comes before the emotional affective component and refers to a process that converts signals and stimuli which have to do with threat and danger into a risk assessment of personally becoming a victim of crime (Oppelaar and Wittebrood, 2006). The *behavioural component* comes after the emotional affective component and refers to the behavioural reactions of fear, such as avoidance and defensive behaviour.

Until now the measurement of fear of crime has almost exclusively focused on the cognitive component (perception of risk) and the behavioural component (avoidance behaviour) of fear of crime. The emotional affective component has been less studied. Some authors take the view that it is impossible to measure the emotional affective component because of the lack of a concrete fear-stimulating situation. They use 'feelings of insecurity' as an alternative (Covington and Taylor, 1991; Taylor and Covington, 1993; Ward, LaGory and Sherman, 1986). Others are of the opinion that an intensity scale, on which respondents need to indicate the severity of their fear, can bring a solution (Chiricos, Hogan and Gertz, 1997; Ferraro, 1995). In general, traditional measures of fear of crime in surveys fail to address the complexity of this emotional affective component. After all, knowledge of prevalence, frequency and severity of fear are absent or inadequate in traditional standard items and scales, and people mostly memorise and report the most serious extent of their fears instead of the most common or typical (Farrall, Bannister, Ditton and Gilchrist, 1997). Undoubtedly this leads to an overestimation of the everyday experience of fear of crime in a population (Farrall, 2004; Farrall and Gadd, 2004; Gray, Jackson and Farrall, 2008), or (Farrall, 2004, p. 163): *'Are we really prepared to unquestioningly accept that almost a third to two-thirds of the westernized, civilized society are 'fearful' of crime 'some' or 'a lot' of the time?'*

Given the fact that traditional measures of fear of crime in surveys encounter a wave of criticism, alternative question structures that build on these strictures and focus on the emotional affective component of fear should be heartily welcomed in quantitative criminology. According to Farrall (2004), change in the survey tradition of fear of crime is desirable, possible and inevitable. Therefore Farrall had developed an alternative question design, considering two important assumptions: first the questions need to refer to the past year only and second the respondents need to recall any occasions on which they felt fearful in the past year because they thought they might be victimised. If both assumptions are fulfilled, additional information can be gathered about the frequency and the intensity of these feelings. In this study we were motivated to use the alternative question structure as used by Farrall and his colleagues (Farrall, 2004; Farrall and Gadd, 2004; Gray, Jackson and Farrall, 2008).

In this paper we focus on the measurement of 'fear' as the emotional affective component of fear of crime by using a 'fear frequency' and 'fear intensity' scale and we mirror these response rates with regard to more traditional measures of fear of crime by using a single-item question and an 'avoidance behaviour' scale, without suggesting that they are comparable. Second, we assess how similar these different measures of fear of crime are in terms of relating to some well-known covariates of fear, such as perceived sense of community, perceived disorder and previous victimisation. Finally, we assess to what extent different measures of fear of crime and their relation with well-known covariates are susceptible to social desirability.

3   Survey measurement error and hypotheses

Measurement can be described as the systematic assignment of numbers to variables to represent features of persons, objects or events (Vandenberg and Lance, 2002, p. 4). In 'fear of crime' studies, researchers assign scores to respondents on variables expected to 'explain' a substantial part of observed differences in 'fear of crime'. When conducting victim surveys we assume that 'fear of crime' is not directly observable. Fear of crime is a latent concept, i.e. not directly observable, which is made observable through the use of indicators. These indicators are assumed to constitute a valid representation of the underlying concept. A careful selection of indicators is therefore obligatory. In the social sciences in general the construction of valid concepts has been called problematic (Waege, 1997).

Validity problems are known to affect survey results under different survey conditions. A thorough definition of measurement error can be found in Billiet (1997, p. 2): *'Survey measurement error refers to error in survey responses arising from the method of data collection, the respondent or the questionnaire (or other instrument)'*. In the present study we are concerned with two major measurement issues: measurement error that arises as a consequence of the questionnaire and measurement error that arises as a consequence of respondent characteristics.

When examining the '*measurement error that arises as a consequence of the question-naire*' we wonder to what extent the measurement of fear of crime by using an alternative question structure will lead to another descriptive picture when we use a scale such as avoidance behaviour, as has traditionally been used in the Belgian Security Monitor and many other surveys. Frequently, in both political and scientific circles, one wrongly intends to measure 'fear' when using traditional measures such as avoidance behaviour. Because of that it is interesting to confront an alternative question structure that really focuses on the emotional affective component. This question structure was used for the first time by Farrall and Gadd (2004) and the findings were very promising. In short, the authors found that traditional measures of fear of crime seemed to overestimate 'fear' in a population because of a lack of depth in question wording. By using an alternative question structure, which was able to dissect the fear of crime concept in terms of frequency and intensity, Farrall and Gadd were able to prove that the fear of crime in the population was less pronounced than what was concluded so far in many traditional fear of crime studies. This is especially of relevance for studies of the epidemiology of fear, but also empirical tests of theories of fear. Measuring fear of crime in surveys is seldom the ultimate goal, but rather a means to analyse the link with some important covariates. With the question wording differences in mind, it is interesting to observe if different measurement tools of fear of crime do have an impact on correlational validity.

When examining the '*measurement error that arises as a consequence of respondents characteristics*' we are restricted to the study of social desirability as a validity problem. Social desirability can be described as the tendency of a respondent to be less willing to admit attitudes and behaviour of a rather threatening character, i.e. attitudes and behaviours that are less socially acceptable (Pauwels and Pleysier, 2005b). In fear of crime studies this type of measurement error is strongly related to the frequently discussed gender differences in fear of crime. Many studies found that women tend to report far more fear than do men, even though their reported risk of being victimised is lower (Fürstenberg, 1971; Smith and Torstensson, 1997). As a consequence, the question can be asked whether socially desirable responding behaviour plays a part in this finding, better known as the fear of crime paradox. One possible reason for this could be the influence of the socially constructed idea that 'boys don't cry' (Goodey, 1997). The stereotype that associates femininity with vulnerability and masculinity with dangerousness could be so deep-rooted in a community that it leads to socially adaptive responding behaviour (Hollander, 2001). Indeed, some men might report lower levels of fear of crime in surveys because of that masculine ideal, notwithstanding the fact that they are more at risk of being victimised than women. In addition to that it is often seen as socially desirable behaviour for women to talk about their emotions and fears, which could even enlarge the gender differences in fear of crime (De Groof, 2008; Hurwitz and Smithey, 1998; Snedker, 2006).

This validity problem can be actively studied with the help of social desirability scales, i.e. scales that are especially designed to measure this tendency. Very often this tendency is measured by asking respondents about behaviours that almost all of us have at some time committed. A striking example can be found in Sutton and Farrall

(2005). By using a 'lie scale' they tested the social desirability of some answers on fear of crime questions by gender and came to a striking conclusion: *'our results suggest that when men are being perfectly honest, they may actually report higher levels of crime than do women'* (Sutton and Farrall, 2005, p. 219). Previously it has been shown that social desirability somewhat affects covariates of self-reported delinquency in a survey of adolescent offending, but that no substantively different results would have been reported if social desirability was not controlled for (Pauwels and Pleysier, 2005b). Pauwels and Pleysier, however, tested the effect of social desirability on a sample of young adolescents, although it has been argued that social desirability is positively related to greater age. Thus, the effect of social desirability might be larger in general surveys where mainly adults are surveyed. In general, two strategies are reported to deal with social desirability: one is to omit respondents that answer desirably (Rovers, 1997). The other strategy is to statistically control for the effect of social desirability (Pauwels and Pleysier, 2005b). In this study we deal with this form of measurement error by using social desirability as a control variable. When testing the relationship between different fear of crime measurement tools and some important covariates, we statistically control for social desirability.

In short, in this study we focus on the previously identified forms of measurement error. More specifically, four general hypotheses can be distinguished:

(1) Different measures of fear of crime lead to different conclusions in terms of magnitude and incidence.
(2) Measuring 'fear' by different fear of crime components has an impact on the correlations with theoretically important covariates (perceived sense of community, perceived disorder and previous victimisation).
(3) Social desirability disturbs this relationship between different measures of fear of crime and measures of perceived sense of community, perceived disorder and previous victimisation.
(4) Social desirability contributes to gender differences in fear of crime.

4 Assessing the construct validity of 'fear of crime': theoretical framework

Construct validity refers to the networks of relationships that exist between theoretical concepts and empirical constructs. If two constructs that are theoretically related are validly measured then one can expect an empirical correlation between these constructs. In this paper construct validity is restricted to simple correlational validity. To assess the impact of measurement on fear of crime outcomes, concepts from a well-known theoretical framework are used. The concepts that are used to assess the construct validity of different measures of fear of crime have their roots in the early Chicago School (Shaw and McKay, 1942; Taylor and Covington, 1993) and refer to the perception of community social climate. It has been established that community social structural characteristics and sense of community affect one's perception of disorder and that these perceptions are strongly related to 'fear of crime' (Lee and Earnest, 2003; Plank, Bradshaw and Young, 2009). In this article we can only take the perception of neighbourhood 'sense of community' and 'disorder' into account. Another important fear of crime covariate is 'previous victimisation'. The relationship between victimisation and fear of crime has been much discussed and questioned. One possible explanation is that previous victimisation shapes one's perception of disorder and crime, which in turn affects fear of crime. For an overview we refer the reader to Hale (1996) and Vanderveen (2006).

5   The present study

The data used in the present study of fear of crime are not based on a representative sample of respondents but on a survey of professional key informants. Professional key informants are persons that have a great deal of knowledge of social situations in neighbourhoods and can provide additional and more accurate information than the average neighbourhood inhabitant in community surveys on social cohesion and disorder (Pauwels and Hardyns, 2009). Recently, Pauwels (2006) and Pauwels and Hardyns (2009) demonstrated that the technique of 'key informant analysis' could be used to create ecologically reliable and valid measures of community (dis)organisational processes. For this study we decided to question key informants by presenting them different measures of fear of crime to detect possible differences in the responses.

The key informants that meet the criterion of above-average knowledge of local area processes were previously identified in jobs such as social work, local police, local shops (e.g. grocers, newsagents, etc.), local pubs and local policy work. One major difference between the use of surveys of inhabitants and profession-based key informants is the selection procedure employed. Whereas random selection is the criterion used in resident surveys, professional key informants are chosen on the basis of their knowledge about community (dis)organisational processes. Key informants are thus field experts. The key informants were selected on the criterion of self-selection. We are aware this selection method delivers an atypical sample of the population. Therefore replications with representative samples are necessary to exclude specific effects on the results.

A survey of 750 key informants in eighteen postal code areas situated along the Belgian coast was conducted between October and November 2008.[2] This Belgian coast area is characterised by a high touristic activity in the summer months and a relatively high violent crime rate (Hardyns, Van de Velde and Pauwels, 2010). Appendix 2 presents descriptive statistics relating to the demographic and professional background characteristics of the questioned key informants. The survey of key informants was originally designed as a pre-test of a large-scale survey of community social cohesion, collective efficacy and disorder and was meant to assess the reliability and validity of the scale constructs used.[3] On this occasion we decided to include a traditional single-item question from the Belgian Security Monitor, a traditional 'avoidance behaviour' scale and the alternative question structure that actually measures frequency and intensity of 'fear' that was derived from Farral and Gadd (2004) to evaluate the measurement issues that were explained above.

---

2   The survey was conducted in De Panne, Koksijde, Oostduinkerke, Nieuwpoort, Lombardsijde, Westende, Middelkerke, Raversijde, Mariakerke, Oostende, Bredene, Vosseslag, De Haan-Centrum, Wenduine, Blankenberge, Zeebrugge, Heist, and Knokke.

3   The large-scale survey fits in with the project 'Social Cohesion Indicators for the Flemish Region', which started in January 2007 and which is a continuation of the scientific and policy interest in the concept of 'social cohesion'. For further information, please refer to www.socialcohesion.eu

6   Measurement of constructs and reliability

6.1   Fear of crime measures and social desirability

A traditional scale that measures the behavioural component of fear of crime is avoidance behaviour. *'Avoidance behaviour'* is measured by an additive index consisting of three items: *does it happen that... ' 'you avoid certain areas in your neighbourhood because you think they are not safe', 'you avoid opening the door to strangers because you think it is not safe', 'you avoid leaving home after dark because you think it is not safe'?* Cronbach's alpha is 0.67.[4] The alternative question structure aims to measure the emotional affective component of fear of crime with special reference to the frequency and intensity of these feelings. *'Fear frequency'* is measured by an index consisting of four items: *'in the last year how frequently have you felt fearful about the possibility of becoming a victim of ... "crime in general", "car theft", "burglary", "assault"?'* Cronbach's alpha is 0.64. *'Fear intensity'* is also measured by an index of four items. Each 'fear intensity' item follows on a 'fear frequency' item with the question: *'on the last occasion, how fearful did you feel?'* Cronbach's alpha is 0.66. Intensity and frequency of fear as the emotional affective component have never been measured like this in a Belgian study of fear of crime. Therefore it is of utmost importance to consider the reliability coefficients of these constructs. The Cronbach's alpha values do not reach the generally suggested level of 0.70. Possibly the relative low number (three or four) of items in these scales could have influenced this conservative reliability indicator. The factor loadings were, however, satisfactory (0.40 or higher) for all the items in the fear of crime scales, the lie scale, the perceived sense of community scale and the perceived disorder scale.

To measure social desirability we have adopted a 'lie scale' which is part of a well-known psychoticism scale in psychological sciences: the Abbreviated form of the Revised Eysenck Personality Questionnaire (EPQR-A).[5] This scale consists of four scales of six items each (Eysenck, Eysenck and Barrett, 1985; Francis, Brown and Philipchalk, 1992). The lie scale is one of these four scales. The items in this scale refer to dichotomous (0=yes, 1=no) questions on which disagreement is socially desirable but highly unlikely to be true. The *'lie scale'* consists of an additive index of the following five items: *'Were you ever greedy in terms of helping yourself to more than your share of anything?', 'Have you ever blamed someone for doing something you knew was really your fault?', 'Have you ever taken anything (even a pin or button) that belonged to someone else?', 'Have you ever cheated at a game?', 'Have you ever taken advantage of someone?'* The more people respond with 'no' to these questions, the higher their score on the social desirability scale. Cronbach's alpha is 0.58, which is not very high but in line with the relatively low values in previous reliability tests based on this scale (Eysenck, Eysenck and Barrett, 1985; Francis, Brown and Philipchalk, 1992). Given that we are dealing with dichotomous items this was the best reliability parameter we could find.

---

4   A reliable scale consists of a set of items that meet the demands of internal consistency. This can be checked by factor analysis of the observational questions and by computing Cronbach's alpha, one of the most well-known estimators of scale reliability (Tacq, 1992). For each scale in this study, additional exploratory factor analyses were conducted as an extra control for reliability at the respondent level.

5   Using an extended version of this lie scale (EPQ-R) Sutton and Farrall (2005) found some interesting effects from social desirability on the relation between fear of crime and gender. More specifically they found that men, with high scores on the social desirability scale, show some reluctance to admit fear of crime.

Future studies, however, might want to look for better or alternative scales of social desirability behaviour.

6.2 Perceived sense of community, perceived disorder and previous victimisation

Social cohesion is such a broad concept that it is hard to find agreement on how it should be measured (Peper et al., 1999). The collective efficacy dimension of social cohesion is increasingly of interest to scholars in Europe (e.g. Flap and Völker, 2005; Friedrichs and Oberwittler, 2007; Oberwittler, 2001). In this context Flap and Völker (2005) introduced the closely related construct 'sense of community', thereby referring to the social trust component of collective efficacy.[6] To measure 'perceived sense of community' key informants were asked on a three-point scale to what extent they agreed with following statements: 'I feel safe in my own neighbourhood', 'contacts in my neighbourhood are rather positive in general' and 'I am respected in my neighbourhood'. Cronbach's alpha was 0.62. Disorder is, like social cohesion, a rather ambiguous concept. Usually both physical (urban decay) and social nuisances (truants on the streets, public use of drugs and alcohol, etc.) are measured. Following Pauwels (2006) and Pauwels and Hardyns (2010), 'perceived disorder' was measured by asking the key informants how many times they observed: 'adolescents hanging around on street corners', 'a group of adolescents harassing persons to obtain money or goods', 'men drinking beer in public', 'persons selling drugs on the streets', 'somebody being threatened with a weapon on the street', 'fights between adolescents because one adolescent was challenged', 'men urinating in public'. Cronbach's alpha was 0.85. Finally, the key informants were asked whether or not they had been the victim of crime. Here 'previous victimisation' is a count variable that summarises the different offences a person (or a household member of the respondent in case of theft from car and bicycle theft) reports as a victim during the last year. The following offences were included in the survey: (1) 'burglary with theft', (2) 'attempted burglary', (3) 'theft from car', (4) 'bicycle theft' (5) 'physical violence' and (6) 'being threatened with violence'. The original survey items were dichotomous (1 = experienced victimisation, 0 = not experienced victimisation). These six victimisation experiences were counted for each respondent.

7 Results

7.1 Comparing different measures of fear of crime

A classic and much criticised single-item question to measure fear of crime can be found in the Belgian Security Monitor: 'does it happen that you feel unsafe? Is that ... (never, seldom, sometimes, often or always)'? As already mentioned, an operational measure of fear of crime through a single-item measurement instrument lacks theoretical depth and conceptual clarity (Pleysier, 2009). For that reason the use of scale constructs to measure fear of crime has been established for some time in this

---

6　Collective efficacy has been defined as 'social cohesion among neighbors combined with their willingness to intervene on behalf of the common good' (Sampson, Raudenbush and Earls, 1997, p. 918). Social cohesion and mutual trust among members of a community ('social trust') are an absolute condition to foster the willingness to intervene in the common interest of a community ('informal social control').

research tradition. In this study we use the well-known 'avoidance behaviour' scale, which measures the behavioural component of fear of crime. Let us take a look at the percentage of respondents that scores high on the single-item question in the Security Monitor and the avoidance behaviour scale in the survey of key informants, before we examine the alternative measurement instrument that measures the emotional affective component of fear of crime. Table 1 presents the distribution for both the single-item question and the scale construct by gender.

Table 1: Frequency distribution for the single-item question (Security Monitor 2006) and the 'avoidance behaviour' scale (Key Informant Survey 2008)

	Unsafe question (Security Monitor 2006)*			Avoidance behaviour (Key Informant Survey 2008)**		
	M	F	Total	M	F	Total
Never or seldom	67.5%	54.2%	60.5%	66.8%	46.7%	55.9%
	13,825	12,365	26,190	231	188	419
Sometimes, often or always	32.5%	45.8%	39.5%	33.2%	53.3%	44.1%
	6,653	10,423	17,076	115	215	330
	100.0%	100.0%	100.0%	100.0%	100.0%	100.0%
	20,478	22,788	43,266	346	403	749

* Chi-square: 896.37 df: 4 p < 0.001
** Chi-square: 38.34 df: 9 p < 0.001

From Table 1 we can see that for the unsafe question ('does it happen that you feel unsafe?'), 39.5 per cent of the 43,266 respondents questioned report that they sometimes, often or always feel unsafe. Split up by gender, this is 32.5 per cent of the male population and 45.8 per cent of the female population (Chi-square: 896.37 df: 4 p < 0.001). Furthermore for the 'avoidance behaviour' scale we observe that 44.1 per cent of the 749 key informants report they sometimes, often or always report avoidance behaviour. Split up by gender, this is 33.2 per cent of the male population and 53.3 per cent of the female population (Chi-square: 38.34 df: 9 p < 0.001). First of all, it is striking that approximately four out of ten people report they sometimes, often or always feel unsafe or show avoidance behaviour. Second, the difference between men and women is remarkable (13 per cent versus 20 per cent respectively).

In short, from Table 1 it can be seen that the use of both a single-item and the 'avoidance behaviour' scale yield large proportions of respondents who answer positively to these questions. It is not clear from these questions whether such large numbers of respondents really fear crime. One other criticism is that these items do not have any follow-up questions that measure in more detail the frequency and the intensity of the reported avoidance behaviour. Omitting such follow-up questions may lead to a serious overestimation of the fear of crime in a population because people largely memorise and report the most serious extent of their fears instead of the most common or typical (Farrall, 2004; Farrall and Gadd, 2004; Farrall, Bannister, Ditton and Gilchrist, 1997; Gray, Jackson and Farrall, 2008). Nevertheless these traditional crime measures have been used for years and are difficult to change in large-scale surveys, especially because of the advantages of comparability over time. One is tempted to start with a new zero point that would complicate the existing trend analyses. The alternative question structure we discuss in this article, however, meets the criticisms mentioned above. This structure refers to the past year only and tries to recall any occasions on which the respondents felt fearful in the past year because they thought they might be victimised. In addition, information is gathered about the frequency and intensity of these feelings (Farrall, 2004). Let us first present the alternative question structure inspired by the work of Farrall and his colleagues:

Q1: In the past year, have you ever felt fearful about the possibility of becoming a victim of crime? (yes, no, can't remember)
Q2: If yes, how frequently have you felt like this in the last year? (absolute number)
Q3: If yes, on the last occasion, how fearful did you feel? (cannot remember, not very fearful, a little bit fearful, quite fearful, very fearful)

The first question (Q1) is a kind of filter question which aims to measure the 'prevalence' of fear of crime. Only those respondents who admitted their fear were asked the subsequent questions in the questionnaire. With the second question (Q2) one wants to gather information about the 'frequency' of the fear of crime feelings. Respondents were asked to give an approximation of this frequency, which could easily be categorised afterwards. The third question (Q3) probes the 'intensity' of the fear of crime feelings by using the last occasion as a reference point. Table 2 presents the results for this alternative question structure which was used in the key informant survey.[7]

Table 2: The incidence, frequency and intensely fearful experiences of crime in general in the past year (Q1, Q2 and Q3 combined)

	Not fearful			Low fear[1]			High fear[2]			Total		
	M	F	Total	M	F	Total	M	F	Total	M	F	Total
Never in past year	273 81.5%	297 77.3%	570 79.3%	-	-	-	-	-	-	273 81.5%	297 77.3%	570 79.3%
Once	-	-	-	6 1.8%	5 1.3%	11 1.5%	5 1.5%	13 3.4%	18 2.5%	11 3.3%	18 4.7%	29 4.0%
Twice	-	-	-	8 2.4%	11 2.9%	19 2.6%	2 0.6%	7 1.8%	9 1.3%	10 3.0%	18 4.7%	28 3.9%
Three times	-	-	-	6 1.8%	7 1.8%	13 1.8%	5 1.5%	2 0.5%	7 1.0%	11 3.3%	9 2.3%	20 2.8%
Four times	-	-	-	0 0.0%	1 0.3%	1 0.1%	1 0.3%	6 1.6%	7 1.0%	1 0.3%	7 1.8%	8 1.1%
Five and more times	-	-	-	13 3.9%	15 3.9%	28 3.9%	16 4.8%	20 5.2%	36 5.0%	29 8.7%	35 9.1%	64 8.9%
Total	273 81.5%	297 77.3%	570 79.3%	33 9.9%	39 10.2%	72 10.0%	29 8.7%	48 12.5%	77 10.7%	335 100%	384 100%	719 100%

Notes: Total N = 719, as the 31 respondents who replied 'Don't know' to one or more of the questions have been excluded from the analyses. All percentages are of totals of the sample size.
1. Includes those respondents who said that they felt 'not very fearful' or 'a little bit fearful'.
2. Includes those respondents who said that they felt 'quite' or 'very' fearful.

The advantage of such a question design is the chance to combine frequency and intensity scores which can help to formulate more subtle pronouncements about fear of crime. In doing so, Farrall and Gadd (2004) found that 'only' 8 per cent of their sample, which was taken in the UK, frequently (five and more times) experienced high levels

---

7   This table is based on the presentation of the alternative question structure results as shown by Farrall and Gadd (2004).

of fear. In Table 2 'only' 5 per cent of the respondents in the key informant survey frequently (five and more times) experienced high levels of fear. The results of the studies are remarkably identical. Split up by gender, 4.8 per cent of the male population and 5.2 per cent of the female population frequently (five and more times) experienced high levels of fear. The observed differences between men and women when traditional questions are used seem to be strongly reduced when this alternative question structure is applied. Over and above this the significant differences analysed between men and women in the traditional structure were not significant in analysis of the alternative question structure.

In summary, we can argue that publications of fear of crime distributions strongly depend on the measurement instrument, and we cite the huge differences between Table 1 and Table 2. Using a traditional single-item question and the measure of avoidance behaviour as the behavioural component of fear, one cannot but conclude that no less than 40% of the Belgian population experiences some feelings of unsafety or expresses some avoidance behaviour. It may be clear that this results in a severely distorted view on the 'fear of crime': the alternative question structure presents a much more balanced result. Thus when statements are made about the extent of 'fear' in a population on the basis of confusing items that do not measure the emotional affective component of fear of crime, this has misleading consequences in terms of evaluation and decision-making. In other words, studies that wrongly intend to measure the emotional affective component will result in a much more pessimistic view.

7.2 Correlational validity of avoidance behaviour, fear frequency and fear intensity

It may be clear from the previous paragraph that univariate results and thus estimates of the proportion within a given population that experiences fear is seriously overestimated when more traditional measures that usually originate from the well-known large-scale surveys are used. Another concern that methodologists have identified is the question of the correlational validity of fear of crime measures. Especially when theories of fear and victimisation are tested, it is important to assess whether different concepts that are related to fear of crime yield different correlations with variant measures of fear of crime. This paragraph therefore tries to answer the question: do different measures of fear of crime have an impact on the relationship between fear of crime and some supposed related concepts such as perceived sense of community, perceived disorder and previous victimisation?

All correlations shown in Table 3 are statistically significant. It can be seen that avoidance behaviour is especially strongly correlated with perceived sense of community and perceived disorder, whereas the scales measuring frequency and intensity of fear are more strongly correlated with previous victimisation. The fear intensity scale in particular is highly correlated with previous victimisation. This is an important finding that may shed some light on the discussion that has been going on in the literature. It seems that the impact of previous victimisation on fear may well be underestimated when the emotional affective component of fear of crime is confused with avoidance behaviour.

Table 3: Correlations between fear of crime measures, perceived sense of community, perceived disorder and previous victimisation

	Fear frequency	Fear intensity	Avoidance Behaviour
Perceived sense of community	-.24***	-.30***	-.38***
Perceived disorder	.22***	.27***	.30***
Previous victimisation	.36***	.44***	.20***

*** p < 0.001

7.3 Does social desirability affect the relationship between fear of crime and covariates of fear?

Methodologists point to the fact that measurement error arising from respondent characteristics can bias results. Social desirability was identified as one such respondent characteristic that may lead to biased results in surveys. The only strategy that allows for confronting empirical results with the effects of measurement error is to actually measure social desirability and to assess whether this respondent characteristic actually disturbs the relationship between theoretically relevant concepts. Thus, in order to get an impression of the impact of social desirability we calculated the partial correlations between the fear of crime scales and perceived sense of community, perceived disorder and previous victimisation. From Table 4 we can see that there is hardly an effect of social desirability on the empirical correlations between all constructs. All partial correlations are identical to the bivariate correlations shown above in Table 3. This finding is interesting because people often expect that some kinds of measurement errors will have a serious impact on empirical findings. In the case of the key informant survey, this is not the case.

Table 4: Partial correlations between fear of crime measures, perceived sense of community, perceived disorder and previous victimisation, controlling for social desirability

	Fear frequency	Fear intensity	Avoidance Behaviour
Perceived sense of community	-0.23***	-0.28***	-0.37***
Perceived disorder	0.21***	0.27***	0.30***
Previous victimisation	0.36***	0.44***	0.19***

*** p < 0.001

7.4 Social desirability and gender differences in fear

Sutton and Farrall (2005) empirically demonstrated that women score significantly higher on fear of crime scales and the lie scale than men, but additionally presented results that suggested that only for men does social desirability have an effect on fear of crime scores. Therefore controlling for social desirability is necessary to get a better understanding of gender differences in fear of crime. These findings were replicated in our key informant survey with some identical, but also some different, results.

Table 5 reveals the mean differences (and standard deviations) between men and women on different measures of fear and the EPQR-A lie scale as a measure of social desirability. From Table 5 it can be seen that, on average, women have higher scores than men on all constructs. This is consistent with most of the fear of crime studies and with the finding of Sutton and Farrall (2005) in particular. The differences between men and women are significant only in the case of 'avoidance behaviour' and the 'EPQR-A lie scale' (respectively: $F_{(747)} = 34.64$, $p = 0.001$ and $F_{(747)} = 22.65$, $p = 0.001$).[8] Anova tests revealed that in the case of 'fear frequency' and 'fear intensity' the differences between men and women are not significant (respectively: $F_{(748)} = 0.02$, $p = 0.887$ and $F_{(742)} = 0.54$, $p = 0.461$).

Table 5: Means (and standard deviations) for fear of crime measures and the EPQR-A lie scale (n = 748)

	Gender	
	Men	Women
Avoidance behaviour **	3.69	4.38
	(1.26)	(1.85)
Fear frequency (ns)	2.76	2.80
	(4.45)	(4.20)
Fear intensity (ns)	1.81	1.97
	(2.96)	(2.85)
EPQR-A lie scale **	3.23	3.69
	(1.40)	(1.25)

** $p < 0.01$
ns = not significant

Table 6 reveals that in the general sample, without differentiating between men and women, there seems to be no relationship between social desirability and avoidance behaviour, fear frequency and fear intensity. The correlations are negligible and not significant. When, however, the analyses are split up by gender and re-run, some significant correlations emerge. The results reveal that social desirability is inversely related to avoidance behaviour, fear frequency and fear intensity among women, whereas it is positively related only to avoidance behaviour among men. Social desirability is not significantly correlated with fear frequency and fear intensity as far as men are concerned.

These findings are the opposite of what was found by Sutton and Farrall (2005). In that study, the only significant correlations between social desirability and fear of crime were related to men. These were negative correlations. Table 6 shows that in this study significantly negative correlations are related to women, whereas the correlation between social desirability and avoidance behaviour is significantly positive for men. This means that women who have high scores on the lie scale, and thus exhibit socially desirable responses, report lower scores of fear of crime. On the other hand, men with high scores on the lie scale report higher scores of avoidance behaviour. Therefore, on the basis of our key informant survey we cannot agree with the hypothesis of Sutton and Farrall (2005) that masculinity should lead to a suppression of reporting fear of crime feelings. Nor can we agree with the argument that women who are characterised by socially desirable responding behaviour should openly admit fear of crime feelings. Caricatural we could say that instead of 'masculinity', this study found a kind of 'feminism' which leads to more socially desirable responding and thus to a reluctance to report feelings of fear of crime as far as women are concerned. The question of whether this finding is culturally determined or not cannot be answered within

---

8   The reporting style of these Anova tests operates in the following fashion: F(degrees of freedom) = [F score], significance level.

the scope of this study. As already mentioned, the atypical sample of key informants in this study could also have an effect on these results. For that reason it is recommended that this study should be repeated with a representative sample of inhabitants.

Table 6: Correlations between the EPQR-A lie scale and fear of crime measures for the whole sample, for women only, and for men only

	EPQR-A lie scale		
	Whole sample	Women	Men
Avoidance behaviour	0.03 (ns)	-0.11*	0.15**
Fear frequency	-0.07 (ns)	-0.13*	-0.01 (ns)
Fear intensity	-0.07 (ns)	-0.13**	-0.02 (ns)

* p < 0.05 ** p < 0.01
ns = not significant

Table 7: Correlations between gender (0 = male, 1 = female) and fear of crime measures by subgroups of social desirability

	Avoidance behaviour	Fear frequency	Fear Intensity
Gender			
Lie scale ↓↓ low score[1]	0.30***	0.07 (ns)	0.09 (ns)
Lie scale Moderate score[2]	0.21**	-0.10 (ns)	-0.02 (ns)
Lie scale ↑↑ High score[3]	0.04 (ns)	0.01 (ns)	-0.02 (ns)

* p < 0.05 ** p < 0.01 *** p < 0.001
ns = not significant
1. Respondents (47.1%) with low score on the lie scale (0, 1, 2 or 3 'no' answers on the lie scale items)
2. Respondents (23.6%) with moderate score on the lie scale (4 'no' answers on the lie scale items)
3. Respondents (29.3%) with high score on the lie scale (5 'no' answers on the lie scale items)

From Table 7 it can be clearly seen how social desirability affects the empirical relationship between gender and avoidance behaviour. This behavioural component of fear is the only fear of crime measure that significantly correlates with gender (r =.21, p < 0.001). This positive correlation means that women score higher on the avoidance behaviour scale than men (0=male, 1=female). To demonstrate the effect of social desirability on the relationship between gender and avoidance behaviour, we split up the whole sample in three approximately equal groups based on the lie scale; i.e. respondents with low scores, moderate scores and high scores on the lie scale. It can be clearly seen that the positive relation between gender and avoidance behaviour only exists for those respondents who do not answer in a socially desirable way. This result means that women show significantly more avoidance behaviour than men only when respondents with low and moderate scores on the lie scale are taken into account. When we focus exclusively on respondents with high scores on the lie scale there is no significant relationship between gender and avoidance behaviour. Thus, the more respondents show social desirable responding behaviour, the less women and men differ in avoidance behaviour.

8   Conclusion and discussion

Fear of crime is politically popular: it appears to provide governments with a new moral target and a well-established arsenal to attack it. Frequently, however, politicians blow the problem out of proportion from a populist perspective. With a punitive attitude towards crime and criminals they try to convince their electorate. In doing so, they often ignore the actual underlying problems (Chevigny, 2003; Scheingold, 1995). From such a policy-oriented perspective it is perfectly understandable that there is a widespread attention to the issue of fear of crime and the development of surveys measuring fear. From a criminological perspective, however, it remains very important to ask oneself what actually is measured, before drawing premature conclusions. Measurement issues are not studied in as much detail or as frequently as one would expect, given the attention the issue has had in worldwide victimisation surveys.

Although more research is needed in this domain, it is already clear from this small-scale study that more precise measures are necessary to evaluate fear of crime. Measuring fear of crime through more traditional single-item questions or scales that refer to avoidance behaviour – as is by and large commonly done in large-scale surveys that are conducted at the request of the (Belgian) government – is not without danger. Few studies have tackled measurement issues. This is especially true of the Belgian situation, with some exceptions (Pleysier, 2009). Measures that actually measure the frequency and intensity of fear reveal that fewer people really are afraid of crime, contrasting with the picture we receive when the behavioural component 'avoidance behaviour' and the single-item question 'does it happen that you feel unsafe?' are taken into account. Statements about the extent of 'fear' in a population based on confusing items that do not measure the emotional affective component of fear of crime have important consequences and often lead to a much more pessimistic view. In addition, the huge differences between men and women that can be observed when fear of crime is assessed by more traditional measures seem to be strongly diminished when the emotional affective component of fear of crime is measured by an alternative question structure.

Furthermore, this study revealed that measuring 'fear' by different fear of crime components does have an impact on the correlations with theoretically important covariates. Although this study was empirically restricted to the study of perceived sense of community (perceived social cohesion), perceived disorder and previous victimisation, it is already clear that the correlation between previous victimisation and fear is seriously affected by the choices that scholars make when they measure fear of crime. This is a very important finding, especially when one reflects on the amount of discussion that has been going on between scholars on the ambiguous relationship that exists between previous victimisation and fear. In the measurement of avoidance behaviour, correlations with previous victimisation are moderate at best but when one actually measures the emotional affective component, it becomes clear that previous victimisation may have a greater effect than previously established.

The measurement of fear is not the only issue that should be taken into consideration in future studies of fear of crime. The actual measurement of potential confounders is another. Confounders may mask the 'true relationships' that exist between fear of crime and its theoretically derived correlates. This study was rather limited because it only took into account social desirability, whereas other potential validity threats exist (such as acquiescence). We found that social desirability does not affect the relationship between fear of crime measures and perceived sense of community, perceived disorder and previous victimisation, but we were able to demonstrate that social desirability does affect the relationship between gender and fear of crime. In particular, the significantly negative correlations for women's social desirability and all fear of crime measurements deserve attention. The correlation between social desirability and avoidance behaviour for men was significantly positive. In this study we found that women who gave socially desirable answers are characterised by a kind of reluctance to report feelings of fear of crime. Moreover, the observed differences in avoidance behaviour between men and women disappear when socially desirable responding behaviour is at stake. These particular findings of the key informant survey are the opposite of what was found in the similar study by Sutton and Farrall (2005). A possible explanation for these opposite results may be found in the atypical sample of key informants that was used for this study. For that reason it is recommended that this study should be replicated with a representative sample of inhabitants. Otherwise the question could be asked whether this finding could be culturally determined. In that respect more international comparative research could contribute.

Finally, we hope that this study brings the issue of carefully measuring constructs related to fear of crime and the study of potential confounders to the attention of both scholars and policymakers. Our understanding of fear cannot develop unless we pay attention to both measurement errors. Only by combining two strategies, avoiding error and measuring error, one can proceed and gain new knowledge about fear of crime in contemporary society.

9   Bibliography

Billiet, J. (1997). *Methoden van sociaal-wetenschappelijk onderzoek*. Leuven: Acco.

Chevigny, P. (2003). The populism of fear: Politics of crime in the Americas. *Punishment Society*, 5 (1), 77-96.

Chiricos, T., Hogan, M. and Gertz, M. (1997). Racial composition of neighbourhood and fear of crime. *British Journal of Criminology*, 35 (1), 107-131.

Covington, J. and Taylor, R. (1991). Fear of crime in urban residential neighbourhoods: implications of between- and within neighbourhood sources for current models. *The Sociological Quarterly*, 32 (2), 231-249.

De Groof, S. (2008). And my mama said... The (relative) parental influence on fear of crime among adolescent girls and boys. *Youth & Society*, 39(3), 267-293.

Ditton, J. and Farrall, S. (Eds.) (2000). *The fear of crime*. Aldershot: Ashgate.

Eysenck, S.B.G., Eysenck, H.J. and Barrett, P. (1985). A revised version of the psychoticism scale. *Personality and Individual Differences*, 6 (1), 21-29.

Farrall, S. (2004). Revisiting crime surveys: emotional responses without emotions? Or: Look back at anger. *International Journal of Social Research Methodology*, 7 (2), 157-171.

Farrall, S., Bannister, J., Ditton, J. and Gilchrist (1997). Questioning the measurement of the 'fear of crime'. Findings from a major methodological study. *British Journal of Criminology*, 37 (4), 658-679.

Farrall, S. and Gadd, D. (2004). The frequency of the fear of crime. *British Journal of Criminology*, 44 (1), 127-132.

Ferraro, K. (1995). *Fear of crime: interpreting victimization risk.* Albany: State University of New York Press.

Ferraro, K. and LaGrange, R. (1987). The measurement of fear of crime. *Sociological Inquiry*, 57 (1), 70-101.

Flap, H. and Völker, B. (2005). Gemeenschap, informele controle en collectieve kwaden. In B. Völker (Ed.), *Burgers in de buurt. Samenleven in school, wijk en vereniging* (pp.41-71). Amsterdam: Amsterdam University Press.

Francis, L.J., Brown, L.B. and Philipchalk, R. (1992). The development of an abbreviated form of the Revised Eysenck Personality Questionnaire (EPQR-A): its use among students in England, Canada, the USA and Australia. *Personality and Individual Differences*, 13 (4), 443-449.

Friedrichs, J. and Oberwittler, D. (2007). Soziales Kapital in Wohngebieten. In A. Franzen and M. Freitag (Eds.), *Sozialkapital (Special Issue 47 of Kölner Zeitschrift für Soziologie und Sozialpsychologie)* (pp.450-486). Wiesbaden: VS Verlag für Sozialwissenschaften.

Fürstenberg, F. (1971). Public Reaction to Crime in the Streets. *American Scholar*, 40, 601-610.

Gabriel, U. and Greve, W. (2003). The psychology of fear of crime. Conceptual and methodological perspectives. *British Journal of Criminology*, 43 (3), 600-614.

Goethals, J., Ponsaers, P., Beyens, K., Pauwels, L. and Devroe, E. (2002). Criminografisch onderzoek in België. In K. Beyens, J. Goethals, P. Ponsaers and G. Vervaeke (Eds.), *Criminologie in actie* (pp.137-188). Brussel: Politeia.

Goodey, J. (1997). Boys don't cry. Masculinities, fear of crime and fearlessness. *British Journal of Criminology*, 37 (3), 401-418.

Gray, E., Jackson, J. and Farrall, S. (2008). Reassessing the fear of crime. *European Journal of Criminology*, 5 (3), 363-380.

Hale, C. (1996). Fear of crime: a review of the literature. *International Review of Victimology*, 4 (2), 79-150.

Hardyns, W., Van de Velde, M. and Pauwels, L. (2010). *Crime, victimisation and urbanisation in Belgium: a descriptive analysis of frequently occurring crime.* In: Pauwels, L. (Ed.), Social disorganisation, offending, fear and victimisation. Findings from Belgian studies on the urban context of crime. Den Haag: Boom Juridische Uitgevers.

Hollander, J.A. (2001). Vulnerability and dangerousness. The construction of gender through conversation about violence. *Gender & Society*, 15 (1), 83-109.

Hurwitz, J. and Smithey, S. (1998). Gender differences on Crime and Punishment. *Political Research Quarterly*, 51 (1), 89-115.

Lee, M.R. and Earnest, T.L. (2003). Perceived community cohesion and perceived risk of victimization: A cross-national analysis. *Justice Quarterly*, 20 (1), 131-157.

Oberwittler, D. (2001). *Neighborhood Cohesion and Mistrust – Ecological Reliability and Structural Conditions.* Working paper retrieved July 14, 2008, from Max-Planck-Institut für ausländisches und internationales Strafrecht.

Oppelaar, J. and Wittebrood, K. (2006). *Angstige burgers? De determinanten van gevoelens van onveiligheid onderzocht.* Den Haag: SCP.

Pauwels, L. (2006). Ecologische betrouwbaarheid en constructvaliditeit van "sociale desorganisatieconstructen" op basis van de methode van "key informant analysis". *Panopticon*, 27 (3), 34-55.

Pauwels, L. and Hardyns, W. (2009). Measuring community (dis)organizational processes through key informant analysis. *European Journal of Criminology*, 6 (5), 401-417.

Pauwels, L. and Pleysier, S. (2005a). Assessing cross-cultural validity of fear of crime measures through comparisons between linguistic communities in Belgium. *European Journal of Criminology*, 2 (2), 139-159.

Pauwels, L. and Pleysier, S. (2005b). Effecten van antwoordstijlen in etiologisch self-report-onderzoek. Een 'causale modellenbenadering'. *Tijdschrift voor Criminologie*, 47 (1), 42-61.

Peper, B., Spierings, F., de Jong, W., Blad, J., Hogenhuis, S. and van Altena, V. (1999). *Bemiddelen bij conflicten tussen buren. Een sociaal wetenschappelijke evaluatie van experimenten met buurtbemiddeling in Nederland.* Delft: Eburon.

Plank, S.B., Bradshaw, C.P. and Young, H. (2009). An Application of "Broken-Windows" and Related Theories to the Study of Disorder, Fear, and Collective Efficacy in Schools. *American Journal of Education*, 115 (2), 227-247.

Pleysier, S. (2009). *'Angst voor criminaliteit' onderzocht. De brede schemerzone tussen alledaagse realiteit en irrationeel fantoom.* Proefschrift tot het verkrijgen van de graad van doctor in de Criminologische Wetenschappen. Afdeling Strafrecht, Strafvordering en Criminologie, Faculteit Rechtsgeleerdheid, Katholieke Universiteit Leuven.

Pleysier, S., Pauwels, L., Vervaeke, G. and Goethals, J. (2005). Temporal invariance in repeated cross-sectional 'fear of crime' research. *International Review of Victimology*, 12 (3), 273-293.

Pleysier, S., Vervaeke, G. and Goethals, J. (2004). Cross-cultural invariance and gender bias when measuring 'fear of crime'. *International Review of Victimology*, 10 (3), 245-260.

Pleysier, S., Vervaeke, G. and Goethals, J. (2002). Het 'onveiligheidsgevoel' onderzocht. Groeipijnen van een onderzoekstraditie in wording. In K. Beyens, J. Goethals, P. Ponsaers and G. Vervaeke (Eds.), *Criminologie in actie* (pp.189-206). Brussel: Politeia.

Shaw, C. R. and McKay, H. D. (1942). *Juvenile Delinquency and Urban Areas*. Chicago: University of Chicago Press.

Scheingold, S.A. (1995). Politics, Public Policy, and Street Crime. *Annals of the American Academy of Political and Social Science*, 539, 155-168.

Smith, W.R. and Torstensson, M. (1997). Gender differences in risk perception and neutralizing fear of crime. Toward resolving the paradoxes. *British Journal of Criminology*, 37 (4), 608-634.

Snedker, K.A. (2006). Altruistic and Vicarious Fear of Crime: Fear for Others and Gendered Social Roles. *Sociological Forum*, 21 (2), 163-195.

Sutton, R.M. and Farrall, S. (2005). Gender, socially desirable responding and the fear of crime. *British Journal of Criminology*, 45, 212-224.

Tacq, J. (1992). *Van probleem naar analyse, de keuze van een gepaste multivariate analysetechniek bij een sociaal-wetenschappelijke probleemstelling*. Lier: Academische Boeken Centrum.

Taylor, R. and Covington, J. (1993). Community structural change and fear of crime. *Social Problems*, 40 (3), 374-395.

Vanderveen, G.N.G. (2006). *Interpreting fear, crime, risk and unsafety. Conceptualisation and measurement*. Den Haag: Boom Juridische uitgevers.

Waege, H. (1997). *Vertogen over de relatie tussen individu en gemeenschap. Ontwikkeling en validering van meetinstrumenten in het kader van survey-onderzoek*. Leuven: ACCO.

Ward, R.A., LaGory, M. and Sherman, S.R. (1986). Fear of crime among the elderly as a person/environment interaction. *The Sociological Quarterly*, 27 (3), 327-341.

Appendix 1: 'Fear of crime' items in the Belgian Security Monitor

Variable	foc item
v63	Does it sometimes happen that you feel unsafe? Is this...?
v57	Does it sometimes happen that you avoid certain areas in your municipality because you do not consider them safe? Is this...?
v58	Does it sometimes happen that you do not open the door to strangers in the evening or at night because you do not consider it safe? Is this...?
v59	Does it sometimes happen that you hide valuable things at home? Is this...?
v60	Does it sometimes happen that you avoid leaving home if it is dark? Is this...?

(always/often/sometimes/seldom/never)

Appendix 2: Descriptive statistics of sample

	Belgian coast survey (2009) Postal code area level Number of units: 18 Total number of informants: 750	
Background characteristics	**Absolute counts**	**%**
Professional background		
Local shops and catering industry	492	65.6
Social work and medical doctors' surgeries	93	12.4
Local governance	111	14.8
Local police and private security	54	7.2
Total	*750*	*100*
Gender		
Male	346	46.1
Female	404	53.9
Total	*750*	*100*
Age		
18-25	69	9.2
26-35	158	21.1
36-45	222	29.6
46-60	251	33.5
60+	50	6.7
Total	*750*	*100*
Length of stay in the postal code area		
< 1 year	17	2.3
> 1 year & < 5 years	66	8.8
> 5 years & > 10 years	90	12.0
> 10 years	577	76.9
Total	*750*	*100*

Mobility and distance decay at the aggregated and individual level

Stijn Van Daele

1   Introduction

Criminology research can start from a variety of positions, including those identified in the literature (Patricia Brantingham & Brantingham, 1981; Cohen & Felson, 1979) – legislation; the criminal; the victim or target of the crime; and place – the scene of the crime, where offenders and targets meet. As such, place can be called the fourth dimension of crime. The study of place and crime has increased in popularity, primarily because place is six times more predictive for future crime than offender identity (Sherman, 1995, pp. 36-37). Studying place therefore offers criminologists a number of opportunities. Several studies on crime and place have identified a 'distance decay pattern' in which most crimes are committed near the offender's home, and the number of crimes declines as distances increase (see for example Phillips, 1980; Turner, 1969; White, 1932). This concept has two subdivisions, although these are rarely mentioned explicitly. First, most crimes are committed near home (i.e. offenders tend not to travel very far). Second, the distance curve shows decay, with the result that the number of crimes committed gradually reduces as distances increase.

Although distance decay is usually considered in relation to short distances, decline may also occur over long distances – decay refers to a distribution and is independent of absolute distance calculations. The primary concern of this paper is, therefore, whether the number of offences reduces over longer distances on both the aggregate and the individual offender level. As such, this paper contributes to a discussion about the application of the distance decay pattern (Rengert, Piquero, & Jones, 1999; Smith, Bond, & Townsley, 2009; Van Koppen & De Keijser, 1997) and to a better understanding of the distance decay pattern. It does so by splitting distance decay in its two components:

1. near-home offending: do offenders primarily operate within the vicinity of their homes, as many studies have found? (for a literature review, see Canter & Youngs, 2008a, pp. 4-6)
2. gradual decline: is there an even or an uneven distribution? Does the proportion of crimes wane with distance? (Canter & Youngs, 2008a, pp. 7-8)

Using mobility features from criminological literature, we compared mobility in terms of average distance travelled, with the mobility shown by specific types of offender at the aggregate level. We expected more emphasis to be put on crimes further from home for these offenders, compared with the general pattern.

However, the average distance travelled to commit a crime is only a central measure and does not reveal any information on the actual shape of the distance decay curve. To flesh out the picture, therefore, we first explored whether high mobility actually results in different mobility patterns. In other words, does the distance decay curve of offences/offenders that are related to high mobility also show a deviating distance decay pattern, or is there straightforward decay, with the curve only slightly 'stretched' over the longer distance?

Second, we bring distance decay down to the level of the individual offender. Distance decay curves are often explored at the aggregated level (which often includes both near-home offending and decay), or central measures are used to calculate offender travelling behaviour, resulting in a loss of some interesting information (Paul Brantingham & Brantingham, 1984, pp. 222, 227). This is often done because listing a number of individual decay curves hampers a clear overall view of the data. In this paper, instead of focusing on central measures, we measure decay itself – i.e. the distribution of distances travelled, independent of the offenders' mean distances travelled.

2   Distance decay

The principle behind the distance decay pattern is that criminals tend to commit most of their crimes close to home, and commit fewer crimes as distance increases. The greater the distance, the fewer the crimes committed. A distance decay curve can be created by counting and combining the individual distances between an offender's residence and the places where they carried out their crimes (Turner, 1969; Phillips, 1980). Figure 1 illustrates a distance decay curve, as presented by Van Koppen and De Keijser (1997, p. 510). This curve is of a fairly standard shape. Nevertheless, the average distances vary in relation to the nature of the crime.

Figure 1: Example of a distance decay curve (Van Koppen & De Keijser, 1997)

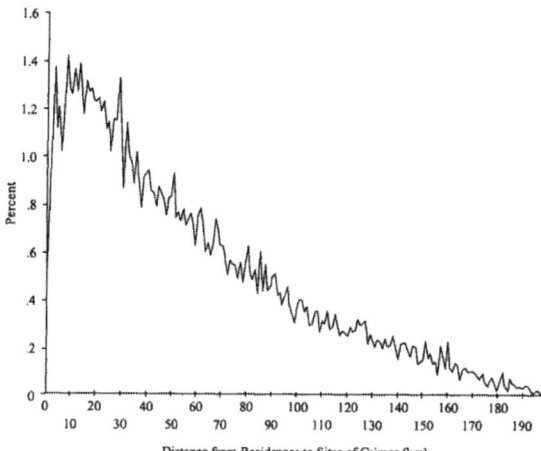

White (1932) observed in his sample an average distance of 2.7km[1] for all types of criminality. He made the observation that crimes against people are committed very

---

1   Many of these studies are US studies and use mile instead of km. For uniform representations, we converted 1 mile into 1,6km.

close to home (1.3km), while crimes against properties occur further away (2.8km). Other researchers have drawn similar conclusions. Phillips (1980, p. 175) found a mean average distance of 2.3km, ranging from 1.1km for assault to 1.7km for burglary, and 4km for petty larceny. Reppetto (1974) observed an average distance for burglary of only 0.8km. Capone and Nichols (1976, p. 209) concluded that average distances differ not only by type of crime, which in their study was robbery, but also in relation to type of target premises. The distances travelled to rob loan companies, for example, averaged 13.4km, while the average distance travelled to rob parking lots was only 3.5km. More recently, Wiles and Costello (2000) found a mean average distance of 3.1km, and in their study on serial arsonists Edwards and Grace (2006, pp. 223-224) found the mean distance travelled from the home base to the crime site was 6.6km. The distance decay curve actually has its origins in other sciences, such as biology, medicine and geography (see for example Mizutani & Jewell, 1998; Snow, 2008; Tobler, 1970), rather than in criminology.

The limited mobility of most offenders can be explained from a variety of perspectives. First, the rational choice perspective would indicate that costs and gains will be weighed against each other (Cornish & Clarke, 1986). The actions that require least effort and result in maximum gains will be taken (see also Zipf, 1949). Travelling further takes more time, effort and cost (Pettiway, 1982, p. 257); for example, travelling longer distances means vehicles need to be used for longer periods of time, increasing the chances of arrest by routine police controls. There are few reasons to choose to travel longer distances if the benefits do not increase accordingly. This could explain why criminal mobility correlates with higher criminal achievements (Morselli & Royer, 2008; Snook, 2004, p. 62).

Second, everyday life influences offender mobility. Criminals will commit their crimes in areas where they also carry out other, non-criminal, activities. Brantingham and Brantingham (1981) highlighted this pattern, and called it 'awareness space'. This is the total space an offender knows. Offenders are likely to perceive criminal opportunities when they are travelling about for other reasons. Most criminals will rarely decide to commit crimes in an area they have never been before, as without any form of reconnaissance they do not know the precise location of opportunities for crime, and have no knowledge of particular risks. Work and recreation are typical examples of non-criminal activities that help criminals to shape their awareness space and identify possible crime sites. Rengert and Wasilchick (1985, pp. 68-71) described these in their study on burglary. Offenders do not tend to commit their crimes when they are actually travelling for non-criminal purposes; instead, they notice opportunities when travelling, and feel safer when they are familiar with an area – which includes getting to know possible escape routes and dead-ends, and developing their risk perception. The authors narrow the concept of 'awareness space' further using the term 'search space', as not every part of the awareness space contains criminal opportunities (Rengert & Wasilchick, 1985, p. 55). Criminals first identify a suitable area that may provide an opportunity for crime, and then narrow their exploration down to potential targets (see also Bernasco & Nieuwbeerta, 2005). The more the space is narrowed, the more concrete criminal planning and operations become. After the principle of least effort, which is part of the rational choice perspective, Goodwill and Alison (2006, p. 408) consider anchoring and familiarity – what we called the influences of everyday life – as the most important issues in explaining criminal mobility.

A third influence is the existence of barriers. Elffers (2004) suggested that physical barriers, such as rivers, seas or forests limit travel options. Other authors have found that social (De Poot, Luykx, Elffers, & Dudink, 2005; Reynald, Averdijk, Elffers, & Bernasco, 2008) and even ethnic (Bernasco & Block, 2009; Gabor & Gottheil, 1984) barriers restrict offender mobility too.

The concept of limited offender mobility, and thus distance decay, is particularly useful in two fields. First, police authorities use it for geographical profiling, where a criminal's anchor point region is deduced from the location of several crimes (Besson, 2004, p. 145; Rossmo, 1995, 2000; Rossmo, Thurman, Jamieson, & Egan, 2008; Van der Kemp & Van Koppen, 2007). Specific computer software has been developed for this purpose, although the actual value of this software and its various functions has been the subject of considerable debate (Alison, Smith, & Morgan, 2003; Canter & Hammond, 2006, 2007; Paulsen, 2006; Snook, Canter, & Bennell, 2002; Snook, Taylor, & Bennell, 2004; Snook, Zito, Bennell, & Taylor, 2005; Van der Kemp & Van Koppen, 2007). Moreover, some offenders act as criminal 'commuters' (Canter & Larkin, 1993). They operate around another anchor point than their residence, making it difficult to create a geographical profile.

Distance decay theory also proves useful in the field of theoretical criminological research. Different offence, offender and target types may influence the distance decay pattern (Bernasco, 2006; Bernasco & Block, 2009; Kocsis, Cooksey, Irwin, & Allen, 2002; Kocsis & Harvey, 1997; Pettiway, 1982; Santtila, Laukkanen, Zappala, & Bosco, 2008). However, the principle behind the distance decay pattern has also been the subject of debate. One discussion relates to the method used. As most papers in this field have been based on official police statistics, it may well be possible that people who commit crimes near home are more likely to be arrested, and thus the role of distance decay would be overestimated (Eck & Weisburd, 1995, p. 16; Laukkanen, Santtila, Jern, & Sandnabba, 2008, p. 233). In other words, criminals who are more mobile may also be more successful and evasive, and are therefore less likely to be arrested (McIver, 1981, p. 43). In addition, not all criminals start their crime trip from home (Wiles & Costello, 2000, p. 40); however, most researchers assume they do, and therefore crime trip calculations may be biased.

A second issue relates to a discrepancy between the individual and the aggregate level. Van Koppen and De Keijser (1997) argued that the assumption of an individual distance decay function is an ecological fallacy. They modelled a number of offenders showing no individual distance decay pattern, resulting in a distance decay pattern on the aggregate level. They were criticised by Rengert, Piquero and Jones (1999) for their interpretation of the ecological fallacy, for neglecting the role of surface calculations over distances, for the interpretation of geographic work on profiling and for the assumption of random target selection. Nevertheless, this did not end the discussion over the role of the distance decay function. By using qualitative offender interviews, Polisenska (2008) put the near-home hypothesis in question, as she found that a number of burglars would travel considerable distances to commit their crimes. Smith, Bond and Townsley (2009, p. 217) also found that the aggregate distance decay function neglects the variation that exists between individual offenders' crime trip distributions.

In the present paper, this second issue is tackled. The aim is to further improve the interpretation of the distance decay pattern. Two main hypotheses are formulated:

1. General distance decay neglects the differing patterns over distinguished groups. As such, larger average mobility may result in different decay patterns as well. Testing this hypothesis, we calculated such patterns for certain sections of offenders. Group divisions were based on a number of basic features and follow a suggestion made by Rengert et al., namely that *"[...] the next step is to identify how distance-decay parameters vary between groups of offenders (i.e., ethnicity, gender, region) and what that says about their offending behavior."* (1999, p. 442)
2. Distance decay is mostly observed at the aggregate level. Assuming the presence of a similar pattern on the level of the individual offender is incorrect. In order to test this hypothesis, we measured individual decay using a method proposed by Smith et al. (2009, pp. 230-232).

3   Method

This study used data on serious property crimes in Belgium drawn from the General National Database, the main nationwide database of the federal police forces, covering the period 2002–2006. We included in our analysis all serious property offences for which the offender was known and was resident in Belgium.

Serious property offences were identified as those with aggravating circumstances. The result of this selection process was data for 72,726 offender–offence combinations. Using police data may potentially bias the result because, as has been said, it is quite possible that offenders who do not travel very far are more likely to get caught (Bruinsma, 2007, p. 485; Canter & Youngs, 2008b, p. 12; Eck & Weisburd, 1995, pp. 15-16; Rhodes & Conly, 1981, p. 177). In this regard, this study suffers from the same flaws as previous analyses of police data.

Because we wanted to calculate the distances between offence location and offender residence we obviously needed to know where offenders lived. As we did not have addresses for offenders who lived abroad, and because it was unlikely that all foreign residences were the starting points for crime trips in Belgium, we only considered those offenders with a known residence in Belgium. Thus, 67,981 cases were included.

Due to the large size of the sample and difficulties encountered with automated detailed geo-coding (see also Wiles & Costello, 2000, pp. 7-8), a simplified way of coding the locations was used. The surface area of Belgium measures approximately 31,000km^2, and is divided into 589 municipalities. This research focused particularly on nationwide offender mobility, as we have observed a considerable proportion of offences being committed outside the home area. Therefore, the Lambert coordinates of the centre of each municipality are used to localise residences and crime places. This means that for every offender starting in a particular municipality or every crime being committed in that area, the same coordinate is used. We admit that this level of detail is quite rough in absolute figures and that small environmental units are preferable (Oberwittler & Wikström, 2009). Yet, this approach fits best in the general framework of this paper, which examines offender mobility and mobile offenders. However, we believe this approach is acceptable, taking into account the fact that other researchers have worked at city level using around 100 geographic subdivisions (see for example Bernasco & Luykx, 2002 for a Western European study; and White, 1932 for one of the pioneer studies in this field), compared with nearly 600 for this paper. Moreover, our aim was to observe general patterns of criminal activity, which can be done at many different units of geography (Swartz, 1999, p. 43). Using these coordinates, Euclidian ('as the crow flies') distances between the home base and crime site were then calculated. In cases where the place of offence and offender residence were located in the same municipality, the distance was estimated by using the formula:

$$d = \frac{\sqrt{S}}{2}$$

in which S measures the surface of the area. This approach has also been used by Bernasco (2006, p. 147). In order to be able to draw easily comparable distance decay curves, we divided the recorded distances into bands of 10km each.

We then used a two-pronged approach. First, a number of aggregated distance decay curves were drawn. In this we were following Rengert et al.'s (1999, p. 442) suggestion that it is important to identify how distance-decay parameters vary between certain groups of offenders. The divisions were related to characteristics that literature suggests influence travel patterns:

- Multiple offending. Experienced offenders are found to travel further than other offenders (Barker, 2000; Beauregard, Proulx, & Rossmo, 2005, p. 587; Gabor & Gottheil, 1984). As our data set contained no information on prior convictions or arrests, we considered those offenders most experienced when they had committed 10 or more crimes in the period under consideration. This notion has been used before (Elffers, 2003; Ferwerda, Kleemans, Korf, & Van der Laan, 2003; Ferwerda, Versteegh, & Beke, 1995; Smith, et al., 2009, p. 224; Snook, et al., 2005). Although this approach is suitable when taking a general perspective, the possible impact of the law of small numbers forced us to only use these offenders in the second stage of analysis. Thus, this criterion merely provides information on the possible bias it generates when calculating individual decay patterns.
- Co-offending. Co-offenders are found to be more mobile than other offenders. They are likely to commit their offences near the residence of one offender, causing the other(s) who live elsewhere to travel longer distances (Bernasco, 2006, p. 147). Co-offending may increase the offenders' awareness space and enable them to travel further (Patricia Brantingham & Brantingham, 1981; Gabor & Gottheil, 1984; Tremblay, 2004, p. 22).
- Eastern European offenders. Previous research has shown that Eastern European offenders tend to be more mobile compared with other offenders, particularly when they commit offences in Western Europe (Ponsaers, 2004; Van Daele, 2008; Van Daele, Vander Beken, & De Ruyver, 2008), although in Eastern European countries offenders generally appear to travel further to commit crimes (Polisenska, 2008).
- Older offenders. Young offenders tend to commit more impulsive, opportunistic offences, and therefore travel shorter distances (Deakin, Smithson, Spencer, & Medina-Ariza, 2007, p. 54; Gabor & Gottheil, 1984, p. 270). Their choices are also affected by their reduced transport options (for example not having a driving licence). In this paper we considered offenders older when they were aged 30 or over at the time of offence.

- Attractive targets. Mobile offenders often travel to attractive targets and richer neighbourhoods (Deakin, et al., 2007, p. 65). Normally, targets in deprived areas are more at risk than those in affluent areas (Johnson & Bowers, 2004, p. 238). However, mobile criminals are more likely to travel to attractive targets further afield (Bernasco & Block, 2009, p. 98; Johnson & Bowers, 2004, p. 243; Maguire, 1982, pp. 19-20). Mawby described this as 'rich pickings' (2001, p. 72). These offenders typically use arterial roads and major highways to reach affluent areas (Paul Brantingham & Brantingham, 1984, p. 357; Fink, 1969; Hakim, Rengert, & Shachmurove, 2000, p. 12; Kleemans, 1996, p. 192; Laukkanen, et al., 2008, p. 232; Maguire, 1982, pp. 41-42; Rossmo, 2000, p. 214). In order to measure the attractiveness of target areas, we used an affluence index designed by the Belgian National Institute for Statistics, in which every area with a wealth index above a mean of 100 is considered to be attractive.

First we analysed whether our data confirms that offenders with these features showed increased mobility. We also considered whether this results in different travel patterns. To do this we analysed the aggregated distance decay curves of criminal actions containing these features, to see whether they deviate from the distance decay curve.

After comparing these aggregated distance decay curves, we took things further. On the individual level, most studies use mean distances to describe offending patterns. However, the distance decay curve would be more useful for this type of analysis (Smith, et al., 2009, p. 220). One problem is how to measure decay patterns. Decay is often represented as a curve, which provides a whole range of information but leaves no room for further calculation; we therefore used a quantification of decay. This approach follows a method proposed by Smith, Bond and Townsley (2009, pp. 230-232), and uses the skewness scores of each individual decay curve. Skewness measures asymmetry, which is exactly what we are looking for in this approach. If skewness has a positive value, the right tail of a distribution is longer than the left. As distance decay illustrates most offences being committed near home (i.e. on the left side of the distribution), we expected most cases to be located on the left, and thus we expected distance decay to correspond with significant high skewness scores. In order to judge the significance of the individual decay curves, we worked with the Z-scores of this skewness, which are calculated through dividing the skewness by its standard error. Only if this value differs more than 2 standard deviations from zero (greater than 1.96 or lower than −1.96) is the skewness significant.

Using one measure means that further calculations can be carried out and comparisons can be made. However, it is also likely to result in a loss of information, as patterns contain richer information than just single values (Paul Brantingham & Brantingham, 1984, p. 222; Rossmo, 2000, p. 101). Because this paper attempts to compare the individual decay patterns of people from different groups that have been distinguished on the aggregate level, this loss of information may be regrettable but it does not hamper the outcome.

Skewness and their corresponding z score measure relative decay. They only consider the shape of the distance decay curve, not the distances that are observed within the curve. Thus, offender A who commits 8 offences at 2km and 2 at 10km from home will have a skewness z-score of 2.59, the same score as offender B who commits 8 offences at 4km and 2 offences at 20km from home.

The skewness score cannot be calculated for all offenders. In order to rule out small-number coincidences, only offenders who have committed 10 or more offences were taken into account. The first 5 offences are indicative for the range in which offenders operate (Barker, 2000, pp. 64-65), so it was imperative that this criterion is maintained. In order to outweigh overgeneralization, we only included offenders who had committed at least 10 offences from this stage on. This avoided a possible bias that could be generated by the law of small numbers. Given the similar curve that multiple offending generated in the first stage of the analysis (see below), we did not expect to see differences by implementing this condition.

Second, skewness estimates could not be calculated for those offenders who committed all their crimes in the same area. Those who commit crimes in the area where they live have been labelled 'marauders'; if it is another area they are 'commuters' (see Canter & Larkin, 1993). The nature of our dataset and the rough geographic variable allowed for no further specification within the boundaries of a particular municipality.

As with the first stage, we compared the individual decay results between our chosen groups with the general results for Belgium. This enabled us to compare aggregated and individual decay, and addressed the question of whether mobility and decay are intertwined or relatively independent from each other.

4   Results

We first considered the mean distances travelled at the aggregate level, observing whether our chosen features result in higher mobility on the aggregate level. In general we noticed a mean distance travelled of 19.1km. This is substantively higher than the results found in literature. However, as most other studies have been executed on city level, not country level, these studies exclude travelling across city borders. Calculating the median distance, we found an average of 7.2km, indicating that our first calculation was certainly influenced by a number of extreme values (we observed a number of crime trips over 250km, which is more or less the maximum distance that can be travelled in Belgium). We then calculated the mean and median distance travelled for each chosen feature. (see Table 1).

Table 1: Distances travelled and 'mobility features'

Mobility characteristics	N	Mean distance (km)	Median distance (km)
Multiple offenders	18,755	24.4	11.1
Co-offenders	41,590	20.1	7.2
Eastern European offenders	7,078	34.7	22.1
Offenders age 30+	16,496	22.1	8.9
Attractive targets	31,569	20.1	8.6
All	*67,981*	*19.1*	*7.2*

Although correlations between travelled distance and these features were negligible, absolute figures indicated a slightly higher mobility for all features except one. This first step showed that these features did result in higher offender mobility than the average for Belgium, though the difference was only slight for some features. A next step, however, was whether these higher central values of mobility (mean and median) were also accompanied by different distance decay curves.

In order to keep the curves comparable, we divided the distances travelled into bands of 10km. The first band contained all crimes committed at less than 10km from an offender's residence, the second included the crimes committed from 10 to 19km, and so on. As very few offences were committed over 100km from an offender's residence, we only included distances of less than 100km. Because the used branches are not detailed, we overcame some of the interpretation problems that may be caused by working only with municipality coordinates. Moreover, because even such rough distinction shows a lot of variation and corresponding distance decay curves, the measure – despite its roughness – is adequate for the framework of this paper.

Figure 2: Aggregated distance-decay patterns

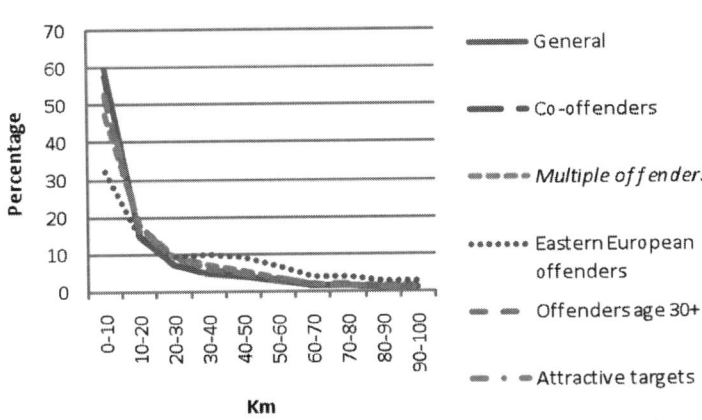

Figure 2 illustrates that the pattern of most offender groups followed a distance decay pattern, regardless of their higher mobility. One pattern deviated from this – Eastern European offenders tended to be less likely to offend within the 10km range from home (32%, compared with percentages of at least 47%, but mostly over 50%, for the other groups). They also committed many more offences over 20km from their residence. Thus, the distance decay pattern for Eastern European offenders appears to be less straightforward than it is for other offenders.

This exercise shows that particular groups of offenders are indeed more mobile, as indicated by previous research, but that higher mobility rarely implies different decay curves. Apart from Eastern European offenders, most offence patterns showed a quite steep level of decay, even with their higher level of mobility. Eastern European offenders show an existing but modest decay pattern, with fewer offences committed within 10km and more at distances over 20km from the residence.

This approach uses offender–offence combinations as the unit of analysis. Although it has also been used by other authors (see for example Hodgson & Costello, 2006; Kleemans, 1996), we admit it is a slightly unusual approach. Most studies use either the offence or the offender as the unit of analysis. If the offender alone is analysed, features that are linked to the offence are excluded, and vice versa. Working with offender–offence combinations allows characteristics of both the offender and the offence to be included in the analysis.

Although the presence of a distance decay pattern is widely accepted within criminological research, it is often observed at the aggregate level and there is an ongoing discussion about whether it can also be observed at the offender level (Rengert, et al., 1999; Smith, et al., 2009; Van Koppen & De Keijser, 1997). Taking into account the observed variations on the aggregated level, we compared these results with an individual measure of decay. It is important to bear in mind that this only measures decay and does not test the near-home hypothesis. Thus, offenders can be quite mobile, still following a decay pattern, or can commit most offences in a limited area but without any particular decay. We could calculate mean skewness, but this again eliminates individual variations.

Co-offending and the choice of attractive targets refer to offence characteristics and not offender features. We therefore defined offenders as co-offenders when they committed more than half of their crimes with other offenders. When the mean attractiveness score of a target area is higher than the overall mean – higher than 100 – we considered offenders to be heading for attractive targets.

Table 2: Individual decay

	Valid	Missing	Mean	Skewness Z			
				> 1.96	> 0 and < 1.96	< 0 and > –1.96	< –1.96
All multiple offenders	765	116	2.10	49.4	24.7	14.1	11.8
Co-offenders	485	70	2.04	48.5	26.6	15.5	9.5
Eastern European offenders	122	6	1.67	38.5	31.1	19.7	10.7
Offenders age 30+	155	10	1.84	45.8	25.8	18.1	10.3
Attractive targets	365	53	1.94	46.0	25.5	16.2	12.3

On the individual level, only a minority of offenders show straightforward distance decay, meaning they have a skewness z-score above 1.96. Depending on which offenders are considered, only 38.5–49.4% show distance decay in the strict sense. For about 25%, we found decay, but could not determine whether this was significant. For another 14–20% we found a distance increase, but again could not determine its significance. Nevertheless, we found that 10% of the offenders showed a significant negative skewness, meaning that they follow a distance increase instead of a distance decay pattern.

For 15.2% of the offenders the skewness could not be determined, because they committed all offences within the same area. For the respective subtypes of offenders, this ranged from 4.9% for Eastern European offenders to 14.5% for offenders heading for attractive targets. This also implies that Eastern European offenders and offenders aged 30 or older have a wider spread of target areas than co-offenders and offenders heading for attractive targets.

5   Discussion

At the aggregate level, we noticed quite a large variation in mean distance travelled, ranging from 19.1km in general to 34.7km for Eastern European offenders. This resulted in differences in distance decay patterns. However, apart from Eastern European offenders, where distance decay deviated slightly more (fewer offences were committed at distances of less than 10km), decay variations appear to be more or less in line with the general distance decay pattern. Thus, at the group level, near-home offending and decay are intertwined to a large extent, as higher mobility leads to slightly less decay. In other words, longer distances travelled result in a distance decay pattern that is slightly stretched to the right. At least, this is what we observed at the aggregate level.

At the individual level, the pattern appears to be less straightforward. By using the skewness z-scores of offenders' decay curves, we calculated the individual level of decay. The result is a relative measure, only focusing on the distribution of decay itself: skewness is independent from mean distance travelled and, therefore, from near-home offending.

Although most offenders tend to commit their crimes near home, this does not always translate into a decay curve at the individual level. About half of the offenders showed a pattern of distance decay at the individual level. Moreover, for every 5 offenders following a clear distance decay pattern, 1 follows a significant distance increase pattern. For Eastern European offenders and offenders systematically heading for attractive targets, this odds ratio becomes 4 to 1.

As z-scores significance is subject to the standard error, we wondered whether individual distance decay variations are partially due to the skewness standard error. The standard error for $z > 0$ is similar to that for $z < 0$ (respective means of 0.56 and 0.57; assumed equality $p<0.05$). For significant and insignificant z-scores (z falling outside or inside the -1.96 to 1.96 span), however, we observe significant differences (respective means of 0.55 and 0.58; $p<0.01$). This shows that z-score significance may indeed be biased by the standard error of the individual distributions.

As the standard error is strongly correlated with the number of crimes committed ($r=0.797$, $p<0.01$), we also controlled for additional bias by taking into account the number of crimes individual offenders committed. We found that offenders with significant distance decay committed more crimes on average than those who followed significant distance increase patterns (respective means of 23.5 and 21.0). Yet these differences were not significant ($p=0.22$). Neither standard error differences nor differences in the average number of committed crimes could falsify our finding that for every 5 offenders showing a distance decay pattern, there is at least 1 who shows a distance increase pattern.

Within our subgroups, co-offenders tend to be most likely to follow distance decay. This is rather unexpected. After all, offending perpetrated in a group setting often cannot be explained by simply the sum of individual rationality (Tillyer & Kennedy, 2008, p. 81). Co-offenders are likely to commit their crimes near the residence of one offender (Bernasco, 2006, p. 147). This would mean that for crimes that are committed near the residence of offender A, we would not expect straightforward decay for offender B. If all co-offended crimes were committed by two offenders, this would mean that at least half of the offenders are likely to show no decay pattern, and as some crimes are committed by more than two offenders, this percentage should be even higher. Before any hasty conclusions are drawn, however, consider the following three points.

First, skewness estimates could not be calculated for 14.4% of the co-offenders. Together with those offenders heading for attractive targets, this is substantially higher than it is for Eastern European offenders (4.9%) and offenders aged over 30 (6.5%). Thus, co-offenders who commit all offences in the same area, which could well be the resident area of one of the co-offenders, are excluded from the analysis.

Second, co-offenders may live near each other, making their decay patterns resemble each other, particularly at our level of analysis. For 53% of the crimes that were committed by two or more offenders, at least two offenders lived in the same area. Thus, for over half of the co-offended crimes, the anchor point of one offender was the same as for at least one other offender. In these cases, the fact that co-offenders would tend to commit their crimes near the anchor point of one participant does not influence the individual decay pattern.

Third, co-offending networks are by no means fixed and not all co-offenders are multiple offenders. In fact, only 23.6% of co-offended crimes were committed by criminals who were classed as multiple offenders. Individual decay patterns were only calculated for multiple offenders. As these offenders are more experienced than others, they may have more discretionary power in the choice of target area, and therefore may choose to offend near their own residence. This behavioural model could apply for 76.4% of the co-offended crimes.

Taking into account these three points, we believe the relatively large proportion of individual decay among co-offenders is not contradictory to previous findings.

Because of their relativity, skewness z-scores by no means contradict the fact that the majority of crimes are committed near home. On the contrary, their main value lies in cutting decay loose from near-home offending. Also, like mean distance, this is only one value. It reveals some of the information that is lost when discussing mean distance, but does not provide as much information as a distance decay curve. Unfortunately, this appears to be the price to be paid for working with comparable and quantifiable values. In addition, it only takes into account information on the crimes, and neglects the environmental backcloth (see Patricia Brantingham & Brantingham, 2004). Intervening opportunities (Elffers, Reynald, Averdijk, Bernasco, & Block, 2008; Stouffer, 1960), awareness space (Patricia Brantingham & Brantingham, 1981), search space (Rengert & Wasilchick, 1985) and incorrect anchor point perception (Wiles & Costello, 2000) are not considered. Yet this is also the case for most journey-to-crime research using mean distance travelled.

In this paper we used Belgian municipalities as a unit of analysis. This related to the nationwide data used, the emphasis on mobile offenders within the broader framework this paper is situated in, and the importance of municipalities from theoretical, methodological and policy perspectives (Hardyns & Pauwels, 2009, pp. 169-171). However, this also implies a loss of information at smaller levels of offender mobility. For a number of offenders, skewness estimates and skewness z-scores could not be computed, because they have committed all offences in the same area. 73.3% of them committed the offences in their home area. A further division of geographic locations could clarify the decay patterns of these offenders as well. Future research could therefore work with exact address locations in order to calculate distances more precisely. This could rule out any possible bias due to the level of analysis.

This paper considered variations in individual decay patterns. It is important, particularly from a theoretical perspective, that we have established that these differences do exist. By providing evidence that supports the conclusion of Smith et al. (2009) that individual differences in distance decay are often lost if analysed only at the aggregate level, we hope to contribute to the discussion about whether distance decay can be applied at the level of the individual offender.

Future steps could contribute to the journey-to-crime research from both a theoretical and a policy perspective. From a theoretical perspective, it is important to explore whether an increase in the distance travelled to commit a crime is accompanied by a directional bias. This would enable the principles of geographic profiling to be either questioned or supported. If decay variations are combined with directional bias, establishing a geographical profile would become difficult for this type of offender. If, however, distance increase curves show limited directional bias, this would mean that even for these offenders, offence locations are mainly spread around the anchor point. In that case, such offenders could be geographically profiled in the traditional way.

From a policy perspective, added value could be found in investigating why offenders decide to travel further, even if they know of and have experience of targets nearby. Is this the result of successful preventive measures close to the residence, forcing offenders to travel more, or is it merely influenced by locations further afield being more attractive or less risky? A next step could therefore be to explore whether decay is a function of target features, offender characteristics and the environmental backcloth.

6   Conclusion

This paper explored two main issues. We were interested in mobile offending and wanted to establish whether high mobility is accompanied by different mobility patterns. We therefore selected some features of mobile offending from literature and found that these did result in higher average distances travelled in our sample. Yet higher mobility does not necessarily lead to different distance decay distributions: slightly fewer offences appear to be committed within 10km of home, but nevertheless a clear distance decay pattern was observed. Only in the case of Eastern European offenders did this turn out not to be the case: the percentage of offences committed by Eastern European offenders near to their homes was found to be considerably lower than for other offenders, and they committed more crimes at distances of 20–40km from their residence than did other offenders. Although some decay was still observed, it was less straightforward for this subgroup.

Establishing distance decay patterns reveals two main issues. On the one hand, it turns out that higher mobility does not always lead to notably different distance decay patterns. Thus, high mobility does not necessarily equal different mobility. On the other hand, even some basic group divisions, such as nationality, can result in deviating distance decay patterns. Even at the aggregate level, taking the universality of the distance decay pattern for granted is questionable, which is confirmed by the skewness z-scores for the 'Eastern European' and the 'aged above 30' subgroups. Variations do indeed exist, and can reveal interesting information if group divisions are well chosen in respect of the crime being studied.

Additionally, we attempted to translate aggregated distance decay patterns into individual distance decay patterns, by calculating individual skewness z-scores. The results were given for each offender type and revealed that fewer than half of the offenders followed a significant distance decay pattern. At the other end of the spectrum we found a distance increase pattern among over 10% of offenders.

Although distance decay can clearly be seen at the aggregated level, at the level of the individual offender it is less apparent. Eastern European offenders differed most from the general pattern in the first analysis. At the offender level, they are less likely to show significant decay, but tend to compensate this with non-significant decay patterns.

The results appear to confirm those of Smith et al. (2009), namely that substantial variation is hidden when only aggregated distance decay patterns are considered. At the aggregated level, the pattern of decay turns out to be a consequence of near-home offending. At the offender level, this is no longer the case. Skewness z-scores indicate that decay is not the simple result of offender mobility. Skewness may therefore function as a measure to estimate the travel patterns of an individual offender. It is important to clarify whether one is measuring distances travelled (i.e. investigating near-home offending), or decay patterns, and to keep them both separated, as the connection they show on the aggregated level cannot easily be seen on the level of the individual offender.

7 Bibliography

Alison, L., Smith, M., & Morgan, K. (2003). Interpreting the accuracy of offender profiles. *Psychology Crime & Law, 9*(2), 185-195.

Barker, M. (2000). The criminal range of small-town burglars. In D. Canter & L. J. Alison (Eds.), *Profiling property crimes* (pp. 57-73). Aldershot: Ashgate.

Beauregard, E., Proulx, J., & Rossmo, K. (2005). Spatial patterns of sex offenders: theoretical, empirical, and practical issues. *Aggression and Violent Behavior, 10*(5), 579-603.

Bernasco, W. (2006). Co-offending and the choice of target areas in burglary. *Journal of Investigative Psychology and Offender Profiling, 3*(3), 139-155.

Bernasco, W., & Block, R. (2009). Where offenders choose to attack: a discrete choice model of robberies in Chicago. *Criminology, 47*(1), 93-130.

Bernasco, W., & Luykx, F. (2002). De ruimtelijke spreiding van woninginbraak: een analyse van Haagse buurten. *Tijdschrift voor Criminologie, 44*(3), 231-246.

Bernasco, W., & Nieuwbeerta, P. (2005). How do residential burglars select target areas? A new approach to the analysis of criminal location choice. *British Journal of Criminology, 45*(3), 296-315.

Besson, J.-L. (2004). *Les cartes du crime*. Paris: PUF.

Brantingham, P., & Brantingham, P. (1981). Notes on the geometry of crime. In P. Brantingham & P. Brantingham (Eds.), *Environmental criminology* (pp. 27-54). Beverly Hills: Sage.

Brantingham, P., & Brantingham, P. (1984). *Patterns in crime.* New York: Macmillan.

Brantingham, P., & Brantingham, P. (2004). Environment, routine and situation: toward a pattern theory of crime. In R. Clarke & M. Felson (Eds.), *Routine Activity and Rational Choice* (pp. 259-294). New Brunswick: Transaction Publishers.

Bruinsma, G. (2007). Urbanization and urban crime: Dutch geographical and environmental research. In M. Tonry (Ed.), *Crime and Justice in the Netherlands* (Vol. 35, pp. 453-502). Chicago: Chicago University Press.

Canter, D., & Hammond, L. (2006). A comparison of the efficacy of different decay functions in geographical profiling for a sample of US serial killers. *Journal of Investigative Psychology and Offender Profiling, 3*(2), 91-103.

Canter, D., & Hammond, L. (2007). Prioritizing burglars: comparing the effectiveness of geographical profiling methods. *Police Practice and Research, 8*(4), 371-384.

Canter, D., & Larkin, P. (1993). The environmental range of serial rapists. *Journal of Environmental Psychology, 13*(1), 63-69.

Canter, D., & Youngs, D. (2008a). Geographical offender profiling: applications and opportunities. In D. Canter & D. Youngs (Eds.), *Applications of geographical offender profiling* (pp. 3-24). Aldershot: Ashgate.

Canter, D., & Youngs, D. (2008b). Geographical offender profiling: origins and principles. In D. Canter & D. Youngs (Eds.), *Principles of geographical offender profiling* (pp. 1-18). Aldershot: Ashgate.

Capone, D., & Nichols, W. (1976). Urban structure and criminal mobility. *American Behavioral Scientist, 20*(2), 199-213.

Cohen, L., & Felson, M. (1979). Social change and crime rate trends: a routine activity approach. *American Sociological Review, 44*(4), 588-608.

Cornish, D., & Clarke, R. (Eds.). (1986). *The reasoning criminal : rational choice perspectives on offending.* New York: Springer.

De Poot, C., Luykx, F., Elffers, H., & Dudink, C. (2005). Hier wonen en daarplegen? Sociale grenzen en locatiekeuze. *Tijdschrift voor Criminologie, 47*(3), 255-268.

Deakin, J., Smithson, H., Spencer, J., & Medina-Ariza, J. (2007). Taxing on the streets: understanding the methods and process of street robbery. *Crime Prevention and Community Safety, 9*(1), 52-67.

Eck, J., & Weisburd, D. (1995). Crime places in crime theory. In J. Eck & D. Weisburd (Eds.), *Crime and place* (pp. 1-33). Monsey: Criminal Justice Press.

Edwards, M., & Grace, R. (2006). Analysing the offence locations and residential base of serial arsonists in New Zealand. *Australian Psychologist, 41*(3), 219-226.

Elffers, H. (2003). Veelplegers of vaakplegers? *Tijdschrift voor Criminologie, 45*(2), 119-126.

Elffers, H. (2004). Decision models underlying the journey to crime. In G. Bruinsma, H. Elffers & J. De Keijser (Eds.), *Punishment, places and perpetrators: developments in criminology and criminal justice research* (pp. 182-195). Cullompton: Willan Publishing.

Elffers, H., Reynald, D., Averdijk, M., Bernasco, W., & Block, R. (2008). Modelling crime flow between neighbourhoods in terms of distance and of intervening opportunities. *Crime Prevention and Community Safety, 10*(2), 85-96.

Ferwerda, H., Kleemans, E., Korf, D., & Van der Laan, P. (2003). Veelplegers. *Tijdschrift voor Criminologie, 45*(2), 110-118.

Ferwerda, H., Versteegh, P., & Beke, B. (1995). De harde kern van jeugdige criminelen. *Tijdschrift voor Criminologie, 37*(2), 138-152.

Fink, G. (1969). Einsbruchstatorte vornehmlich an einfallstrassen? *Kriminalistik, 23*, 358-360.

Gabor, T., & Gottheil, E. (1984). Offender Characteristics and Spatial Mobility – an Empirical-Study and Some Policy Implications. *Canadian Journal of Criminology-Revue Canadienne De Criminologie, 26*(3), 267-281.

Goodwill, A., & Alison, L. (2006). The development of a filter model for prioritizing suspects in burglary offences. *Psychology Crime & Law, 12*(4), 395-416.

Hakim, S., Rengert, G., & Shachmurove, Y. (2000). *Knowing your odds: home burglary and the odds ratio*: Carress working paper.

Hardyns, W., & Pauwels, L. (2009). The geography of social cohesion and crime at the municipality level. In M. Cools, S. De Kimpe, B. De Ruyver, M. Easton, L. Pauwels, P. Ponsaers, G. Vande Walle, T. Vander Beken, F. Vander Laenen & G. Vermeulen (Eds.), *Contemporary issues in the empirical study of crime* (pp. 157-178). Antwerp: Maklu.

Hodgson, B., & Costello, A. (2006). The prognostic significance of burglary in company. *European Journal of Criminology, 3*(1), 115-119.

Johnson, S., & Bowers, K. (2004). The burglary as clue to the future: The beginnings of prospective hot-spotting. *European Journal of Criminology, 1*(2), 237-255.

Kleemans, E. (1996). *Strategische misdaadanalyse en stedelijke criminaliteit*. Enschede: Universiteit Twente.

Kocsis, R., Cooksey, R., Irwin, H., & Allen, G. (2002). A further assessment of 'circle theory' for geographic psychological profiling. *Australian and New Zealand Journal of Criminology, 35*(1), 43-62.

Kocsis, R., & Harvey, I. (1997). An analysis of spatial patterns in serial rape, arson, burglary: the utility of the circle theory of environmental range for psychological profiling. *Psychiatry, Psychology and Law, 4*(2), 195-206.

Laukkanen, M., Santtila, P., Jern, P., & Sandnabba, K. (2008). Predicting offender home location in urban burglary series. *Forensic Science International, 176*(2-3), 224-235.

Maguire, M. (1982). *Burglary in a dwelling: the offence, the offender and the victim.* London: Heinemann.

Mawby, R. (2001). *Burglary.* Cullompton: Willan Publishing.

McIver, J. (1981). Criminal mobility: a review of empirical studies. In S. Hakim & G. Rengert (Eds.), *Crime spillover* (pp. 20-47). Beverly Hills: Sage.

Mizutani, F., & Jewell, P. (1998). Home-range and movements of leopards (Panthera pardus) on a livestock ranch in Kenya. *Journal of Zoology, 244*(2), 269-286.

Morselli, C., & Royer, M.-N. (2008). Criminal mobility and criminal achievement. *Journal of Research in Crime and Delinquency, 45*(1), 4-21.

Oberwittler, D., & Wikström, P.-O. (2009). Why small is better: advancing the study of the role of behavior contexts in crime statistics. In D. Weisburd, W. Bernasco & G. Bruinsma (Eds.), *Putting crime in its place: units of analysis in geographic criminology* (pp. 35-59). New York: Springer.

Paulsen, D. (2006). Connecting the dots: assessing the accuracy of geographic profiling software. *Policing-an International Journal of Police Strategies & Management, 29*(2), 306-334.

Pettiway, L. (1982). Mobility of robbery and burglary offenders: ghetto and nonghetto spaces. *Urban Affairs Quarterly, 18*(2), 255-270.

Phillips, P. (1980). Characteristics and typology of the journey to crime. In D. Georges-Abeyie & K. Harries (Eds.), *Crime: a spatial perspective* (pp. 167-180). New York: Columbia University Press.

Polisenska, A. V. (2008). A qualitative approach to the criminal mobility of burglars: questioning the 'near home' hypothesis. *Crime Patterns and Analysis, 1*(1), 47-60.

Ponsaers, P. (2004). Rondtrekkende dadergroepen: rationele Nederlandse criminologen en irrationele criminelen in Vlaanderen. *Tijdschrift voor Criminologie, Jubileumuitgave,* 15-23.

Rengert, G., Piquero, A., & Jones, P. (1999). Distance decay reexamined. *Criminology, 37*(2), 427-446.

Rengert, G., & Wasilchick, J. (1985). *Suburban burglary: a time and a place for everything.* Springfield: Thomas.

Reppetto, T. (1974). *Residential crime.* Cambridge: Ballinger.

Reynald, D., Averdijk, M., Elffers, H., & Bernasco, W. (2008). Do social barriers affect urban crime trips? The effect of ethnic and economic neighbourhood compositions on the flow of crime in The Hague, The Netherlands. *Built Environment, 34*(1), 21-31.

Rhodes, W., & Conly, C. (1981). Crime and mobility: an empirical study. In P. Brantingham & P. Brantingham (Eds.), *Environmental Criminology* (pp. 167-188). Beverly Hills: Sage.

Rossmo, K. (1995). Place, space and police Investigations: hunting serial violent criminals. In J. Eck & D. Weisburd (Eds.), *Crime and place* (pp. 217-235). Monsey: Criminal Justice Press.

Rossmo, K. (2000). *Geographic profiling*. Boca Raton: CRC Press.

Rossmo, K., Thurman, Q., Jamieson, J. D., & Egan, K. (2008). Geographic patterns and profiling of illegal crossings of the southern U.S. border. *Security Journal, 21*(1-2), 29-57.

Santtila, P., Laukkanen, M., Zappala, A., & Bosco, D. (2008). Distance travelled and offence characteristics in homicide, rape, and robbery against business. *Legal and Criminological Psychology, 13*(2), 345-356.

Sherman, L. (1995). Hot spots of crime and criminal careers of places. In J. Eck & D. Weisburd (Eds.), *Crime and Place* (35-52 ed., pp. 35-52). Monsey: Criminal Justice Press.

Smith, W., Bond, J., & Townsley, M. (2009). Determining how journeys-to-crime vary: measuring inter- and intra-offender crime trip distributions. In D. Weisburd, W. Bernasco & G. Bruinsma (Eds.), *Putting crime in its place: units of analysis in geographic criminology* (pp. 217-236). New York: Springer.

Snook, B. (2004). Individual differences in distance travelled by serial burglars. *Journal of Investigative Psychology and Offender Profiling, 1*(1), 53-66.

Snook, B., Canter, D., & Bennell, C. (2002). Predicting the home location of serial offenders: a preliminary comparison of the accuracy of human judges with a geographic profiling system. *Behavioral Sciences and the Law, 20*(1-2), 109-118.

Snook, B., Taylor, P., & Bennell, C. (2004). Geographic profiling: the fast, frugal and accurate way. *Applied Cognitive Psychology, 18*(1), 105-121.

Snook, B., Zito, M., Bennell, C., & Taylor, P. (2005). On the complexity and accuracy of geographic profiling strategies. *Journal of Quantitative Criminology, 21*(1), 1-26.

Snow, J. (2008). Excerpt from: on the mode of communication of cholera (originally from 1855). In D. Canter & D. Youngs (Eds.), *Principles of geographical offender profiling* (pp. 33-39). Aldershot: Ashgate.

Stouffer, S. (1960). Intervening opportunities and competing migrants. *Journal of Regional Science, 2*(1), 1-26.

Swartz, C. (1999). The spatial analysis of crime: what social scientists have learned. In V. Goldsmith, P. McGuire, J. Mollenkopf & T. Ross (Eds.), *Analyzing crime patterns: frontiers of practice* (pp. 33-46). Thousand Oaks: Sage.

Tillyer, M., & Kennedy, D. (2008). Locating focused deterrence approaches within a situational crime prevention framework. *Crime Prevention and Community Safety, 10*(2), 75-84.

Tobler, W. (1970). A computer movie simulating urban growth in the Detroit Region. *Economic Geography, 46*(2), 234-240.

Tremblay, P. (2004). Searching for suitable co-offenders. In R. Clarke & M. Felson (Eds.), *Routine Activity and Rational Choice* (pp. 17-36). New Brunswick: Transaction Publishers.

Turner, S. (1969). Delinquency and distance. In T. Sellin & M. Wolfgang (Eds.), *Delinquency: Selected Studies* (pp. 11-26). New York: Wiley.

Van Daele, S. (2008). Organised property crimes in Belgium: the case of the 'itinerant crime groups'. *Global Crime, 9*(3), 241-247.

Van Daele, S., Vander Beken, T., & De Ruyver, B. (2008). Rondtrekkende dadergroepen: een empirische toets. *Panopticon, 29*(4), 25-39.

Van der Kemp, J., & Van Koppen, P. (2007). Finetuning geographical profiling. In R. Kocsis (Ed.), *Criminal profiling: international perspectives in theory, practice, and research* (pp. 347-364). Totowa: Humana.

Van Koppen, P., & De Keijser, J. (1997). Desisting distance decay: On the aggregation of individual crime trips. *Criminology, 35*(3), 505-515.

White, C. (1932). The relations of felonies to environmental factors in Indianapolis. *Social Forces, 10*(4), 498-509.

Wiles, P., & Costello, A. (2000). *The 'road to nowhere': the evidence for travelling criminals.* London: Home Office.

Zipf, G. (1949). *Human behavior and the principle of least-effort.* New York: Hafner.

Exploring the role of exposure to offending and deviant lifestyles in explaining offending, victimisation and the strength of the association between offending and victimisation

Gerwinde Vynckier[1]
Lieven Pauwels

1   Introduction

During the past decades, an increase in victimisation studies can be observed. Research predominantly reveals that not all segments of population share the same risk of becoming a victim. The risk of becoming a victim differs by background characteristics, such as gender, ethnic background, status and age (Hindelang et al., 1978; Garofalo, Siegel & Laub, 1987; Davies et al., 2003; van Noije & Wittebrood, 2007; Walklate, 2007a). Although studies have been fostering, there is surprisingly little known about the causes of individual level variation in the risk of becoming a victim. Victimisation seems to vary by background variables that are also related to self-reported offending (Pauwels & Pleysier, 2008). What is known about the prevalence of victimisation among minors, is mostly derived from studies in which the extent of victimisation is only a restricted part (for example Smith et al., 2001; MORI, 2004; Armstrong et al., 2005; Vettenburg et al., 2007). Surveys on victimisation among young adolescents reflect surveys which measure victimisation among adults (for example Aye Maung, 1995; Wood, 2005). In contrast to e.g. the Netherlands, where there has been some independent studies on the prevalence of victimisation of adolescents at school (Mooij, 1994 & 2001; Debarbieux & Blaya, 2001; Lawrence, 2007), in Belgium, levels of victimisation are not studied as extensively. However, also in Belgium we have a partial picture of victimisation levels of adolescents: the federal victim survey, which is known as the 'Security Monitor', draws a picture of personal and household victimisation among Belgian citizens aged 15 years or older[2]. Besides that, the 'Jop-monitor' gives an idea of victimisation among 14-25 year olds[3]. Nevertheless, information on levels of victimisation among young adolescents (aged 12-13 years old) is seriously lacking. Based on data drawn from a school survey in Sint-Niklaas among 1554 pupils of the first grade of the secondary compulsory educational system in Belgium, this study aims at partially filling this gap. As the school context is very important for adolescents

---

1   Gerwinde Vynckier holds a Ph. D. fellowship of the Research Foundation – Flanders (FWO)
2   The Security Monitor probes the needs of the population concerning security and police and gives some statistical information about security and victimisation. See for example Pauwels & Pleysier, 2008.
3   The first results of this monitor were based on a questionnaire by a representative group of 2503 youngsters from 14 till 25 years old which took place in 2005-2006 and focused on the conditions, convictions and conducts of young people. See Vettenburg, Elchardus & Walgrave, 2007.

from a contextual perspective (Bronfenbrenner, 1979), this study emphasises personal victimisation at school. As victimisation surveys and self reports are helpful in learning about the experiences of students in and around schools (Gottfredson, 2001), we present bivariate contingency tables and describe levels of victimisation at school by levels of self-reported offending and further analyse these bivariate relationships by using three-way tables, i.e. by studying these bivariate relationships by different categories of lifestyles. In the explanatory part of this contribution, the applicability of an integrated self-control/lifestyle-exposure model of victimisation is carefully assessed. Often, it has been assumed that there exists an overlap between victimisation and offending, and that theories of victimisation therefore partially "mirror" or reflect theories of offending. The extent in which this really is the case, is assessed in the explanatory part of this study. We try to achieve this goal by answering the following research questions: 1) How strong are the effects of concepts derived from theories of victimisation and offending, such as exposure to offending at school, propensity to offend and lifestyle risk, on victimisation and offending? 2) How strong is the bivariate and partial correlation between victimisation and offending? 3) How strong is the independent effect of offending on victimisation and the independent effect of victimisation on offending, controlling for exposure to offending at school, propensity to offend and lifestyle risk? By answering these questions the results contribute to knowledge on victimisation and offending of young adolescents and the empirical assessment of theories of victimisation as a mirror of theories of offending. Theories on lifestyle (Hindelang, Gottfredson & Garofalo, 1978) and routine activity (Cohen & Felson, 1979) were originally developed to explain victimisation. In short, in this study we question to what extent such theories can really explain individual differences in victimisation of young adolescents in a similar way as they can explain individual differences in offending while further fuelling the discussion on the study of association between victimisation and offending.

2   Theoretical backdrop

2.1   The role of lifestyle / routines in explaining individual differences in victimisation and offending

The integrated lifestyle exposure theory (Hindelang et al., 1978) is probably the best known classic theory that explicitly explains individual differences in personal victimisation. This theory has been developed to explain why the distribution of victimisation is unequal and varies by background characteristics. According to the theory, individuals are differentially involved in leisure activities, vocational activities and professional activities due to their different backgrounds. Such background characteristics that refer to age, race, income and marital status bring about differences in role expectations (what to expect in life) and constraints (what can be achieved given a certain structural reference). Individuals are restricted by their backgrounds in realising their goals. Individuals react and thus adapt differentially to the structural conditions they are part of. Such adaptations bring about observable differences in lifestyles. In this classic model lifestyle is seen as an indicator of a person's daily routines. The theory focused in its original version on vocational, professional and leisure related activities. Such activities set the stage for exposure to offending in at-risk situations. The routine activity theory is closely related to the lifestyle model of victimisation and states that offenders and targets meet in absence of capable guardians. Differences in lifestyles cause individuals to be differentially exposed to the environmental settings they encounter. Furthermore, differences in lifestyles lead to different associations with significant others. The classic lifestyle exposure theory can be visualized as follows:

Figure 1 – Visualisation of the classic lifestyle exposure theory

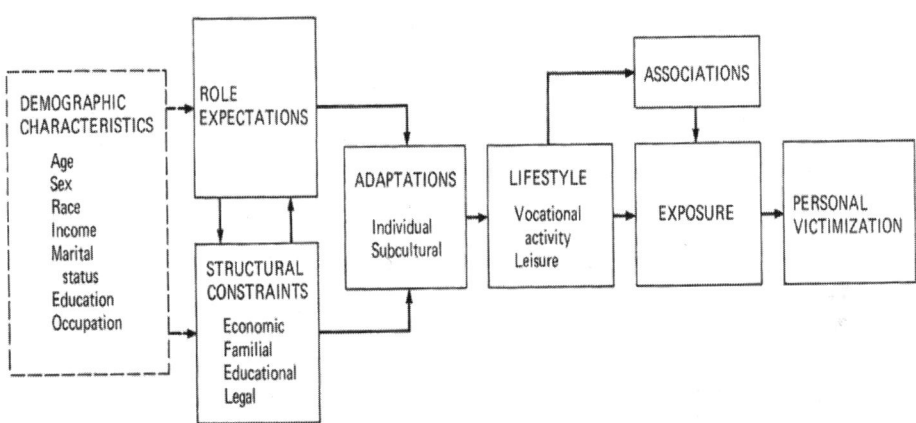

© Hindelang et al., 1978

The lifestyle model has some strong and weak points. The fact that exposure to environmental settings is taken as one situational concept that contributes to our understanding of victimisation is considered to be a strong point. The fact that the model is rather vague in explaining how individual differences in lifestyles are brought about is a weak point. The classic model overemphasized structural and cultural differences, thereby ignoring the role of individual characteristics, such as morality and self-control.

Recent developments within theories of offending increasingly point to the importance of crime inducing settings. In doing so, they very similarly point to the role of lifestyles and daily routines (Wikström & Butterworth, 2006). Previous studies have mostly studied lifestyles in a restricted meaning, referring exclusively to vocational and leisure activities (e.g. Nofzinger & Kurtz, 2005; Osgood & Anderson, 2004; Vazsonyi et al., 2002). Some contemporary scholars use the concept of lifestyle risk broader and consider also associations as a dimension of lifestyle (Wikström & Butterworth, 2006; Svensson & Pauwels, 2008). One contemporary theory that explicitly emphasizes the importance of lifestyle risk or situational risk, within a framework of structural relationships between causal variables, is the Situational Action Theory (SAT) developed by Wikström (2005). The SAT stresses the importance of the simultaneous study of individual characteristics and situational characteristics in explaining offending. More specifically, the theory stresses the interaction between individuals and settings to explain why some adolescents perceive offending as an action alternative, ultimately causing them to choose to offend. Thus, different contemporary versions of lifestyle theories exist, and they all have in common that they consider lifestyles as an indicator of exposure, while actual exposure to risky settings is generally not measured.

2.2   Victimisation, offending and exposure to offending at school

One factor that influences the chances of becoming a victim is the extent to which potential victims appoint themselves to potential offenders. Ecological proximity to violence is an important determinant of victimisation (Hindelang et al., 1978; Cohen et al., 1981; Miethe & Meier, 1994; Sampson & Lauritsen, 1990). Young adolescents spend much time at school. Garofalo, Siegel & Laub (1987) wrote about the school as *'probably the most predominant location for the routine daily activities of adolescents, at least for 9 months of the year'* (Garofalo, Siegel & Laub, 1987; 329) and Felson described the school as *'the heart that pumps adolescents about in society at 3:00 P.M. and then pulls them back the next morning. On Friday Evening, it pumps them out for the week-end; it then draws them back on Monday morning'* (Felson, 1998: 109). The school context can itself be a place where potential victims are appointed to potential offenders. Schools concentrate large numbers of teenagers in one place for long periods of time (Gottfredson, 2001). So the school is not only the major institutionalized sphere for adolescent interaction and social recognition but *'as an arena of intense social interaction'* (Bjarnason et al., 1999: 108), the school can also be a breeding ground for delinquent and subcultural peer groups and, consequently, for victimisation and offending (Fiqueira-Mcdonough, 1986; Pauwels, 2009). Piquero and Hickman (2003) found a positive and significant relation between the number of hours spent on campus and general victimisation. Similarly, rates of delinquency and victimisation vary between schools and other social units that young people belong to (Smith et al., 2001). Besides that, the dependency of others makes that adolescents have less choice than adults over their environments and over whom they associate with. At school, young adolescents may be exposed to motivated offenders and (potential) delinquent peers but because of their limited autonomy they cannot simply change schools or quit (Finkelhor, 1995; Hashima & Finkelhor, 1999). So the question of the effect of differential school composition (i.e. school social structure) with regard to the frequency of offending is of equally great importance as well as the question of the effect of differential school composition with regard to the frequency of becoming a victim as going to school has consequences for both offending and victimisation (Felson, 1998). Schreck, Miller, and Gibson (2002) found that having potential offenders at school, such as students involved in gangs, and having peers who drink alcohol, smoke, or use drugs increased the risk of victimisation within the school setting.

2.3   Low self-control, offending and victimisation

Gottfredson and Hirschi (1990) pointed to the importance of self-control as a key mechanism in understanding offending and victimisation. In their well-known book "A general Theory of Crime" the authors assessed evidence that linked offending and victimisation to low self-control. The importance of low self-control is further developed in Wikströms aforementioned situational action theory: both morality and low self-control determine whether individuals tend to see crime as an alternative and choose that option, but low self control enhances the chance that offending is chosen as an option, especially for people with low morality.[4] A number of scholars have found empirical evidence that low self-control is related to offending (e.g. Gibbs et al. 2003; Grasmick et al. 1993; Pratt & Cullen 2000; Turner & Piquero, 2002). In contrast to the large body of evidence that low self-control is associated with higher levels of offending, fewer studies have assessed victimisation as being one of these negative consequences. Schreck (1999) extended self-control theory through integration with a lifestyle model to explain why individuals that have low levels of self-control are at increased risk for developing a deviant lifestyle and thus for experiencing higher levels of victimisation. Additionally, other scholars have suggested to extend the lifestyle-exposure model and have pointed to the fact that self-control theory can easily be integrated with lifestyle theory in explaining (violent) victimisation (Nofziger, 2007; Childs & Gibson, 2009). Self-control or propensity to offend seems to be one key mechanism in explaining individual differences in lifestyle risk that operates similarly by gender and ethnic background (Pauwels & Svensson, 2009)[5].

3   Modelling the relationship between victimisation and offending

One can find different explanations for the association between victimisation and offending. The idea that individuals mostly interact with people who share similar characteristics, is the most typical explanation for the relationship between offending and victimisation: the relation between victimisation and offending can primarily be explained by the fact that offenders and victims have a common lifestyle; the risk of victimisation is in direct proportion with the amount of the characteristics they share with perpetrators (Hindelang et al. 1978). This explanation is based on the assumption that some daily routines can increase the chance to offend as well as the chance of becoming a victim, which is called the '*principle of homogamy*' by Hindelang et al. (1978). Thus, in that case, controlling for common characteristics, such as common demographic background variables, deviant lifestyles and low self-control would make the relationship between offending and victimisation spurious (Wittebrood & Nieuwbeerta, 1997 & 1999).

A second explanation for the association between offending and victimisation is that offending actively contributes to the chances of becoming a victim. In that case, offending is seen as one causal mechanism in explaining victimisation '*because of the motives, vulnerability, or culpability of people involved in these activities*' (Jensen & Brownfield, 1986: 87). Delinquent lifestyles expose offenders to rough environments and risky situations while introducing them to other (potential) offenders, which in turn increases their chance of becoming a victim (Lauritsen et al., 1991; Wittebrood & Nieuwbeerta, 1997 & 1999; Bjarnason et al., 1999; Shaffer & Ruback, 2002).[6] Smith et

---

4   While this interaction effect has been found elsewhere (Svensson and Pauwels, 2008), it will not be considered in the subsequent analyses. The explanatory part of the present study is restricted to evaluating main effects of concepts derived from an integrated model, including explanatory variables that are met in both victimisation and offending studies.

5   In the present study, we use the concept of propensity to offend rather than self control. This is legitimate because of the strong correlation between low self control and low levels of morality (0.6, p < 0.001).

6   Van Noije and Wittebrood (2007) point out that offenders, who do spend time more regularly with other offenders than non-delinquents, are in fact, more at risk of becoming victimized.

al. (2001) speak about incidents of delinquency and victimisation being *'the outcome of an interactive process, in which both the eventual victim and the eventual delinquent play a part'* (Smith et al., 2001: 76). The idea that lifestyles are causally related to victimisation is reflected in sayings such as "he who lives by the sword shall die by the sword".

Most scholars seem to consider offending as one causal mechanism that explains victimisation, while the relationship between both characteristics can be non-recursive, i.e., there may exist reciprocal relationships or feedback loops between offending and victimisation. The association between victimisation and offending can also be explained by the fact that repeated trauma of violence-exposed children can lead to processes of anger potentially causing them to participate in high-risk behaviours or associate with offenders (Bailey & Whittle, 2004). Offenders may become the target of a victim's revenge, when seeking compensation for previous suffering, or victimisation can in its own turn lead to offending when the victim's feeling of being a victim of injustice, may induce vengeance. Despite all these different interpretations we are restricted in this empirical test by one theoretical model. This is discussed in the next paragraph.

4   The present study

The present study describes and explains (repeated) victimisation among young adolescents. The present study presents cross-tabs of (repeated) victimisation by (repeated) offending and further describes this pattern by different types of lifestyles (deviant or not deviant). The study then turns to the main questions set out in this study. The first explanatory question is how well can the integrated self-control/lifestyle-exposure model explain individual differences in victimisation and offending? To answer this question several regression models are presented and evaluated. The second explanatory question is how strong is the association between victimisation and offending, when controlling for common causes that have been identified in the literature? To answer this question some partial correlation analyses and regression models are discussed. It is hypothesised that exposure to offending at school is associated with higher levels of victimisation and offending, independent of the composition of schools. Further it is assumed that demographic background characteristics are rather indirectly related to offending and victimisation, and that *propensity to offend* and *deviant lifestyles* are the main mechanisms that explain the observed relationship between exposure to offending, background characteristics, offending and victimisation. The present study builds upon previous studies by testing an integrated theoretical model of exposure to offending at school, propensity to offend and deviant lifestyles on an urban sample of young adolescents. Unlike previous studies (e.g. Reep and Oudhof, 2009), we were able to use direct measures of exposure to offending at school, propensity to offend and deviant lifestyles and conduct the analyses from a multilevel framework. We explicitly hypothesize that the contextual effect of exposure to offending at school will be by and large moderated by propensity to offend and lifestyle risk, two key constructs in both the lifestyle exposure theory and the contemporary situational action theory. Ultimately, we critically evaluate the relation between victimisation at school and self-reported offending by simultaneously studying the direct effect of victimisation on offending, controlling for, propensity to offend and lifestyle risk and the direct effect of offending on victimisation at school, controlling again for those two background characteristics, propensity to offend and lifestyle risk.

Figure 2 – the integrated self-control/lifestyle-exposure model

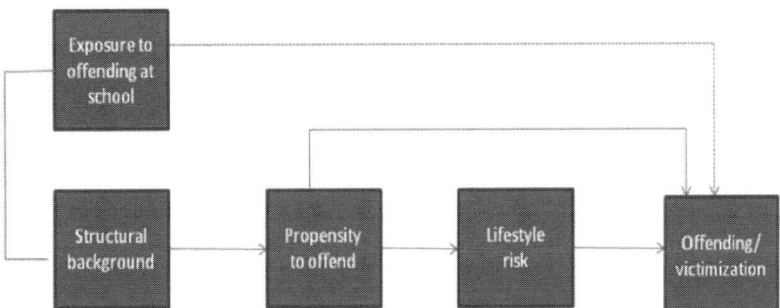

5   Data

The Sint-Niklaas school survey is a cross-sectional school survey using 1,554 young adolescents (effective sample size) in 10 schools in the first grade of the secondary compulsory education schooling system (out of a total of 12 schools). This population is on average 13 years old when entering the first grade and 14 years old when leaving the first grade. Sint-Niklaas belongs to the group of 13 urbanized municipalities of Flanders and has approximately 70,000 inhabitants. The survey took place in the academic year 2007-2008. All schools were addressed in writing to ask for participation in the survey. The original goal of the Sint-Niklaas survey was twofold: on the one hand, estimating personal levels of victimisation at school and comparing these levels by school type (See De Bruyn, 2008) while on the other, retesting some hypotheses derived from Wikströms situational action theory (Pauwels, 2007). The survey was conducted by De Bruyn under the supervision of one of the co-authors of this contribution. Due to her professional status as a school teacher, De Bruyn realized a very satisfying response rate. A detailed evaluation of the unit response rate in the Sint-Niklaas survey is rather difficult, but we could make an estimation using information on the number of students enrolled at the time of the survey. The population in the participating schools consisted of 2074 students during the academic year 2007-2008. 1,554 students enrolled in these schools have participated in the survey. Therefore the unit response rate is estimated 74.9 percent. Despite the fact that some selection can be expected to be present, this is a rather satisfying number. The overall response rate of students attending school in Sint-Niklaas (in the first grade of secondary compulsory educational system) is 50.69%[7]. The survey was conducted using a questionnaire for self-administration that could be placed into an envelope. Confidentiality was guaranteed.

---

7   The unit response rate is the number of adolescents that participated in the survey, divided by the amount of pupils that where participating.
     The overall response rate is the number of adolescents that participated in the survey, divided by all adolescent students attending school in Sint-Niklaas in the first grade.

5.1 Variables

5.1.1 Dependent variables

Two key dependent variables are used in the current study. *Self-reported offending* is measured by summarizing positive responses into a number of thirteen dichotomous offence-items (ever-prevalence of specific items). The offence-items used in the questionnaire are: damaging public goods around school, graffiti around school, verbally threatening somebody around school, verbally threatening somebody at school, participating in a fight at school, participating in a fight around school, hitting somebody around school, hitting somebody at school, forcing somebody to give money or goods at school, forcing somebody to give money or goods around school, using a weapon to take something from somebody at school, using a weapon to take something from somebody around school and trying to get in somewhere to steal something. *Self-reported victimisation at school* is measured by summarizing the number of different offences one became a victim of. The offences that were surveyed were blackmailing, being put under pressure, bullying by other pupils in the class room, bullying by other students, being threatened with a weapon, vandalism on a bicycle, bicycle theft, theft in general and physical violence.

5.1.2 Independent variables

Lifestyle risk is measured as a combined index that measures how often adolescents hang around street corners or parks, how many delinquent peers they have and how frequent they get drunk at the weekend. The measure is a combined risk score summing the risk (+1), balanced (0) and protective (-1) scores of each dimension (see Wikström & Loeber, 2000 for details). The definition and measurement of lifestyle risk used in the present study are based on a previous study by Wikström and Butterworth (2006). Lifestyle risk is thus thought of as a multidimensional phenomenon, indicating where adolescents spend their leisure time (hanging around in unstructured environments such as street corners), what kind of risky activities they engage in (consuming alcohol) and whom they are spending their leisure time with (delinquent peers)[8]. Each of these aspects can from a theoretical point of view be studied as independent characteristics, but because they are correlated dimensions, Pauwels (2007) argued that the multidimensional approach is of interest if the researcher wants to avoid that adolescents are classified as "high risk" too early. *Propensity to offend* is an additive index of two previously validated sub scales, namely a delinquency tolerance scale and the low self-control scale (Pauwels, 2007). Delinquency tolerance and low self-control are strongly related to each other (r=.58 p <0.001) and are considered to be two important dimensions of one's propensity to continuously see the commitment of acts of crime as a behavioural alternative (Wikström & Butterworth, 2006; Pauwels & Svensson, 2009). The delinquency tolerance scale is an additive index and Alpha is 0.802. A high

---

8   Although arguments to study the different dimensions of lifestyle risk separately are valid, we chose to make one scale because these three dimensions are quite strongly correlated and because we argue that one can only speak of lifestyle risk if one is at risk at the different dimensions that are a part of one's lifstyle. Thus, in the present study we are more interested in the overall risk, rather than to study each risk dimension separately.

value on this measure indicates a high level of delinquency tolerance. Low self-control, on the other hand, is an additive index mainly based on the items used and developed by Grasmick et al. (1993). This scale measures whether an individual has the capability to resist temptation and provocation. High values on this measure indicate low levels of self-control. Seven items are included in the present study. Alpha is 0.817. The overall Alpha value is 0.863. *School level exposure* to offending is measured by aggregating the individual level life style risk measure at the school level and can be seen as a valid indicator of exposure to offending. The questions used to create the scales can be found in the appendix.

Four other variables were used as statistical controls.[9] These variables are demographic background characteristics that are known correlates of both offending and victimisation. *Gender* refers to the biological sex of the respondents rather than the concept of "gender", as used in gender studies. *Family structure* refers to single-parent households (coded 1) versus multi-parent households (coded 0). *Immigrant background* is measured by dividing both samples into subgroups of adolescents whose both parents are born in Belgium (coded 0) versus those adolescents who have at least one parent with an immigrant background (coded 1). *Disadvantage* is measured by combining specific answers to two survey questions: one question that asks whether the family owns a car (no=rough indicator of disadvantage) and two, whether the respondent sometimes hears at home that there are insufficient financial resources to supply in daily living (yes=rough indicator of disadvantage) (see Rovers, 1997 for a similar measure). *To be kept down a class* is a dichotomy that measures if a pupil had to redo at least one school year (coded 1 if this was the case and 0 if this was not the case). The descriptive statistics for these variables are displayed in table 1.

Table 1: Descriptives

Variables	Mean	SD	Min	Max
Dependent				
Victimisation	2.15	2.05	0.00	10.00
Offending	1.48	1.92	0.00	13.00
Independent				
Gender	0.47	0.50	0.00	1.00
Immigrant background	0.19	0.39	0.00	1.00
Family structure	0.25	0.43	0.00	1.00
Disadvantage	0.04	0.19	0.00	1.00
To be kept down a class	0.25	0.43	0.00	1.00
Propensity to offend	28.10	9.25	11.00	55.00
Lifestyle risk	-1.23	1.50	-3.00	3.00
Exposure to offending	1.53	0.54	0.81	2.47

5.2   Analysis plan

Herein we start by presenting crosstabs that show frequencies of victimisation in the general sample and split up by involvement in offending. We then reanalyze the same analysis by level of lifestyle risk. Furthermore we examine whether exposure to offend-

---

9   We acknowledge that these variables are attributes and not causes of offending and victimisation, but from a methodological point of view it is important to control for confounders in survey-based studies.

ing, propensity to offend and lifestyle risk are related to both offending and victimisation. As the respondents are grouped in 10 schools and as exposure to offenders is measured at the school level, multilevel modelling is used.[10] Multilevel modelling is concerned with detecting contextual influences of setting characteristics (or higher level variables) on behavioural and attitudinal outcomes at lower levels and with distinguishing true contextual effects from compositional effects, or consequences of segregation or selection. For example when students with similar motivations or attitudes are grouped in highly selective schools or colleges, this grouping effect can have several consequences. In one case we usually refer to "causal influence" of variables at a higher level on outcome variables measured at the lowest level, independent of the composition of the higher level units. The variables at the lower level should be included to control for confounding effects of compositional effects. The effect of a variable at the higher level, independent of the composition, is called the true contextual effect. In school studies the contextual effect represents the situation where school characteristics have a direct effect and can be equated with 'genuine ecological effects'. The explanatory analyses are conducted as follows. Firstly, block wise random intercept regression models are used to test the independent effects of (1) exposure to offending, (2) demographic background characteristics, (3) propensity to offend, and (4) lifestyle risk on both victimisation and offending. Secondly, bivariate and partial correlation analysis is applied to evaluate the strength of the association between offending and victimisation. Thirdly, multiple random intercept regression models are used to evaluate the independent effect of offending on victimisation and the independent effect of victimisation on offending, controlling for the other variables in the model.

6   Results

6.1   Victimisation by levels of offending

In this paragraph we describe levels of victimisation in the general sample and by levels of offending. To describe the prevalence of victimisation the original victimisation and offending variables were categorized. Victimisation was categorized into the following categories: never, once and more than once. Offending was categorized into the following categories: never, reported 1-3 different acts, reported 4 or more different acts of crime. Such categories give a first impression of the seriousness of the respondents' involvement in victimisation by involvement in offending. We expect higher levels of victimisation among those adolescents that report most offences. Next, we discuss levels of victimisation by offending, but further split the descriptive table by levels of lifestyle risk

10   A guideline rule argues that multilevel modelling can best be applied when sufficient level-2 units are available. Sufficient is often interpreted as a minimum of 25 level-2 units. In this study the level-2 observations are restricted to 10 schools. Despite the fact that this situation is far from ideal, it remains important to obtain correct standard errors. We argue that, from a theoretical and methodological point of view, the situation of performing multilevel modelling is still more correct than treating level-2 characteristics as level-1 characteristics.

Table 2 – Victimisation by levels of offending

	Total	Committed no acts	Committed 1-3 acts	Committed ≥ 4 acts
Never victim	25,9% (376)	41,9% (262)	17,0% (106)	4,0% (8)
Victim once	20,8% (302)	24,6% (154)	20,3% (127)	10,4% (21)
Victim more than once	53,3% (774)	33,5% (210)	62,7% (392)	85,6% (172)

From table 2 it can be derived that in general only 25,9 percent of the respondents has never become the victim of one of the offences listed in the questionnaire. Victimisation seems to be widespread among young adolescents. However, when levels of victimisation are split up by offending, it becomes clear that repeated victimisation is especially high among those that are most frequently involved in offending. There is a clear association between victimisation and offending (Pearson Chi-square = 228,38; df = 4; p <.01). 41,9 % of the persons who never committed an act of crime did also never become a victim; of the persons who did commit an act of crime most of them did also become a victim more than once during lifetime: 62,7 % of the young adolescents who reported 1-3 acts and 85,6 % of those who reported four or more acts of crime.

6.2 Victimisation by offending by level of lifestyle risk

As we are interested in the effect of lifestyle risk on both offending and victimisation, we split up the findings concerning the association between offending and victimisation by lifestyle risk (table 3). Lifestyle risk was dichotomized using the median lifestyle risk score.

Table 3 – Victimisation by offending by level of lifestyle risk

		Total	Committed no act	Committed 1-3 acts	Committed ≥ 4 acts
Low lifestyle risk	Never victim	30,7% (321)	43,2% (244)	17,5% (75)	3,8% (2)
	Victim once				
	21,5% (225)	24,4% (138)	19,9% (85)	3,8% (2)	
	Victim more than once	47,8% (500)	32,4% (183)	62,6% (268)	92,5% (49)
High lifestyle risk	Never victim	13,5% (49)	34,8% (16)	15,4% (27)	4,2% (6)
	Victim once	18,2% (66)	21,7% (10)		
	21,7% (38)	12,7% (18)			
	Victim more than once	68,3% (248)	43,5% (20)	62,9% (110)	83,1% (118)

Levels of victimisation vary by levels of offending, as has been described above, but the association is stronger for those with a low score lifestyle risk (Pearson Chi-square = 145,8; df = 4; p <.01 for young adolescents with a low lifestyle risk; Pearson Chi-square = 38,31; df = 4; p <.01 for those with a high lifestyle risk).

In general, 30,7 percent of those respondents with a low lifestyle risk never have been the victim of the aforementioned offences, while this percentage is much lower among those that have an above median score on lifestyle risk. In both groups, the percent of respondents that have become victimized more than once increases by their level of involvement in offending.

6.3 Do exposure to offending at school, propensity to offend and lifestyle risk have independent effects on victimisation and offending?

In this paragraph we present the results of a series of block-wise hierarchical regression models that have victimisation and offending as dependent variables. We start by evaluating the model of victimisation.

Table 4 – Hierarchical regression models for self-reported victimisation at school

	Block 0 B / Beta (S.E.)	Block 1 B / Beta (S.E.)	Block 2 B / Beta (S.E.)	Block3 B / Beta (S.E.)	Block 4 B / Beta (S.E.)
Intercept (B0)	2.13 (0.18)**	1.46 (0.12)**	1.49 (0.10)**	1.60 (0.11)**	1.63 (0.11)**
Level-1					
Gender		0.78/0.19 (0.11)**	0.77/0.18 (0.11)**	0.63/0.15 (0.11)**	0.60/0.14 (0.11)**
Immigrant background		0.31/0.06 (0.13)*	0.30/0.05 (0.13)*	0.25/0.04 (0.13)(*)	0.23/0.04 (0.13)(*)
Disadvantage		0.05/0.00 (0.26)	0.03/0.00 (0.26)	-0.03/-0.00 (0.26)	-0.03/-0.00 (0.26)
One parent family		0.55/0.11 (0.11)**	0.55/0.11 (0.11)**	0.46/0.09 (0.11)**	0.46/0.09 (0.11)**
Being kept down a class		0.37/0.07 (0.12)**	0.34/0.07 (0.12)**	0.22/0.04 (0.12)(*)	0.16/0.03 (0.12)
Propensity to offend				0.04/0.22 (0.01)**	0.03/0.16 (0.01)**
Lifestyle risk					0.18/0.13 (0.03)**
Level-2					
Exposure to offending at school			0.15/0.11 (0.05)*	0.15/0.10 (0.06)*	0.14/0.10 (0.06)(*)
Varcomp L2	0.32**	0.08**	0.04**	0.05**	0.05**
Varcomp L1	3.87	3.62	3.62	3.39	3.33
Deviance	6275.10	6045.60	6041.98	5812.03	5633.51
Pseudo-R squared	0%	12%	13%	18%	20%
ICC	7.63%	2.16%	1.09%	1.45%	1.47%

* p < 0,05   ** p < 0,01   *** p < 0,001

In *block 0*, an empty random intercept model is presented, allowing us to study the percentage of the variance in victimisation at school that is situated at the school level. From this analysis it becomes clear that there exists substantial and significant variation in victimisation at school: the ICC (intra-class correlation) is 7,63 %. An intra-class coefficient of 7,63 % means that 7,63 % of the observed variance at level one is situated at the higher level. Multilevel modelling is essentially about partitioning or unfolding variance.[11] In *block 1*, background characteristics are entered into the equation allowing us to control for the socio-demographic composition of schools. The standardized regression parameters of gender, immigration background, family structure and being kept down a class are statistically significant related to victimisation,

---

11   The intra-class correlation coefficient (ICC) is defined as the share of the between-group variance of the sum of between- and within-group variance and is independent of the number of observations.

but these standardized effects are in most cases rather small.[12] The squared multiple correlation is about 12 %.[13] The observed differences between schools are partially due to the differential school composition of pupils: this can be seen from the ICC that has shrunk substantially to a value of 2,16 %. In *block 2,* exposure to offending as a school level characteristic is entered into the equation to assess the existence of a real contextual effect of exposure to offending at the school level on individual differences in victimisation at school. In line with routine activity theory, exposure to offending at school is significantly related to individual differences in victimisation at school, independent of one's background characteristics. The squared multiple correlation is now 13 %. In *block 3,* the independent effect of propensity to offending is assessed. Propensity to offending has the strongest independent effect on victimisation at school and reduces the effects of background characteristics. The squared multiple correlation has increased to 18 %. The effect of being kept down a class is reduced to unacceptable levels of significance ($p > 0.10$). According to Hirschi and Gottfredson (1990), propensity to offending should have an independent effect on individual differences in victimisation at school. We explicitly hypothesised that propensity to offend would moderate the effect of exposure to offending at the school level. This does not seem to be the case, the effect of exposure to offending does still exist and hardly decreased. In *block 4,* lifestyle risk is entered into the equation. Both lifestyle risk and propensity to offend have substantial and significant effects on victimisation. Together, all variables into the equation explain 20 % of the variance in victimisation.

Table 5 – Hierarchical regression models for self-reported offending

	Block 0 B / Beta S.E.	Block 1 B / Beta S.E.	Block 2 B / Beta S.E.	Block3 B / Beta S.E.	Block 4 B / Beta S.E.
Intercept (B0)	1.52 (0.16)**	0.69 (0.09)*	0.72 (0.08)**	0.97 (0.08)**	1.04 (0.07)**
Level-1					
Gender		1.08/0.28 (0.10)**	1.04/0.27 (0.10)**	0.74/0.19 (0.09)**	0.72/0.18 (0.08)**
Immigrant background		0.56/0.11 (0.12)**	0.55/0.11 (0.12)**	0.45/0.09 (0.10)**	0.41/0.08 (0.10)**
Disadvantage		0.16/0.01 (0.25)	0.14/0.01 (0.25)	0.00/0.00 (0.21)	0.41/0.00 (0.10)**
One parent family		0.30/0.06 (0.11)**	0.29/0.06 (0.11)**	0.10/0.02 (0.09)	0.02/0.00 (0.09)
Being kept down a grade		0.40/0.09 (0.11)**	0.37/0.08 (0.11)**	0.20/0.04 (0.09)*	0.10/0.02 (0.09)
Propensity to offend				0.10/0.50 (0.00)**	0.07/0.34 (0.00)**
Lifestyle risk					0.41/0.32 (0.02)**
Level-2					

---

12   We consider a standardized coefficient that does not exceed the absolute value of 0.1 of little statistical importance.

13   Multilevel modelling for continuous dependent variables does not yield squared multiple correlations as in done in OLS-regression models, but we computed a pseudo squared multiple correlation for each model, using the methodology suggested by Snijders and Bosker (1999).

	Block 0 B / Beta S.E.	Block 1 B / Beta S.E.	Block 2 B / Beta S.E.	Block3 B / Beta S.E.	Block 4 B / Beta S.E.
Exposure offending at school			0.09/0.07 (0.04)*	0.09/0.07 (0.04)*	0.05/0.03 (0.03)
Varcomp L2 Varcomp L1	0.27** 3.40	0.02** 3.09	0.01 3.09	0.01 2.21	0.01 1.95
Deviance	6134.25	5866.50	5868.46	5253.87	4946.82
Pseudo-R squared	0%	15%	16%	40%	47%
ICC	7.35%	0.64%	0.32%	0.45%	0.51%

* p < 0,05   ** p < 0,01   *** p < 0,001

Let us now compare these findings with the situation whereas self-reported offending is the dependent variable. If victimisation mirrors offending then it is expected that similar results are found. In *block 0*, an empty model is presented, allowing us to study the percentage of the variance in self-reported offending that is situated at the school level. Surprisingly, 7,35 % of the variance between adolescents is situated at the school level. Previous multilevel studies of offending in Belgium yielded lower ICCs in empty models (Pauwels; 2007 & 2008). In *block 1*, background characteristics are entered into the equation allowing us to control for the socio-demographic composition of schools. After controlling for demographic background variables it seems that the variance that is situated at the school level has disappeared almost completely. Background variables explain 15 % of the variance in offending. Only 0,64 % of the variance in offending is situated at the school level, when controlling for background variables. Although the ICC coefficient is still significant, this finding suggests that the effect of the school context should not be overestimated. It seems that almost all variance in offending in Sint-Niklaas at the school level is due to the differential compositions of schools. Previous studies in Antwerp revealed higher ICC levels when controlling for the demographic composition of schools (Pauwels, 2007). Gender, immigrant background, family structure and being kept down a class have positive independent effects on offending, but the standardized coefficients are rather moderate. Disadvantage does not have any significant effect. In *block 2*, exposure to offending as a school level characteristic is entered into the equation to assess the existence of a real contextual effect of exposure to offending at the school level on individual differences in self reported offending. Although the variance at the school level is small, this variance can be explained by exposure to offending at school. This means that exposure to offending at school leaves its mark on all pupils, but its impact should not be overestimated in comparison with other mechanisms. The squared multiple correlation is now 16 %. In *block 3*, the independent effect of propensity to offending is assessed. According to Hirschi and Gottfredson (1990) as well as Wikström (2004) propensity to offending should have an independent effect on individual differences in self reported offending. We explicitly hypothesised that propensity to offend would moderate the effect of exposure to offending at the school level. Propensity to offend does seriously moderate the effects of all variables in the equation with the exception of exposure to offending and has the strongest direct effect on offending. The standardized coefficient is substantial and the squared multiple correlation is now 40 %, which is a very large increase. In *block 4*, lifestyle risk is entered into the equation. This model explains 51 % of the variance in offending. Lifestyle risk moderates the effect of propensity to offending on individual differences in offending. Both variables have similar substantial effects.

From these analyses we cannot but conclude that there exist some similarities in the effects of exposure, background variables, propensity to offend and lifestyle risk on victimisation and offending. The key theoretical concepts derived from the integrated self-control/lifestyle-exposure model are significantly related to both victimisation and offending, but nevertheless, the differences between both models are rather remarkable. The squared multiple correlations are much larger when offending is the dependent variable. Based on this study, we can therefore conclude that it would be wrong to see the effects of exposure, background variables, propensity to offend and lifestyle risk on victimisation as a mirror of the impact of these effects on offending. In fact, that would be a reduction of reality. Nevertheless, exposure to offending, propensity to offending and lifestyle matter in the explanation of both dependent variables, they just seem to fit offending better than victimisation. While exposure and propensity to offend have been suggested in both theories of victimisation and offending, our findings suggest that the variables used in the present study seem to be more appropriate to test theories of offending. Although the lifestyle perspective originated from theories of victimization, it is interesting to see that it proves to be of importance in explaining differences in offending.

6.4   How strong is the bivariate association between victimisation and offending?

We have found that victimisation can only partially be explained by variables from the integrated theoretical framework. Let us now turn to the question how strong offending and victimisation are correlated. We do so not with the intention to verify whether victims can be "blamed" but because it is not possible to understand juvenile victimisation without also understanding the association or linkage with patterns of juvenile offending (Lauritsen et al., 1991). We are merely interested in the association between victimisation and offending as interpersonal dynamics (Fattah, 1992).

Table 6- Bivariate and partial correlation between victimisation and offending

Control variables	Pearson 's R between victimisation and offending
None	0.44***
Demographic background variables	0.37***
Demographic background variables and propensity to offend	0.30***
Demographic background variables, propensity to offend and lifestyle risk	0.274***

$* p < 0,05$ $** p < 0,01$ $*** p < 0,001$

From the partial correlations it becomes clear that the correlation between self-reported victimisation and self-reported offending is rather strong and significant. The positive bivariate correlation of 0,44 does not disappear when controlling for background variables. Therefore, common background variables cannot be used to explain the observed correlation between both concepts. When including propensity and lifestyle risk the observed correlation between offending and victimisation remains positive and statistically significant (0, 27, p > 0.001). One can only observe that the partial correlation between victimisation at school and offending can to a certain extent be due to background variables, propensity to offend and lifestyle risk. Therefore the hypothesis of lifestyle being the most important common cause of the observed relation between victimisation and offending is falsified in this study.

Finally, we try to answer following questions in the present study: how large are the independent effects of offending on victimisation and of victimisation on offending and what happens to the effects of all other variables from the integrated model? This becomes clear from the next paragraph.

Table 7- Final models of victimisation and offending

	Dependent: Victimisation	Dependent: Offending
Intercept (B0)	1.80 (0.10)**	1.14 (0.06)**
Level-1		
Gender	0.33/0.08 (0.11)**	0.59/0.15 (0.08)**
Immigrant background	0.05/0.01 (0.13)	0.36/0.07 (0.09)**
Disadvantage	-0.01/-0.00 (0.26)	-0.03/0.00 (0.19)
One parent family	0.42/0.08 (0.11)**	-0.04/0.00 (0.08)
Being kept down a class	0.19/0.04 (0.12)	0.01/0.00 (0.09)
Propensity to offend	0.01/0.03 (0.01)	0.06/0.30 (0.00)**
Lifestyle risk	0.08/0.05 (0.04)(*)	0.28/0.22 (0.03)**
Offending	0.38/0.36 (0.03)**	--
Victimisation	--	0.21/0.22 (0.01)**
Level-2		
Exposure offending at school	0.12 (0.06)(*)	0.01/0.00 (0.03)
Varcomp L2 Varcomp L1	0.04 3.08	0.00 1.7
Deviance	5411.87	4618.08
Pseudo-R squared	26%	54%
ICC	0%	0%

* p < 0,05   ** p < 0,01   *** p < 0,001

The analyses conducted here learn us that only gender, family structure and offending have direct effects on victimisation, while the effects of lifestyle risk and propensity to offend are rendered insignificant when controlling for offending. Offending probably fully mediates the effects of propensity and lifestyle risk on victimisation, yet, there is a lot to be explained, as the squared multiple correlation is only 26 percent. The analyses reveal that victimisation has a direct effect on offending, but the effects of the variables from the integrated model hardly change: propensity to offend and lifestyle risk still have very similar independent effect on offending and the squared multiple correlation is 54 %. As this analysis is cross-sectional, it is not possible to derive the correct causal order. Nevertheless, these findings are important.

7   Discussion and conclusion

This study contributed to our understanding of victimisation within an urban sample of young adolescents in a Belgian urbanized area. From this study we conclude that victimisation at school is a rather common experience among young adolescents, with about half of the sample having experienced victimisation of at least two offences. Clearly, repeated victimisation is much more common among those that have committed offences themselves, and this finding was reproduced in both subgroups based on their lifestyles.

We questioned to what extent the explanation of victimisation could be seen as a mirror of the explanation of offending. Therefore, we used a recently integrated theory of self-control/lifestyle-exposure to find out if the variables from this integrated theory could equally explain individual differences in offending and victimisation. Although the integrated model in its original form has been developed to explain victimisation, the model at the very most only explains 20 % of the observed differences in victimisation, while the same model does explain about 50 % of the observed differences in offending. This is a very big difference which causes to question the similarity in explaining between offending and victimisation as the self-control/lifestyle-exposure theory is far better suited to explain observed differences in offending.

The (integrated) lifestyle-exposure theories and their (limited) explanation of risk of victimisation have been criticized on different topics by several authors. For example, some kinds of victimisation that occur in private space like incest and domestic violence cannot be explained by these theories (Finkelhor & Hashima, 2001; Finkelhor, 2007). Besides that, these theories analyse only partially the role of human action and structural constraints in victimisation (Spalek, 2006; Walklate, 2007b). There still exists a lacuna in our knowledge of understanding mechanisms which may increase the risk of victimisation. Furthermore, when concluding that only 20 % of the observed differences in victimisation may in fact be explained by concepts such as exposure to offending and propensity to offending and lifestyle, it should be avoided to reason in circles. By that we mean that it is dangerous to argue that all victims of crime are in general characterized by some specific demographic features and that they "because of these features" consequently have the highest chances to become victim of crime "just because they have those characteristics". One should avoid to fall back into what is called 'the guilty conscience of victimology' by blaming the victim (Van Dijk 2008) or by awakening the impression some victims provoke their victimisation (Brants, 1982; de Beer & Fiselier, 1989; Wittebrood & Nieuwbeerta, 1999)[14]. It is not our intention to blame adolescent victims who were involved in delinquent activity as also these victims still do not deserve their victimisation (Lauritsen et al., 1991).

---

14   Victim blaming: 'any allusion to the victim's role in criminal transactions' (Rock, 2002: 12).

In short, it remains important to recognize the idea that adolescent victimisation is predicted by involvement in offending, but only partially. Thus, it is ambiguous and even dangerous to generalize all victimisation among all victim groups as a consequence of offending: to do so would disregard the generally involuntary and high-risk nature of adolescent environment (Finkelhor, 2007).

This study has, apart from the above mentioned theoretical caveats, some important restrictions. The current study has a cross-sectional design. It is known that the cross-sectional design is not an appropriate one in testing any causal relationships between variables. And this applies to the study of the association between offending and victimisation. To establish causal relationships through any survey-design might be problematic, because self-reported offending and victimisation both may be differentially subject to issues of telescoping.

The association between victimisation and offending may vary by offence type, i.e. may be stronger for violent crime then for property crime. In this study we used only one measure of victimisation and offending.

The original questionnaire was not designed to allow such a detailed comparison. Nevertheless, future studies that combine questions on offending and victimisation should consider the problems we encountered in this study. This would make the interpretation more straightforward. The association between offending and victimisation is not exclusively due to exposure to offending but may also be due to the fact that offenders are more psychologically distressed and therefore more vulnerable or that delinquents are more aggressive and therefore more likely to arouse antagonism (Finkelhor & Asdigian, 1996).

Studying the relationship between victimisation and offending should not rely exclusively on self-report studies as it is known that more serious offences are underreported and self-reported victimisation can be confound by measurement error due to telescoping. Despite these problems, we have observed a strong association between victimisation and offending. However, the strength of the association between offending and victimisation becomes clear when both concepts are used as dependent variables, while controlling for the same independent variables: offending has the only direct effect on differences in victimisation while victimisation has some effect and the effect of variables from the integrated lifestyle perspective still remain. To conclude, we must clearly state that not all victims are linked to acts of crime. There are also victims who are not involved in offending (Finkelhor & Hashima, 2001; Finkelhor, 2007). The correlation cannot be completely ascribed to similar background characteristics. The relation even holds when controlling for propensity to offend and lifestyle risk. So firstly, this suggests once again that there are also other factors relevant to explain the risk of victimisation which mean that the explanation of victimisation needs to move beyond lifestyle concepts (Singer, 1981; Finkelhor & Asdigian, 1996; Wittebrood & Nieuwbeerta, 1999; Purugganan et al., 2000). Secondly, when we conclude that the effects of life style risk and propensity to offend on victimisation become insignificant when controlling for offending, this means that offending thus mediates the effects of the aforementioned concepts, offending might explain why those at risk in terms of low self-control and lifestyles become victimised, but not the other way around. Our study may trigger other scholars to rethink theories that aim at explaining both offending, victimisation and the well documented relationship between offending and victimisation. Clearly, there is a need for a more profound and in-depth theoretical research about the explanation of the association between victimisation and offending in general, not only at school, as was the case in the present study. Only by testing of theories, one can increase insight in the mechanisms that are at work when explaining offending, victimisation and the association between the both. Finally, we recommend to reanalyze studies like this one for different types of offences, this is especially important because the context of victimisation might be more versatile then the context of adolescent offending.

8 Bibliography

Armstrong, D., Hine, J., Hacking, S., Armaos, R., Jones, R. Klessinger, N. & France, A. (2005), *Children, risk and crime: the On Track Youth Lifestyles Surveys*, Home Office Research Study 278, Home Office Research, Development and Statistics Directorate.

Aye Maung, N. (1995), *Young people, victimisation and the police: British Crime Survey findings on experiences and attitudes of 12 to 15 year olds*, Home Office Research Study No. 140, London, Home Office.

Bailey, S. &Whittle, N. (2004), "Young people: victims of violence", *Current Opinion in Psychiatry*, Vol. 17, pp. 263-268.

Bjarnason, T., Sigurdardottir, T. J. & Thorlindsson, T. (1999), "Human Agency, Capable Guardians, and Structural Constraints: A Lifestyle Approach to the Study of Violent Victimization", *Journal of Youth and Adolescence*, 28, pp. 105-119.

Brants, C. (1982). "Definiëring van slachtofferschap: een kritische verkenning van de grenzen der victimologie", *Tijdschrift voor Criminologie*, 24, 325-340.

Bronfenbrenner, U. (1979), *The Ecology of Human Development*, Cambridge/Mass.: Harvard University Press.

Childs, K. & Gibson, C. (2009). "Low self-control, lifestyles, and violent victimization: Considering the mediating and moderating effects of gang involvement" Paper presented at the annual meeting of the American Society of Criminology, Royal York, Toronto (http://www.allacademic.com/meta/p32216_index.html (27-04-2009))

Cohen, L. E. & Felson, M. (1979), "Social change and crime rate trends: A routine activity approach", *American Sociological Review*, Vol. 44, pp. 588-608.

Cohen, L., Kleugel, J. & Land, K. (1981), "Social inequality and predatory criminal victimization: An exposition and test of a formal theory, *American Sociological Review*, Vol. 46, pp. 505-524.

Davies, P., Francis, P. & Jupp, V. (2003), "Victimology, Victimisation and Public Policy" in Davies, P., Francis, P. & Jupp, V. (Eds.), *Victimisation: theory, research and policy*, Houndmills: Palgrave, pp. 1-27.

Debarbieux, E. & Blaya, C. (2001) (Eds.), *Violence in Schools: Ten European Approaches*, Paris: ESF.

de Beer, P.W.G. & Fiselier, J.P.W. (1989), *Voorrang aan slachtoffers van criminaliteit,* Arnhem: Gouda Quint bv, 61 p.

De Bruyn, P. (2008), *Vrijetijdsbesteding en slachtofferschap op school bij jongeren: een verkennend onderzoek te Sint-Niklaas,* Onuitgegeven verhandeling tot het behalen van de Graad van Master in de Criminologische Wetenschappen, Gent: UGent.

Fattah, E. A. (1992), "Victims and Victimology: the Facts and the Rhetoric" in Fattah, E.A. (Ed.), *Towards a Critical Victimology.* Houndmills: The Macmillan Press, pp. 29-56.

Felson, M. (1998), *Crime & Everyday Life,* Thousand Oaks: Pine Forge Press.

Finkelhor, D. (1995), "The victimization of children: A developmental perspective." *American Journal of Orthopsychiatry* Vol. 65, pp. 177-193.

Finkelhor, D. (2007), "Developmental Victimology: The Comprehensive Study of Childhood Victimizations" in Davis, R.C., Lurigio, A.J. & Herman, S. (Eds.), *Victims of Crime,* Los Angeles: Sage Publications, pp. 9-34.

Finkelhor, D. & Asdigian, N.L. (1996), "Risk Factors for Youth Victimization: Beyond a Lifestyles/Routine Activities Theory Approach", *Violence and Victims,* Vol. 11, pp. 3-19.

Finkelhor, D. & Hashima, P.Y. (2001), "The Victimization of Children and Youth: A Comprehensive Overview" in White, S.O. (Ed.), *Handbook of Youth and Justice,* New York: Kluwer Academic Publishers Group, pp. 49-78.

Fiqueira-McDonough, J. (1986), "School context, gender, and delinquency", *Journal of Youth and Adolescence,* Vol. 15,pp. 79–88.

Garofalo, J., Siegel, L. & Laub, J. (1987), "School-Related Victimizations Among Adolescents: An Analysis of National Crime Survey (NCS) Narratives", *Journal of Quantitative Criminology,* Vol. 3, pp. 321-338.

Gottfredson, D.C. (2001), *Schools and Delinquency,* Cambridge: Cambridge University Press.

Grasmick, H.G., Tittle, C.R., Bursik Jr., R.J. & Arneklev, B.J. (1993), "Testing the core empirical implications of Gottfredson and Hirschi's General Theory of Crime", *Journal of Research in Crime and Delinquency,* Vol. 30, pp.5-29.

Gibbs, J., Giever, D., & Higgins, G. (2003). "A test of Gottfredson and Hirschi's generally theory using structural equation modeling", *Criminal Justice & Behavior,* 30, pp. 441- 458.

Hashima, P.Y. & Finkelhor, D. (1999), "Voilent Victimization of Youth Versus Adults in the National Crime Victimization Survey", *Journal of Interpersonal Violence,* Vol. 14, pp. 799-820.

Hindelang, H.J., Gottfredson, M.R. & Garofalo, S. (1978), *Victims of Personal Crime: An Emperical Foundation for a Theory of Personal Victimization,* Cambridge: Ballinger Publishing Company.

Hirschi, T. & Gottfredson, M. (1990), *A general Theory of Crime*, Stansford: Stansford California Press.

Jensen, G.F. & Brownfield, D. (1986), "Gender, lifestyles, and victimization: beyond routine activity", *Violence and Victims*, Vol. 1(2), 85-99.

Lauritsen, J.L., Sampson, R.J. & Laub, J.H. (1991), "The link between offending and victimization among adolescents", *Criminology*, 29, pp. 265-292.

Lawrence, R. (2007), *School crime and juvenile justice* (2nd edition). *New York/Oxford: Oxford University Press*.

Miethe, T.D. & Meier, R.F. (1994), *Crime and Its Social Context: Toward an Integrated Theory of Offenders, Victims, and Situations*, Albany, NY: State University of New York Press.

Mooij, T. (1994), *Leerlinggeweld in het voortgezet onderwijs. Sociale binding van scholieren*, Nijmegen: ITS.

Mooij, T. (2001), *Veilige scholen en (pro)sociaal gedrag. Evaluatie van de veilige school-campagne in het voortgezet onderwijs*, Nijmegen: Instituut voor Toegepaste Sociale Wetenschappen, Katholieke Universiteit Nijmegen.

MORI (2004), *MORI Youth Survey 2004*, Youth Justice Board for England and Wales.

Nofziger, S., & Kurtz, D. (2005), "Violent Lives: a Life Style Model Linking Exposure to Violence to Juvenile Violent Offending", *Journal of Research in Crime and Delinquency*, 42, pp. 3-26.

Nofziger, S. (2007), "Integrating Self-Control and Lifestyle Theories to Explain Juvenile Violence and Victimization", Paper presented at the annual meeting of the American Society of Criminology, Atlanta Marriott Marquis (http://www.allacademic.com/meta/p200176_index.html 2009-02-03))

Osgood, D. W. & Anderson, A.L. (2004), "Unstructured Socializing and Rates of Delinquency", *Criminology*, Vol. 42, pp. 519-549.

Pauwels, L. (2007), *Buurtinvloeden en Jeugddelinquentie: een Toets van de Sociale Desorganisatietheorie*. Den Haag: BJU.

Pauwels, L. (2008), "Geweld in groepsverband onder Antwerpse Jongeren: de rol van schoolcontext en leefstijl", *Tijdschrift voor Criminologie*, 50 (1), pp. 3-16.

Pauwels, L. (2009), "Disentangling neighbourhood and school contextual variation in serious offending: assessing the effect of ecological disadvantage" in: Cools, M, De Kimpe, S., De Ruyver, B., Easton, M., Pauwels, L., Ponsaers, P., Vande Walle, G., Vander Beken, T., Vander Laenen, F. & Vermeulen, G. (Eds), *Governance of Security Research Papers Series I, Contemporary Issues in the Empirical Study of Crime*, Antwerpen: Maklu, pp. 179-210.

Pauwels, L. & Pleysier, S. (2008), "Crime Victims and insecurity surveys in Belgium and the Netherlands" in Zauberman, R. (Ed.), *Victimisation and Insecurity in Europe. A Review of Surveys and their Use*, Brussel: VUBPRESS, pp. 39-64.

Pauwels, L. & Svensson, R. (2009), "Individual Differences In Adolescent Life Style Risk By Gender And Ethnic Background: A Test in Two Urban Samples", *European Journal of Criminology*, Vol. 6 (1), pp. 5-23.

Piquero, A. R. & Hickman, M. (2003), "Extending Tittle's control balance theory to account for victimization", *Criminal Justice and behavior*, Vol. 30, pp. 282-301.

Pratt, T. C. & Cullen, F. T. (2000), "The empirical status of Gottfredson and Hirschi's general theory of crime: A meta-analysis", *Criminology*, Vol. 38, pp. 931-964.

Reep, C. & Oudhof, K. (2009), "Wie kwaad doet, kwaad ontmoet? Over de samenhang van daderschap en slachtofferschap", *Tijdschrift voor Criminologie*, 51 (3), pp. 228-245.

Rock, P. (2002), "On Becoming a Victim" in Hoyle, C. & Young, R. (Eds.), *New Visions of Crime Victims*, Oxford: Hart Publishing, pp. 1-22.

Rovers, B. (1997), *De buurt, een broeinest?, Een onderzoek naar de invloed van de woonomgeving op jeugdcriminaliteit*, Nijmegen: Ars Aequi Libri.

Purugganan, O. H., Stein, R.E.K., Johnson Silver, E. & Benenson, B.S. (2000), "Exposure to Violence Among Urban School-Aged Children: Is it Only on Television?", *Pediatrics*, Vol. 106, pp. 949-953.

Sampson, R. J. &. Lauritsen, J.L. (1990), "Deviant lifestyles, proximity to crime, and the offender-victim link in personal violence", *Journal of Research in Crime and Delinquency*, Vol. 27, pp. 110-133.

Schreck C (1999), "Criminal victimization and low self-control: and test of a general theory of crime", *Justice Quarterly*, Vol. 16 (3), pp. 633-654.

Schreck, C., Richard, J., Wright, A. & Miller, J. M. (2002), "A Study of Individual and Situational Antecedents of Violent Victimization", *Justice Quarterly*, Vol. 19, pp. 159-180.

Shaffer, J. N. & Ruback, R. B., (2002), "Violent Victimization as a Risk Factor for Violent Offending Among Juveniles", *Juvenile Justice Bulletin* (http://www.ncjrs.gov/html/ojjdp/jjbul2002_12_1/contents.html (12-03-2009))

Singer, S. I. (1981), "Homogeneous victim-offender populations: a review and research implications", *The Journal of Criminal Law & Criminology*, Vol. 72, pp. 779-788.

Smith, D. J., McVie, S., Woodward, R., Shute, J., Flint, J. & McAra, L. (2001), *The Edinburgh Study of Youth Transitions and Crime: Key findings at ages 12 and 13*, 209 p. (http://www.law.ed.ac.uk/cls/esytc/findreport/wholereport.pdf (18-03-2009))

Snijders, T., & Bosker, R. (1999). *Multilevel Analysis: an introduction to basic and advanced multilevel modeling*. London: Sage Publications.

Spalek, B. (2006), *Crime Victims: theory, policy and practice*, Houndmills: Palgrave Macmillan.

Svensson, R. & Pauwels, L. (2008), "Is a risky lifestyle always "risky"? The interaction between individual propensity and lifestyle risk in adolescent offending: A test in two urban samples", *Crime & delinquency,* Online first published on September 24, 2008, DOI:10.1177/0011128708324290.

Turner, M. G. & Piquero, A. R. (2002), "The stability of self control", *Journal of Criminal Justice,* Vol. 30, pp. 457-471.

van Noije, L. & Wittebrood, K. (2007), "Veiligheid" in Bijl, R., Boelhouwer, J. & Pommer, E. (Eds.), *De sociale staat van Nederland 2007,* Den Haag: Sociaal en Cultureel Planbureau, pp. 213-244.

Vazsonyi, A.T., Pickering, L.E., Belliston, L.M., Hessing, D. & Junger, M. (2002), "Routine activities and deviant behaviors: American, Dutch, Hungarian, and Swiss youth", *Journal of Quantitative Criminology,* Vol. 18, pp. 397-422.

Vettenburg, N., Elchardus, M. & Walgrave, L. (2007), *Jongeren in cijfers en letters. Bevindingen uit de JOP-monitor 1,* Leuven: LannooCampus.

Walklate, S. (2007a), *Imagining the victim of crime,* Berkshire: Open University Press, 189 p.

Walklate, S. (2007b), *Understanding Criminology: Current theoretical debates,* Maidenhead: Open University Press.

Wikström, P-O.H. (2004), "Crime as alternative. Towards a cross-level situational action theory of crime causation" in McCord, J. (Ed.), *Beyond empiricism: Institutions and intentions in the study of crime,* Advances in criminological theory, Vol. 13, New Brunswick: NJ: Transaction.

Wikström, P-O. H. (2005), "The social origins of pathways in crime: Towards a developmental ecological action theory of crime involvement and its changes" in Farrington, D.P. (Ed.), *Integrated developmental and life course theories of offending,* Advances in criminological theory, Vol. 14, New Brunswick: NJ Transaction.

Wikström, P-O.H. & Butterworth, D.A. (2006), *Adolescent crime: Individuals differences and lifestyles,* Willan Publishing.

Wikström, P-O.H. & Loeber, R. (2000), "Do Disadvantaged Neighborhoods Cause Well-Adjusted Children To Become Adolescent Delinquents? A Study of Male Juvenile Serious Offending, Individual Risk and Protective Factors, and Neighborhood Context", *Criminology,* Vol. 38 (4), pp. 1109-1142.

Wittebrood, K. & Nieuwbeerta, P. (1997), "Wie misdaden pleegt, kan klappen verwachten. De invloed van daderschap en leefpatronen op de kans slachtoffer van geweld te worden", *Tijdschrift voor Criminologie,* 4, pp. 341-356.

Wittebrood, K. & Nieuwbeerta, P. (1999). "Wages of sin? The link between offending, lifestyle and violent victimization", *European Journal on Criminal Policy and Research,* 7, pp. 63-80.

Wood, M. (2005), *The victimisation of young people: findings from the Crime and Justice Survey 2003,* London, Home Office.

Appendix 1 – Measurement of scale constructs

Life-style risk:	
Combination of risk scores (Wikström and Loeber, 2000): (risk= Highest quartile =1, average= centrality = 0, protective = lowest quartile= -1) Summation of risk scores on peer delinquency, frequency of hanging out on streets and frequency of alcohol use	
Propensity to offend: (Delinquency tolerance and self-control combined)	
Delinquency tolerance:	
rules are made to be broken	
ok to break rules, as long as not get caught	
fighting ok when provoked	
if honest ways to achieve sth fail, then use dishonest ways	
Self-control:	
I often do things without thinking first	
when angry, others can better stay away from me	
I make fun when I can, even if I get into trouble afterwards	
when I am angry, I'd rather hit then talk	
I say what I think, even if its not smart	
I often do what I want to	
get angry very fast	

"Safety: everybody's concern, everybody's duty"?

Questioning the significance of 'active citizenship' and 'social cohesion' for people's perception of safety.

Evelien Van den Herrewegen[1]

1   Introduction

"Safety: everybody's concern, everybody's duty"[2] is a catchphrase launched by the King Baudouin Foundation and is an appeal to policymakers and the public to tackle 'unsafety' in a local manner and in collaboration with citizens[3]. As such, the King Baudouin Foundation suggests that the active involvement of citizens is vital to the governance of safety. Active citizens are believed to behave in a more responsible manner, which in turn would result in feelings of safety. Active citizens are furthermore sensitive to the safety of others. The concept of active citizenship implies that individuals interact with one another so that mutual trust and feelings of social connectedness among citizens can emerge. 'Active citizenship' and 'social cohesion' are thus closely interwoven, and both concepts are believed to be essential to people's 'perception of safety'.

In this paper, we will question the assumption that 'social cohesion' and 'active citizenship' are prerequisites for people to feel safe. First, we will outline that the emergence of 'social cohesion' and 'active citizenship' in the governance of safety is the result of the current conceptualization of 'fear of crime'. Fear of crime is perceived as a product of concerns and doubts about one's position and identity in late modernity (part 1). Next, we will discuss current theories that suggest that these concerns and doubts can be countered by re-embedding people into the community. In this context, civic engagement and social integration are viewed as the new tools to improve people's wellbeing, which in turn would result in more positive perceptions of safety (part 2). However, we will question these alleged positive linkages between 'perception of safety', 'active citizenship' and 'social cohesion' and point out that the three concepts might even counteract (part 3). Finally, an alternative perspective is suggested that acknowledges the importance of 'identity' in understanding people's perception of safety. This perspective first and foremost recognizes that people's identity is not limited to their social integration and involvement in the local community. Consequently,

---

1   The author likes to thank K. Verfaillie and three anonymous reviewers for their pertinent remarks on an earlier version. However, responsibility for the contents is exclusively mine.
2   Veiligheid: een zaak van iedereen, een taak voor iedereen.
3   http://www.kbs-frb.be/pressitem.aspx?id=177526&LangType=2067 [20/01/2009]

research or policy initiatives that focus on 'social cohesion' and 'active citizenship' are perhaps not fully addressing the complexity of people's perception of safety.

2   Perception of safety: genealogy, aetiology and autonomization

2.1   Genealogy and aetiology of 'fear of crime'

Historically the interest for people's fears or perceptions concerning crime and unsafety is relatively new. In his book 'Inventing fear of crime' Murray Lee (2007) situates its origins in North-America in the 1960s. He systematically illustrates that 'fear of crime' was not a phenomenon waiting out there to be discovered, but that its emergence was a result of interactions between political willingness to act on behalf of the 'people' (especially the growing interest in victims) and developments in social scientific enquiries (especially the victim crime survey). Both developments initiated a feedback loop that helped to sustain and intensify the interest in fear of crime.

Since its 'discovery' in 1965, research on 'fear of crime' has been driven by the search for causation: 'what causes fear of crime?' and 'how can we control this fear of crime?' (Lee, 2007). There is, however, still no scientific consensus about the main features and causes of 'fear of crime'. This quest for the causes of 'fear of crime' is even hardened by a strange paradox that emerged from the very first victim surveys that contained questions measuring people's attitudes and opinions about safety such as: 'How safe do you feel walking in your neighbourhood at night?' In brief, the paradox states that people with the lowest risk of criminal victimization, exhibit the highest fear (e.g. women and elderly), whereas people with a higher risk of victimization have less fear (e.g. young men) (Vanderveen, 2006). This discrepancy between the objectively measured risk of victimization (e.g. calculated by means of crime statistics) and the subjectively measured fear of crime (by means of victim crime surveys) lead to a debate about the rationality of people's fear of crime and instigated the question whether 'fear of crime' was still a legitimate research object or an appropriate focus for policy initiatives. Nonetheless, despite this legitimization crisis, research and policy never ceased trying to understand and control this 'fear of crime', or more broadly 'perceptions of safety'. On the contrary, 'fear of crime' evolved as an important niche within the criminological domain[4], as well as in political circles.

In order to answer the 'fear of crime' paradox, researchers broadened their scope and included non-criminal factors to explain feelings of unsafety. The inclusion of non-criminal factors led to a plethora of variables that are believed to explain, in a direct or indirect manner, people's feelings and worries about crime. Researchers who synthesized forty years of etiological 'fear of crime' research, distinguish *four broad dimensions in the theoretical perspectives on fear of crime*: "vulnerability", "victimization experience", "the environment" and "psychological factors". We will briefly discuss

---

4   In his renowned review Chris Hale (1996) noted that over 200 reports dealt with the subject 'fear of crime'. In 2000 Ditton and Farrall did an online research and located 837 entries. In 2007 Lee dragged up 242.000 'fear of crime' entries using the Google search engine. Now, in 2009, googling the term "fear of crime" (with double quotation marks) discloses 480.000 hits and in Google Scholar "fear of crime" reveals 24.300 links. In this sense 'fear of crime' is measuring up with other prominent criminological phenomena such as "hooliganism" (18.400 hits), "money laundering" (41.700 hits), "organized crime" (64.900 hits), "violent crime" (94. 600 hits).

these four dimensions, but for a more thorough review we refer to Hale (1996), Ditton & Farrall (2000) but also Vanderveen (2006) and Pleysier (2009).

The first two dimensions, "vulnerability", "victimization experience", include variables discovered and analyzed in the early days of fear of crime research. In the first dimension, the thesis is that certain populations are physically (e.g women, elderly) and/or socially (e.g ethnic minorities, long term unemployement) more vulnerable to crime and fear of crime. The second dimension explores the direct or indirect victimization (e.g friends, relatives) and its consequences for people's fear of crime. In the third dimension the physical and/or social organization of one's neighbourhood is considered to be detrimental for people's perception of safety.

The final dimension in the aetiology of 'fear of crime' is relatively new and is perceived as the new avenue for researchers to pursue. The assumption is that in order to account for 'fear of crime', a symbiosis of sociological and socio-psychological factors is needed. As such, the fourth dimension integrates the theses of the three other dimensions and adds the wide range of socio-psychological factors. (Jackson, 2004; Pleysier, 2009). In this dimension, 'fear of crime' is not solely considered as a direct emotional reaction to crime or other deviant behavior, but as a manifestation of a broader sense of non-well-being. As such, 'fear of crime' has an 'experiential' component that refers to everyday experiences with crime (victimization) and the lack of resources to cope with these experiences (vulnerability and social disorganization). Yet, additionally, there is also a 'expressive' component in which 'fear of crime' is a result of individual's attitudes and opinions about society as a whole.

In sum, the perpetual search for the causes of people's fear of crime and the acknowledgement that a focus on crime-related experiences is insufficient to account for or to control people's fear of crime, stimulated research and policy to approach 'fear of crime' as something which is not necessarily crime related. Particularly the fourth dimension in the aetiology of 'fear of crime' influenced this disentanglement process.

2.2   Autonomization of ' fear of crime'

The autonomization of 'fear of crime' is a process that is visible in the other dimensions of the aetiology of fear of crime, as people's divergent and seemingly irrational responses to crime were explained by taking into consideration non-crime related individual characteristics and/or environmental factors. However, it is precisely in the fourth dimension that 'fear of crime' totally disentangles itself from crime experience, and is considered to be an expression of people's attitude and concerns about society. In this sense, the term 'perception of safety' is preferred because it captures more accurately the idea that people's reaction to safety is the outcome of an interpretative process in which a variety of feelings, rationalizations, information resources, experiences, and so forth might come into play.

This autonomization process and the evolution of 'fear of crime' towards 'perception of safety' are important developments, rooted in very influential sociological theories that describe our society as 'reflexive modernity' (Beck, 1992), 'liquid modernity' (Bauman, 2000) or 'radicalized modernity' (Giddens, 1990). These sociological theories are not only relevant for the autonomization process but also for the incorporation of 'social cohesion' and 'active citizenship' as important concepts to understand and control 'perception of safety'.

2.2.1 Perception of safety as a late modern concept

Although their analyses stem from divergent backgrounds, sociologists such as Ulrich Beck, Zygmunt Bauman and Anthony Giddens, criticize adherents of 'postmodernity' that presume the ending of the modernization process and the dawning of a new era. Contemporary modernity, they argue, rather involves a continuation or even a radicalization of the modernization process: as such *"we are witnessing not the end but the beginning of modernity – that is, of a modernity* beyond *its classical industrial design."*(Beck, 1992: 10). In this sense, the term 'late modernity' is preferred as it neatly encapsulates the continuity of the modern ideas and projects. However, this continuing process of modernization not only liberates and creates opportunities but also produces new risks and uncertainties.: *"Continuity becomes the 'cause' of discontinuity. People are* set free *from the certainties and modes of living of the industrial epoch (...) The system of coordinates in which life and thinking are fastened in industrial modernity – the axes of gender, family and occupation, the belief in science and progress – begins to shake, and a new twilight of opportunities and hazards comes into existence."* (Beck, 1992: 14-15).

This new found freedom of choice is not without risk because it also involves a rising awareness of the individual's responsibility for the consequences and the limits that these decisions entail. In the words of Bauman (2000: 37-38): *The yawning gap between the right of self-assertion and the capacity to control the social settings which render such self-assertion feasible or unrealistic seems to be the main contradiction of fluid modernity (...)"*. In this sense, the gap between individual freedom combined with a heightened risk awareness and the lack of readymade answers or formulas to control these risks, can lead to a state of loss and uncertainty that triggers feelings of unease and anxiety. Therefore, late modernity is characterized by a prominent need for security and safety: *"The driving force in the class society can be summarized in the phrase:* I am hungry! *The movement set in motion by the risk society, on the other hand, is expressed in the statement:* I am afraid! *The commonality of anxiety takes the place of the commonality of need."* (Beck, 1992:49).

While Beck still questions whether and how 'anxiety' and 'safety' can bind people, Bauman decisively argues that safety-issues are stimulating and directing collective action and policy. His thesis is that people's and the state's incapacity to control the future (uncertainty) and their lack of resources to deal with risks (insecurity) are channelled into concerns about the safety of one's body, family and property. However, this preoccupation with 'safety' is unlikely to ease people's mind because the roots of uncertainty and insecurity are left intact. (Bauman, 2000). Nonetheless, because of a lack of tools to tackle uncertainty and insecurity, people as well as policy makers seem to focus their attention on sources of fear that are identifiable and assignable. Not unexpectedly, the focus shifts to people who are unlike 'us': *"Strangers are unsafety incarnate and so they embody by proxy that insecurity which haunts your life. In a bizarre yet perverse way their presence is comforting, even reassuring: the diffuse and scattered fears, difficult to pinpoint and name, now have a tangible target to focus on, you know where the dangers reside and you need no longer take the blows of fate placidly."* (Bauman, 2001: 145).

Inspired by these new sociological insights, criminologists further analyzed in what sense this late-modernity has an impact on the control of crime and people's perception of safety. The best known criminological author is David Garland[5] who states in his widely appraised book 'Culture of Control' (2001) that contemporary citizens are caught up in a 'crime complex', i.e. *"high crime rates are regarded as a normal social fact and crime-avoidance becomes an organizing principle of everyday life. (...) A high level of 'crime consciousness' comes to be embedded in everyday social life and institutionalized in the media, in popular culture and in the built environment."* (Garland & Sparks, 2000: 16). As a consequence of this 'crime complex' many citizens exhibit high levels of fear and *"take on the identity of (actual or potential) crime victims and think and act accordingly"* (Garland, 2001: 164). According to Garland, this institutionalization also explains the fear of crime paradox: *"Our attitudes to crime – our fear and resentments, but also our common sense narratives and understandings – become settled cultural facts that are sustained and reproduced by cultural scripts and not by criminological research or official data."* (Garland, 2001: 164).

Also Hope & Sparks (2000) identify 'fear of crime' as a result of the disintegration of society: *"'Fear of crime' thus intersects with the larger consequences of modernity, and finds its lived social meaning among people's senses of change and decay, optimism and foreboding in the neighbourhoods, towns, cities and wider political communities in which they live and move. Sometimes the question of fear seems chronically enmeshed with the dynamics of de-traditionalisation and an accompanying sense of disruption of formerly settled moral and customary orders."* (Hope & Sparks, 2000: 5). As such 'fear of crime' is a dialectic between on the one hand, people's concern about risks and uncertainties in their local community and everyday life, and on the other hand, their worry about social and cultural transformations on the national and global level (Hope & Sparks, 2000; Loader, Girling, & Sparks, 1998; Pleysier, 2009; Sparks, Girling, & Loader, 2001).

Additionally some researchers connect with Bauman's thesis and conceive the '(fear of) crime' discourse not only as a way to express late modern uncertainty, but also as a way to cope with these anxieties. Unlike the late modern risks, crime and crime-related issues function as a relative familiar domain with identifiable victims and blameable culprits that are manageable and potentially controllable. As such, Hollway & Jefferson, conclude that crime *"serve[s] unconsciously as a relatively reassuring site for displaced anxieties which otherwise would be too threatening to cope with."* (Hollway & Jefferson, 1997: 264). According to this theory 'fear of crime' is a projection of a formless and ambiguous feeling of uneasiness and uncertainty about one's position and identity in society.

In sum, in sociology as well as in criminology, it is recognized that in order to understand and explain people's 'fear' and 'fear of crime' we need to go beyond the obvious safety related issues (e.g. terrorism, crime, paedophily, food-poisoning, toxic waste, ...) and also take into account other anxieties that are triggered by threats to our certainty (e.g. financial and economical crisis) and security (e.g. breakdown of the welfare state). Empirically, this recent insight is operationalized with a diversity of latent constructs such as political impotence, anomy, alienation, intolerance, distrust, ... (Jackson, 2004; Pleysier, 2009). By conceptualizing 'perception of safety' as the outcome of a process of loss or the lack of a stable identity, caused by late modern uncertainties, the way is paved for concepts such as 'active citizenship' and 'social cohesion'

---

5   Although we also need to acknowledge the work of Jock Young (1999): 'The Exlcusive Society: Social exclusion, Crime and Difference in Late Modernity'.

as the prime tools to re-embed and integrate people into society, and thus eliminate their insecurities.

2.2.2 Subjective safety as a legitimate policy objective

Not only in research, but also politically 'fear of crime' emerged as an autonomous policy object, i.e. a problem that needed to be addressed independent of crime fighting. This disentanglement process was already noticeable in the early years of fear of crime research.

Initially the main goal of crime surveys was to measure people's victimization in order to evade the dark number of official (police) figures. But the reasons to include questions measuring people's attitudes and feelings about crime were more politically tinted. In their report[6] "The Challenge of Crime in a Free Society" published in 1967 the American Presidential Commission on Law Enforcement already noticed that fear of strangers could have severe implications for the social order and trust in society.

> "(...) the most dangerous aspect of fear of strangers is its implication that the moral and social order of society are of doubtful trustworthiness and stability. (...) The tendency of many people to think of crime in terms of increasing moral deterioration is an indication that they are losing their faith in their society. And so the costs of the fear of crime to the social order may ultimately be even greater than its psychological costs to individuals." (The Challenge of Crime in a Free Society. A Report by the President's Commission on Law Enforcement and Administration of Justice, 1967: 51).

In a policy report of 1976 that introduced crime surveys in the Netherlands, it was argued that governments that neglected people's opinions, feelings and emotions about crime, might encourage people to take the law into their own hands (vigilantism) and as such endanger public safety and governmental authority (Vanderveen 2006). In short, from the early days of its 'discovery', 'fear of crime' was already conceived as more than just a signal that people were concerned or worried about crime. Fear of crime was perceived as a symptom of a disintegrating society in which people's trust in each other and in government was declining. Consequently, "[t]he pervasive instrumental role that the (crime) statistics and crime surveys played, including the items on 'fear of crime' became clear; statistics are thought to enable politicians and policy makers to 'count & control' or to 'explain & tame'. (Vanderveen, 2006: 203).

These days this reasoning is still very much present in criminal policy. It is generally acknowledged that reducing 'perception of unsafety' is as important as fighting crime and antisocial behavior. In resonance with the objectively measured risk of victimization and the subjectively measured perception of safety, a distinction is made between objective and subjective forms of safety. The former is caused by criminal acts that have to be tackled primarily by the official safety institutions (police and justice). Subjective safety, however, is conceived as more problematic and harder to control because it is perceived as caused by "a declining of social ties, alienation, individualism and a lack of cultural identity."[7] ("Federal Plan of Security and Detention", 2000: 20, au-

---

6 http://www.ncjrs.gov/pdffiles1/nij/42.pdf [9-04-2009]
7 "...het wegvallen van de sociale band, de vereenzaming, ook het individualisme en het ontbreken van culturele identiteit spelen een rol.".

thor's translation). In this sense, policy is referring to recent criminological theories that consider 'perception of unsafety' as a manifestation of late modern uncertainty. Similar to the Presidential report of 1967, subjective safety is perceived as detrimental for trustiness in society but also as potentially dangerous for social cohesion and democracy as people no longer participate in community life and are more susceptible to populist and extreme-right parties and their tough approach on crime. Therefore, it is argued, in addition to fighting crime, policy needs to restore social trust by enhancing social connectedness and improving social integration. ("Federal Plan of Security and Detention", 2000: 15 and 21-22).

By introducing the term 'subjective safety' criminal policy recognizes that in order to improve people's perception of safety more is needed than preventing and fighting crime. In this way, 'social cohesion' and 'active citizenship' become important elements in the governance of people's perception of safety. In the next part, we will further describe how both concepts develop as two prime sources to enhance people's perception of safety.

3 Sources of perception of safety: Social cohesion and Active citizenship:

In the previous part we discussed the autonomization of 'perception of safety' and its conception as a product of concerns and doubts about one's position and identity in late modernity. In this second part, we describe the most prominent theories and strategies that advocate the importance of 'social cohesion' and 'active citizenship' as the prime sources to counter these late modern uncertainties, insecurities and unsafety.

3.1 Communitarianism: civic engagement as a prerequisite of social cohesion

According to Bauman (2000), communitarianism is an all-too-expectable reaction to the contemporary social processes that on the one hand are liberating individuals but on the other are deepening their need for security. In essence, communitarianism promises to defy this imbalance by re-embedding people into a community.

One of its most prominent advocates is the American sociologist Robert Putnam. In his book 'Bowling Alone' (2000) he claims that since the 1970s people's *social capital* is declining, i.e. people are less active in social networks and groups that share common norms and values and are marked by mutual trust and support. Because of this decline in social capital, people's social connectedness and civic engagement have been severely damaged. This collapse of social networks and the disintegration of civil society are detrimental for the individuals' integration and welfare as well as for society's economic position and prosperity. According to communitarians the crisis in social cohesion can be stopped by revitalizing social capital, i.e. restoring safe and secure communities by enhancing people's involvement in civil society.

In this sense active citizenship is conceived as a prerequisite of social cohesion. It is assumed that active citizens do not only preserve their rights and liberties ascribed to them as members of a particular state, but that a virtuous citizen is also concerned about the common good. An active citizen does not only obey and follow, or act critical and controlling, but is also actively involved in protecting, sustaining and constructing society. (Carton & Pauwels, 2005). As such, *"citizenship is about much more than the passive membership of a particular political entity. To be a citizen in the fullest sense,..., you have to be active. It is about a willingness to get involved and make a contribution to both political debate and social action."* (Brannan, John, & Stoker, 2006: 994). The assumption, moreover, is that activating citizens and enhancing civic engagement are essential to face today's crisis's and problems: *"Generating civicness is perceived as a panacea for numerous previously intractable social, economic and political problems: social exclusion, community cohesion, crime, democratic deficit, political apathy and disillusionment, and unresponsive and underperforming public services."* (Brannan et al., 2006: 1005).

In his book Putnam argues that it is the government's responsibility to enhance this civic engagement by stimulating civilians to organize and to participate in voluntary associations. Putnam's theory therefore resonates on the policy level, particularly with policymakers who advocate the *third way approach* in which it is emphasized that dealing with social problems is no longer solely the task of public institutions and that private companies and individual citizens need to recognize and take up their responsibility. Therefore Putnam's theory coheres with a new 'governance' model in which *"state, private and non-governmental organisations and citizens themselves form partnerships to attack problems in new ways."*(Brannan et al., 2006: 994).

3.2   Social disorganization and Community Safety

The analysis of communitarianism also pervades criminology and criminal policy. It is assumed that without social cohesion *"a society (...) would be one which displayed social disorder and conflict, disparate moral values, extreme social inequality, low levels of social interaction between and within communities and low levels of place attachment."* (Forrest & Kearns, 2001: 2128).

In criminology, it is mainly the social disorganization theory that stipulates the importance of social cohesion in reducing crime and fear of crime. In brief, the theory states that disruptions in the social organization of a setting weakens the informal social control mechanisms and as a result deviant behaviour and incivility are not or minimally restrained, which leads to a rise in crime (Sampson & Groves, 1989; Ralph B. Taylor, 1996) and fear of crime (Hale, Pack, & Salked, 1994; Marlowitz, Bellair, Liska, & Liu, 2001; Ralph. B. Taylor & Covington, 1993). Recently, this theory broadened its conceptualization of social cohesion and stresses that research needs to consider the quantity as well as the quality or the 'collective efficacy' of social ties. The concept of 'collective efficacy' refers to a set of common norms and values that sustain informal social control mechanisms. A community with inadequate collective efficacy levels is characterized by a decline in social trust and a reluctance of its members to intervene on behalf of the common good (Sampson, Morenoff, & Gannon-Rowley, 2002; Sampson & Raudenbush, 1999; Sampson, Raudenbush, & Earls, 1997).

By defining social cohesion as 'collective efficacy' and the willingness of people to intervene, social disorganization theory is, intentionally or unintentionally, affirming the concept of 'active citizenship', i.e. that good citizens act on behalf of the common good. As such it is assumed that civic engagement is essential for the social organization of a community and thus for its members' perception of safety: for, it is argued, people who care about their community and its members, act responsible and intervene if this community or one of its members is put into danger. Therefore, it makes sense to relate social disorganization theory with active citizenship, as the latter set in motion the mechanisms of informal social control that are deemed to reduce crime and people's perception of unsafety.

The relationship between social cohesion, active citizenship and (perception of) safety is, however, more explicitly asserted by the concept of 'community safety'. Community safety is *"an approach which seeks to enable local communities or neighbour-hoods to develop the protective capacity to reduce or eliminate the risks of crime and disorder (...) To be successful and sustainable, this capacity must include relationships of trust, mutuality and inter-dependence between community members and between those members and the local agencies of crime control."* (Prior, 2005: 360). As such, more so than other traditional crime prevention strategies, community safety is reliant for its effectiveness on processes of civil renewal (Prior, 2005). The prospect is that a civic attitude and conduct will encourage positive and discourage negative behaviour (Brannan et al., 2006). However, it is argued, that this civicness has to be renewed because late modern processes are eroding the basic moral principles of citizenship and therefore the self-evident guidelines of good conduct. As such, the state is acting to stimulate civicness, willingly or unwillingly. (Boutellier, 2007).

Consequently, as public policy is defining contemporary crime and disorder in terms of *"the breakdown of informal control, moral decline and a collapse in social capital"* (Crawford, 2006: 958), a new 'community governance' model is fostered in which statutory, voluntary and commercial organizations are casted into novel community safety partnerships and security networks (Crawford, 2002; Hughes, 2002). The idea is that *"[t]he reinvigoration of 'community' (...) facilitates informal social control mechanisms which prevent crime. Strong communities can speak to us in moral voices, allowing the policing by communities rather than the policing of communities"* (Crawford, 2004: 513). The slogan 'Safety, Everybody's Duty" must be situated in this context and is an appeal to all of us, individual citizens included, to be 'partners against crime' (Crawford, 2004), which brings Crawford to the observation that *"[w]here once the public was told to 'leave it to the professionals', now they are enjoined to active participation in a 'self-policing society'"* (Crawford, 2004: 67).

According to Lee (2007), 'fear' in this context, is a greater asset than 'crime' as it is better equipped to responsibilise people to take preventive safety measures: *"The fearing subject is a responsibilised active citizen whose civic duty includes keeping one's self and one's belongings safe."* (Lee, 2007: 141). Late modern citizens are inundated by messages to act responsible: however this 'governance-through-fear' is not only enhanced by official governmental functionaries, but is also advocated by non-governmental or non-profit organizations as well as by privately organized companies and institutions.

In sum, security is no longer perceived as the state's monopoly, but as a responsibility of every citizen: *"Security is a citizen's right and everyone's duty[8]."* ("Federal Plan of Security and Detention", 2000: 15-16, author's translation).

3.3 Local approach to a diffuse, relative and global phenomenon

In research and policy it has been argued that enhancing 'active citizenship' is beneficial to 'social cohesion' and as such both are valuable resources to improve people's perception of safety. So far, we have been reluctant to geographically locate these

---

[8] "Veiligheid is een recht van de burger en een plicht van iedereen".

sources of perception of safety. Although the term 'community' might suggest a preference for a local approach, theoretically the concepts of 'social cohesion' and 'active citizenship' are not necessarily limited to a particular geographical setting. However, in research and in policy a local approach is emphasized. Moreover, it is argued that it is mainly disadvantaged urban areas that lack the resources to produce and sustain social cohesion. (Forrest & Kearns, 2001).

This local approach is elicited by situating perception of safety as a late modern product with diverse causes and different shapes that change over time and according to location. Consequently, by defining perception of safety as a *diffuse and relative phenomenon*, it makes sense to assume that there are no universally applied methods to deal with perception of safety and so, it is argued, improving perception of safety *"requires a differentiated approach which is tailor-made for the local situation"* (Lasthuizen, van Eeuwijk, & Huberts, 2004: 218). In particular the neighbourhood approach is further solicited. Especially the social disorganization theory considers the social organization of people's residential setting as a primary source of safety as well as unsafety. It is argued that *"[t]o create and maintain a safe, manageable and predictable society, we shall have to start within the community, in the public's direct residential areas. The neighbourhood is the place where it all starts for the public and therefore it should be the starting point for actions that really matter."* (Lasthuizen, van Eewijk, 2004: 218). By the same token, 'active citizenship' is rather seen as a bottom-up process and coherent with a local approach in which government stimulates citizens to participate in meetings and organise their own initiatives to improve the liveability and safety of their residential area (van Caem, 2008; van Ostaaijen & Tops, 2007). As such, civic engagement is not so much related to making global statements or participating in (trans)national organisations, but is rather perceived as an attitude that must be cultivated and manifested locally.

When perception of safety is described as a diffuse and relative phenomenon for which only custom-made and locally sensitive methods are adequate, it is not a surprise that in most European countries, crime and insecurity are increasingly defined as problems for local policy and local intervention (van der Vijver & Terpstra, 2004). Thus, although the fourth dimension of the aetiology of 'fear of crime' situates 'perception of safety' as a product of global forces, the (policy) response is geographically limited to the third dimension, i.e. the local environment, and more particularly the residential neighbourhood.

4   All for safety, safety for all?

The slogan "Safety: Everybody's concern, everybody's duty" assumes that safety is a common interest and promises that by uniting and working together we, i.e. government, private associations and individual citizens, can overcome the problems and dangers that are threatening our families and communities. In this sense, being (social cohesion) and acting (active citizenship) together seems a natural reaction to safeguard social order and people's perception of safety. In this third part, however, we question the concepts of 'social cohesion' and 'active citizenship' as the prime sources to enhance 'perception of safety'. Is social cohesion beneficial and necessary to people's perception of safety? Is active citizenship required to enhance social cohesion and thus people's perception of safety?

4.1 Assumptions about social cohesion

The main assumption about social cohesion is that it functions as a natural barrier against crime and other deviant behaviour. Is this the case, however? It seems that there are some dubious assumptions about social cohesion and its relationship with safety.

4.1.1 Social cohesion is in crisis

The greatest misapprehension about social cohesion is that it is in dire straits. This apprehension is mainly based on a nostalgic notion about the past and the idea that people then were more connected and showed more solidarity with each other. However, is this longing for the 'days of yore' warranted and are we not nostalgic about a society that never existed? Social cohesion, it is argued, is not a predefined and stable concept but is very much context related, i.e. its form and content varies according to time and scale. As such, loyalty to and connection with a certain neighbourhood or group can be detrimental for solidarity on other levels such as the city or national level. Therefore, to claim that social cohesion is in crisis, depends upon what timeframe and/or spatial scale one is examining. (Blokland, 2005; Forrest & Kearns, 2001).

4.1.2 Social cohesion is a local process

Consequently, recognizing that social cohesion is relative to time and space,, implicates that its manifestation is not limited to a certain geographical space, such as a neighbourhood. Processes of globalization and individualization are extending people's social networks and interactions. As such, the residential area is but one of the many contexts where social cohesion emerges. However, this does not entail that the local is no longer important for socialization and social identity. By and large, identification with extra-local connections and groups is increasing, but to what extent 'location matters' depends upon the individual's use of the neighbourhood (Blokland, 2005) and his/hers lifestyle and life phase (Forrest & Kearns, 2001).

4.1.3 Social cohesion is positive

The third great misapprehension is that social cohesion is unambiguously a good thing. This assumption is first of all based on the misunderstanding that social cohesion can only be possessed by mainstream groups and organizations set up by obeying citizens and recognized by government. However, social cohesion can also be a characteristic of criminal organizations (cfr. Mafia) and non-conforming organizations that question the current *status quo*. These upstream organizations also have social networks based on trust, mutual support, shared values and informal control mechanisms, and as such they do not differ from mainstream forms of social cohesion, except for there disobedience to the current laws and authorities. (Mayer, 2003).

These non-conformists can raise anxiety, protest and conflict and can therefore be menacing for the social order and people's perception of safety. However, there are also some criminological examples of neighbourhoods where upstream forms of social cohesion are safeguarding people's safety. In these settings the presence of criminal gangs did not increase people's fear, as their activities were not directed to the residents. Moreover, by some, their presence was appreciated because they interfered when internal conflicts or outsiders tended to disturb the neighbourhood. As such, these gangs were perceived as upholders of peace in a neighbourhood where official forms of control were missing or mistrusted. (Crawford, 1999; I. Taylor, Evans, & Fraser, 1996; Triplett, Sun, & Gainey, 2005; Walklate, 1998a, 1998b, 2001).

The positioning of social cohesion as something unquestionably good, is in fact a result of the communitarian analysis in which there is no reference to the notion of 'power'. In Putnam's analysis, social capital is perceived as something beneficial for the wellbeing of every member of society. However, the manifestation and preservation of a social network depends upon the power the group has in relation to other groups and individuals in society. (Bolt & Torrance, 2005). Furthermore, it is argued, other forms of capital, i.e. economical and cultural capital, are essential to the emergence and existence of social cohesion. (Portes, 1998). By not recognizing the power element, communtarians never considered or neglected the negative consequences of social cohesion. However, any form of social cohesion, mainstream or upstream, is not without risk. Forming a group is not only about defining who is part of the group, but is also about deciding who is outside the group. As such group formation and cohesion can be detrimental for those without a membership card and lacking alternative resources to attain and defend their interests: *"[Social cohesion] can be about discrimination and exclusion and about a majority imposing its will or value system on a minority."* (Forrest & Kearns, 2001: 2134). Some go as far as to state that social cohesion is inherently connected with social exclusion and they warn us about the dangers of desiring 'community':

> *"Community therefore contains another fundamental contradiction at its heart. Coveted for its secure sense of belonging and inclusiveness, even its most fragile ephemeral realization hinges upon vigorous exclusion and differentiation. To the extent we can experience it, we do so by expressing the insecurity of our own difference from others and their collective exclusion from our ranks."* (Carson, 2004: 8).

4.1.4 Social cohesion is essential for safety

The final misapprehension that we will discuss here, is the assumption that social cohesion is essential to social control and therefore a prerequisite to safety. By exposing social cohesion as not necessarily positive, we have already illustrated that social cohesion can be detrimental to social order and people's perception of safety as its manifestation can exhibit resentment, exclusion, protest, tension and even conflict between groups and individuals. But we do not only contest its positive consequences, but we also question its necessity to enhance people's perception of safety: is social harmony an essential prerequisite for safety? In research there are ample counterexamples that doubt this assumption.

Research into social cohesion, is too much focused on problem neighbourhoods and as such there is but a partial view on the importance of social networks in a neighbourhood (Forrest & Kearns, 2001). In his research Baumgartner (1988) studied conflict resolution in American suburbs and concluded that social order was established by avoiding conflict and non-intervention. As such, social order was attained, not by familiarity and connection, but by precisely those factors that are assumed to be detrimental for conflict and violence: fragmentation, isolation, indifference and volatility. In well-off residential areas, it is argued, people seem to be more inclined to appeal on formal control mechanisms to resolve conflicts than on informal ones. (Crawford, 1999).

Another extreme example are the 'gated communities' where social interaction between residents is rather limited, but where private companies are paid to make sure that the neighbourhood is secure and safe. Conversely, there are also neighbourhoods where residents are very close and mutual supportive, but have to deal with a lot of crime and incivility. (Crawford, 1999; DeFillipis, 2001; Forrest & Kearns, 2001; Foster, 1995).

In sum we can state that social cohesion conceptualized as individuals bonded together in a strong (local) social network, is not necessarily a prerequisite for more social order and safety.[9] Other factors are thus as, or even more, important to secure a local setting. In his study, Patrick Carr (2003) concluded that effective social control is possible without strong social networks, but that its functioning clearly depends upon the extent to which local individuals or organizations can appeal to political and institutional resources *outside* the neighbourhood. Research and policy tend to overlook the importance of *vertical relationships* in the exertion of social control (Crawford, 1999). However, the extra-local context, i.e. the social connections of residents with the city council, is important to understand a neighbourhood's capacity to deal with criminal and deviant behaviour. (Crawford, 2006).

Next to the strong and vertical relationships of a neighbourhood, the importance of *weak ties* must not be overlooked. Although these occasional and fluid connections are not as supportive as strong ties, they are less provisional and can be very useful if these connections give access to resources that are not provided for by one's own strong social network (Granovetter, 1973, 1983; Henning & Lieberg, 1996)[10]. Moreover, some also value these weak ties because these small and sporadic encounters make the setting and the people in it more predictable and trustworthy (Blokland, 2005; Soenen, 2001). In this sense, the term 'social cohesion' is replaced by 'cognitive cohesion':

"Proximity politics should focus on getting to know each other – on becoming acquaintances, not necessarily best friends. (...) To increase the feeling of safety and 'livability' in neighbourhood requires that the quality of social relations meet certain

---

9   In this article I criticize the assumption that a community's social capital, i.e. the presence of a strong social network, is a prerequisite for social order and safety. The same argument applies for individuals' social capital and their (perception of) safety, i.e. belonging or having access to a social network is not necessarily beneficial for the individual as the network can restrict individuals' freedom or access to resources. Consequently, the relationship between individuals' social capital and their (perception of) safety is not unequivocal or straightforward positive. Cfr. Portes (1998) for a critical assessment about the functions of social capital at the individual and community level.

10   Putnam (2000) himself has also acknowledged this by making a distinction between bonding and bridging ties.

minimum standards. To gain a sense of social safety, we don't have to be friends with each other, but we do need to get along." (Duyvendak, 2004: 33).

4.2 Assumptions about active citizenship

According to communitarianism and adherents of 'community safety', active citizenship is a prerequisite to attain a cohesive and thus safe society. In the following paragraphs we question these assumptions about active citizenship and its relation with social cohesion and safety.

4.2.1 Active citizenship: "Yes, We Can![11]"?

Active citizens are attributed a wide range of good qualities: good citizens are expected to be on guard for themselves as well as for others, to keep themselves informed, to automatically cooperate with the authorities and to feel responsible for the socially weak members of our society. These expectations, however, presume a lot of knowledge and skills that are not evenly distributed throughout society: not every citizen disposes of the necessary economical, social or cultural capital to exert their citizen's rights and duties, i.e. lack the money, relationships and assertiveness to mobilize, to formulate demands and get involved in top-down decisions. (Uitermark, 2007; Uitermark & Duyvendak, 2007).

Additionally, these high expectations are risky, as citizens are sometimes accorded responsibilities that belong in fact to official authorities. As such, there is always the risk of 'blaming the victim'. Idealising active citizenship might turn the citizen into the scapegoat when problems re-emerge or sustain. (Duyvendak, 2004).

Even if citizens are adequately skilled, is it reasonable to expect them to sort out solutions to problems that are perhaps not within their reach? It is not because citizens express problems, that the causes and solutions are citizen related. These problems are often caused by structural problems that cannot be solved by activating citizens. In this sense, an approach that is mainly citizen orientated, is focusing on symptom treatment or risk containment, but is rarely or never offering a long term solution. (Uitermark & Duyvendak, 2008; Uitermark, Duyvendak, & Kleinhans, 2007).

Finally, we question the willingness of citizens to invest in dealing with problems. A much heard frustration and disappointment among official authorities is that citizens are not interested and difficult to motivate. Often only a minority is willing to contribute time and effort to get involved. Most citizens, however, lack the time and vigor to participate in initiatives. Especially residents in problem neighbourhoods are predominantly preoccupied with their own survival and are therefore not interested or able to invest in improving their living conditions with others. (Van den Broeck, 2004).

---

11 Slogan used by President Barack Obama in the 2008 presidential elections.

4.2.2 Active citizenship is essential for social cohesion

Active citizenship is also assumed to induce positive consequences: the activation of citizens is expected to improve individual's social integration and enhance mutual contact between citizens. Accordingly, active citizenship is depicted as the driving force of social cohesion. However, there are some indications that active citizenship is not unambiguously positive for social cohesion and that its implementation can even (further) disrupt social life within the community.

In his research, Maarten Loopmans (2005) assessed two kind of tensions in activated citizens. On the one hand, activated citizens can have a stressful relationship with non-active citizens whose indifference and even adverseness, can be provoking and demoralizing. Consequently, the community (further) disintegrates and clashes are possible. On the other hand, activation can also lead to inner tensions. Activation often raises people's expectations and hopes. But, at the same time, it might also elevate their awareness and alertness to problems. This situation becomes a problem if negotiated solutions and successes tend to be overridden by failures that bring feelings of helplessness and desolation. Consequently activation is no longer considered to be a duty towards the community and on behalf of the common good, but as a way to express their despair and individual frustrations. As such, activation is self-centered and intensifies in-between differences rather than it is bringing people closer together.

This focus on active citizenship is especially tart when it is directed towards alleged socially marginal groups such as immigrants and youngsters in high crime areas. In a judicial formal way these groups are considered to be citizens because legally they are members of society and as such they have rights and obligations. However, in reality these groups have always been considered as non-integrated, mainly because they are structurally deprived, but since the mid-90s their underprivileged status is also or predominantly explained from a cultural perspective (Schinkel, 2007). These groups are outside of mainstream society because they are culturally different and therefore classified as passive citizens that need to be initialized in good citizenship, i.e. one that is fully in line with the dominant culture. In this sense, this kind of activation is not aimed at emancipation and inclusion, quite the contrary: these initiatives are suppressing identity and individuality (Schinkel, 2007) and are more likely intended to discipline and to civilize (Uitermark & Duyvendak, 2008). Consequently, whatever its principles and good intentions, implementing activation initiatives can be detrimental to a community's social connectedness as it is a very delicate and difficult task to realize active citizenship of socially marginalized groups without exposing and possibly stigmatizing their otherness.

4.2.3 Active citizenship is essential for safety

In the final paragraph, we examine the assumption that activating citizens is a required tactic in governing safety. However, some researchers, who examined the implementation of 'community safety' strategies, paint another picture. They question the extent to which the 'community safety' governance model is effectively improving people's involvement and their perception of safety. Moreover, some researchers warn for the negative effects that stimulating activation has on people's perception of safety.

In the second part of this paper, we suggested that government is fostering a multi-agency approach to handle crime and unsafety. This 'community safety' governance model often produces a plethora of projects and approaches that is very confusing for the targeted citizen groups. As such, Van den Broeck concludes that *"[t]he policy network for the local governance of crime becomes a jumble, a Babylon confusion of tongues, which fails to bridge the (communication) gap between citizens and government and which does not produce any lasting results."* (Van den Broeck, 2004: 132). Furthermore, activated citizens are often very disappointed about their degree of involvement and complain that the public hearings are in fact reduced to meetings for announcing top-down decisions or that their attendance and commitment is misused as a pretext to justify and impose some unwished decisions. In addition, activation is not a guarantee for a more democratic decision process as some activated citizens experienced that politically influential citizens had more impact on policymakers. In sum: *"Mere decentralization does not eliminate or 'magic away' the existing 'autocratic', top-down management and leadership styles in the municipal apparatus."* (Van den Broeck, 2004: 132).

Additionally, it is argued, that partnerships are not necessarily more successful in handling crime and unsafety. A partnership entails that responsibility is devolved and dispersed, but as such no partner is ultimately accountable: *"the problem of many hands where so many people contribute that no one contribution can be identified; and if no person can be held accountable after the event, then no one needs to behave responsibly beforehand. As authority is 'shared' it becomes difficult to disentangle and become almost intangible."* (Crawford, 2004: 77). Some researchers even argue that the success of safety initiatives depends on the dedication of merely a handful of motivated people: *"Despite all the talk about partnership crime prevention policies are shaped and determined by quite literally one handful of individuals."* (Foster, 2002: 172). Certain key figures, such as a motivated police officer, an enthusiastic community worker or an engaged citizen, are often more effective than a partnership comprised of diverse powerful organizations. The importance of a number of motivated and well-placed persons is wonderfully illustrated in Ben Rovers' research. Rovers (2007a, 2007b) states that criminal policy is too much focused on improving the environment and/or citizen's knowledge and skills (emancipation, sensitization, informing). A project's success, however, is also influenced by the way the parties concerned (citizens, but as well as initiators and partners) are committed to the project and the extent they can transport this belief to others. Thus, according to Rovers, this *belief effect*, exerted by official parties and individuals, is crucial in the successful implementation of any project aimed at activation in order to improve perception of safety.

Some researchers question the beneficial effects of community safety interventions all together. They argue that such interventions will not bring people together and potentially even harm their perception of safety. David Prior (2005) convincingly states that: *"...whilst a core aim of community safety intervention is an increase in the level of trust within a community, the outcome may often be an increase of suspicion."* (Prior, 2005: 360). As such, in stead of increasing people's trust in each other, safety interventions are increasing distrust and therefore sentiments of unsafety. For example, the main goal of CCTV is to control and monitor behaviour, but its installment is also believed to enhance the perception of safety among residents and other legitimate users of the setting. However, intentionally or unintentionally, CCTV also raises suspicion and distrust as every uncanny conduct or any stranger is to be regarded as a potential hazard. In this sense, *"the objective of increased trust and confidence in everyday life is pursued through the principle of the active and routine suspicion – the presumed distrust – of others."* (Prior, 2005: 361). Another example are social prevention strategies that (re)activate (potential) offenders in order to (re)integrate them into society. These strategies presume that certain social groups threaten the social order and need to be reinitiated, whether by persuasion or coercion, into the dominant attitudes, values, lifestyles and behaviours. These programmes, however, are more likely *"to reproduce and possibly exacerbate the inherent dynamics of social exclusion that exist within communities"* (Prior, 2005: 365) and therefore will confirm people's suspicion and their perception of unsafety. Prior concludes that safety is not a neutral concept because 'community safety' strategies unavoidable propagate certain ideas and norms about safety and order, and consequently define, condemn and eliminate deviant lifestyles and behaviours. Therefore, it is not unsurprising that these community safety strategies do not succeed in enhancing trust and social cohesion.

4.3   Conclusion

In this third part we outlined that active citizenship and social cohesion are not necessarily positively interlinked and are not necessarily a prerequisite for more perception of safety. Moreover we also indicated that civicness and cohesiveness are possibly harmful for people's perception of safety. The reason for these ambivalent relationships is that both concepts, intentionally or not, bring into view differences and contrasts. Consequently, they may affirm and enhance superficial tensions and suspicions between citizens and as such decrease mutual trust and people's sense of connectedness and safety.

Yet, this is not a call to totally discard the input of citizens in safety prevention strategies. However, we do argue that policymakers should be careful to use 'active citizenship' and 'social cohesion' to increase people's perception of safety. Stimulating civicness and/or cohesiveness can be in itself valuable goals, but they are no guarantees for more safety. 'Safety' is in itself not a neutral concept as its interpretation varies according to time, setting and parties evolved. Because safety can be construed in various ways, negotiating a consensus is impossible and even undesirable, as consensus implicates smoothing of differences on behalf of one particular interpretation. But, as we have shown in this part, ignoring or neglecting these differences in opinion and lifestyle is in fact not necessarily beneficial for the social cohesion of a community, nor for the active involvement of certain citizen groups and as such "All for safety" does not equal "Safety for all". Therefore, some argue that we should accept these differences and assert a *conflict model* in which different perspectives and attitudes towards safety are taken into account. (Goris, 2001). In this sense, it is also essential to recognize that people have different forms of perception of safety. In the final part we set out the main contours of an alternative stance in comprehending perception of safety.

5   An alternative stance: identity and perception of safety

In the first part 'perception of safety' was described as a late modern phenomenon, i.e. a product of macro-sociological processes that causes doubts and concerns about one's identity and position in society. In the new dimension of the aetiology of 'fear of crime', perception of safety is therefore considered as a result of a combination of sociological (gender, age, ethnicity, ...) and socio-psychological features (political impotence, alienation, intolerance, distrust, ...). In this sense, the assumption is that people who feel afraid combine certain *identity features* that make him/her more insecure and uncertain, i.e. a high perception of unsafety is more likely to be found with someone who is e.g. physical vulnerable, unskilled, unemployed, resident of a high crime and who tends to be intolerant, a-political, alienated and xenophobic. According to this analysis, it seems sensible to focus on problem neighbourhoods and draw up strategies aimed at informing, sensitizing and enhancing civic involvement and social interaction.

In sum, based on late modern theories it is assumed that perception of unsafety is caused by *a loss or lack of a stable identity*. However, researchers do not seem to agree which identity features are more influential than others. Consequently, 'perception of safety' is still a *black box*: after 40 years we seem to have a fairly complete overview of *what* kind of factors trigger perceptions of unsafety and *what* kind of consequences these perceptions elicit or aggravate, but we do not seem to fully comprehend *why* these factors are sometimes exceedingly important and then again seem to be totally irrelevant. As such, 'perception of safety' is more than ever considered to be a *diffuse and relative phenomenon*, i.e. a multilayered phenomenon with diverse causes and different shapes that changes over time and according to location. If we are not careful, the spectre of illegitimacy re-emerges, i.e. the conception of 'perception of safety' as indefinable, intangible and thus uncontrollable. However, we argue, that its diffuse and relative nature, is not a to be considered as a final result but as a starting point of research. In this final part we shortly suggest an alternative perspective to understand and research 'perception of safety'.

First of all, we acknowledge that perception of safety is a complex phenomenon, but this does not entail that it is not researchable or uncontrollable. Perception of safety is indeed an unstable and dynamic entity (Lupton, 2000) because it is based on personal experiences and individual biography. In this sense people might differ in identifying and selecting situations as unsafe. Safety is a social phenomenon that is constructed through social and cultural processes. Thus people's experiences become meaningful through their social-cultural framework of meaning. Consequently, what people define as unsafe may differ, but just as important are the definitions of unsafety which are shared among individuals. These shared understandings of unsafety are rooted in social and cultural processes. The goal of research is then to examine these social-cultural frameworks that people share in constructing their perception of safety.

Secondly, in line with late modern theories we acknowledge the importance of identity in the construction of perception of safety, i.e. we agree that different aspects of social identity work together in different ways to impact on the nature of 'perception of safety' (Pain, 2001); or that the individuals' unique biographies impact on their identification with fear of crime discourses (Hollway & Jefferson, 1997). However, we do not agree with the way identity is commonly researched. Generally identity is considered as a result of (a combination of) individual features, i.e. it is examined by measuring individual characteristics such as gender, race, age, class or by appraising their place of birth or residence and their membership to certain associations. People's identity, however, is not an individual feature formed in a vacuum, but constructed in relation to others: to understand people's identity we need to research how they are assessing their own position and that of others, i.e. people are defining their identity by continually categorizing others and determining to what category they belong to (Tulloch, 2000).

Consequently, the process of identity formation and social categorization is important to understand people's perception of safety. More concretely we need to research the social categories people draw upon when they are assessing an unsafe situation and how they position themselves and others as potential victims or assaulters. In this sense, perception of unsafety is not a result of a lack or unstable identity, but on the contrary: in describing unsafe situations people are highly aware of their identity and they act accordingly. For example, most older people perceive themselves as (getting more) vulnerable and tend to avoid risky situations such as taking the bus or going out late at night. Moreover, they perceive younger people as acting dangerous and reckless. Consequently, their perception of safety is very much guided by their self-categorization as 'old' and 'vulnerable' and the categorization of young people as threatening. Conversely, young people, especially men, like to perceive themselves as capable of defending themselves and being in control. They tend to perceive gangs of young people, especially those of other subcultures, as threatening. (Tulloch, 2000).

In this alternative stance we acknowledge the dynamic and relative nature of perception of safety. However, we do not consider this as a result of macro-sociological processes that are undermining people's identity. On the contrary, people tend to assess unsafe situations in accordance to their identity and that of others. Identification with neighbourhoods and/or communities might still be essential to some people, but in a cosmopolitan world people tend to have more (important) identities than being a resident of a neighbourhood or being a member of an association. Maybe people complain about unsafe situations in their neighbourhood, but are these situations considered as unsafe because they are menacing the personal living conditions and/or diminishing involvement in society? Or are the perceived unsafe conditions rather fuelled by other identities: for instance parents who worry about their children's safety or such as young women who intuitively act more careful when they are wandering in any street late at night. Therefore, policies that are directed to stimulate social cohesion and civic engagement in local settings might not fully address the problem because their envisioning the wrong identities, stimulating identities that enhance perceptions of unsafety and neglecting the identities that do matter to people. Perception of safety is a complex phenomenon. Therefore, policy makers need to analyze exactly what it means for people to feel unsafe before implementing community safety measures.

6   Bibliography

Bauman, Z. (2000). *Liquid Modernity*. Cambridge: Polity Press.

Bauman, Z. (2001). *Community. Seeking Safety in an Insecure World*. Malden (USA): Polity Press in association with Blackwell Publishers Ltd..

Baumgartner, M. P. (1988). *The Moral Order of a Suburb*. USA: Oxford University Press.

Beck, U. (1992). *Risk Society: Towards a New Modernity*. London: Sage.

Blokland, T. (2005). *Goeie buren houden zich op d'r eigen. Buurt, gemeenschap en sociale relaties in de stad*. Den Haag: Dr. Gradus Hendriks-stichting.

Bolt, G., & Torrance, M., I. (2005). *Stedelijke herstructurering en sociale cohesie*. Utrecht: DGW/NETHUR.

Boutellier, H. (2007). *Nodale orde. Veiligheid en burgerschap in een netwerksamenleving. Inaugurele rede bij de aanvaarding van de Frans Denkers leerstoel Veiligheid & burgerschap aan de Vrije Universiteit vanwege de gemeente Amsterdam en het politiekorps Amsterdam-Amstelland*. Amsterdam: Faculteit der Sociale Wetenschappen, Vrije Universiteit Amsterdam.

Brannan, T., John, P., & Stoker, G. (2006). Active Citizenship and Effective Public Services and Programmes: How Can We Know What Really Works? *Urban Studies, 43*(5/6), 993.

Carr, P., J. (2003). The New Parochialism: The Implications of the Beltway Case for Arguments Concerning Informal Social Control. *American Journal of Sociology, 108*(6), 1249-1291.

Carson, W. G. (2004). Is Communalism Dead? Reflections on the Present and Future Practice of Crime Prevention: Part One. *Australian and New Zealand Society of Criminology, 37*(1), 1-21.

Carton, A., & Pauwels, G. (2005). Burgerschap in Vlaanderen anno 2004. De perceptie van de rol van de burger en de overheid in de weegschaal gelegd, *Vlaanderen gepeild 2005*. Brussel: Administratie Planning en Statistiek. Departement Algemene Zaken en Financiën. Ministerie van de Vlaamse Gemeenschap.

Crawford, A. (1999). Questioning Appeals to Community within Crime Prevention and Control. *European Journal on Criminal Policy and Research, 7*, 509-530.

Crawford, A. (2002). The Governance of Crime and Insecurity in an Anxious Age: the Trans-European and the Local. In A. Crawford (Ed.), *Crime and Insecurity. The Governance of Safety in Europe* (pp. 27-51). Devon: Willian Publishing.

Crawford, A. (2004). The Governance of Urban Safety and the Politics of Insecurity. In K. van der Vijver & J. Terpstra (Eds.), *Urban Safety: Problems, Governance and Strategies* (pp. 65-85). Enschede: Institute for Social Safety Studies. University of Twente.

Crawford, A. (2006). 'Fixing Broken Promises?': Neighbourhood Wardens and Social Capital. *Urban Studies, 43*(5/6), 957-976.

DeFillipis, J. (2001). The Myth of Social Capital in Community Development. *Housing Policy Debate, 12*(4), 781-806.

Ditton, J., & Farrall, S. (2000). *The fear on crime*. Dartmouth: Ashgate.

Duyvendak, J. W. (2004). Neighbourhoods, Cohesion and Social Safety. In K. van der Vijver & J. Terpstra (Eds.), *Urban Safety: Problems, Governance and Strategies (pp. 107-132)*. (pp. 27-35). Enschede: Institute for Social Safety Studies. University of Twente.

Federal Plan of Security and Detention. (2000). Brussels: Department of Justice.

Forrest, R., & Kearns, A. (2001). Social Cohesion, Social Capital and the Neighbourhood. *Urban Studies, 38*(12), 2125-2143.

Foster, J. (1995). Informal Social Control and Community Crime Prevention. *British Journal of Criminology, 35*(4), 563-583.

Foster, J. (2002). 'People Pieces'; the Neglected but Essential Elements of Community Crime Prevention. In G. Hughes & A. Edwards (Eds.), *Crime Control and Community: The New Politics of Public Safety* (pp. 167-196). Devon: Willian Publishing.

Garland, D. (2001). *The Culture of Control. Crime and Social Order in Contemporary Society*. Oxford: Oxford University Press.

Garland, D., & Sparks, R. (2000). Criminology, Social Theory and the Challenge of our Times. *British Journal of Criminology, 40*, 189-204.

Giddens, A. (1990). *The Consequences of Modernity*. Cambridge: Polity Press.

Goris, P. (2001). Community Crime Prevention and the 'Partnership Approach': a Safe Community for Everyone? *European Journal on Criminal Policy and Research, 9*, 447-457.

Granovetter, M. (1973). The Strength of Weak Ties. *The American Journal of Sociology, 78*(6), 1360-1380.

Granovetter, M. (1983). The Strength of Weak Ties: A Network Theory Revisited. *Sociological Theory, 1*, 201-233.

Hale, C. (1996). Fear of Crime: a Review of the Literature. *International Review of Victimology, 4*, 79-150.

Hale, C., Pack, P., & Salked, J. (1994). The Structural Determinants of Fear of Crime: an Analysis Using Census and Crime Survey Data from England and Wales. *International Review of Victimology, 3*, 211-233.

Henning, C., & Lieberg, M. (1996). Strong Ties or Weak Ties? Neighbourhood Networks in a New Perspective. *Scandinavian Housing & Planning Research, 13*, 3-26.

Hollway, W., & Jefferson, T. (1997). The Risk Society in an Age of Anxiety: Situating Fear of Crime. *British Journal of Sociology, 48*(2), 255-266.

Hope, T., & Sparks, R. (2000). *Crime, risk and insecurity : law and order in everyday life and political discourse* London: Routledge.

Hughes, G. (2002). Plotting the Rise of Community Safety: Critical Reflections on Research, Theory and Politics. In G. Hughes & A. Edwards (Eds.), *Crime Control and Community: The New Politics of Public Safety* (pp. 20-45). Devon: Willian Publishing.

Jackson, J. (2004). Experience and Expression. Social and Cultural Significance in the Fear of Crime. *British Journal of Criminology, 44,* 946.

Lasthuizen, K., van Eeuwijk, B., & Huberts, L. (2004). Feelings of Insecurity in Communities: What does really matter? Results of Survey Research in the Netherlands. In K. van der Vijver & J. Terpstra (Eds.), *Urban Safety: Problems, Governance and Strategies* (pp. 205-220). Enschede: Institute for Social Safety Studies. University of Twente.

Lee, M. (2007). *Inventing Fear of Crime. Criminology and the Politics of Anxiety.* Devon: Willan Publishing.

Loader, I., Girling, E., & Sparks, R. (1998). Narratives of Decline. Youth, Dis/order and Community in an English 'Middletown'. *Britisch Journal of Criminology, 39*(3), 388-403.

Loopmans, M. (2005). From Residents to Neighbours: The Making of Active Citizens in Antwerp, Belgium. In J. W. Duyvendak, T. Knijn & M. Kremer (Eds.), *Policy, People, and the New Professional. De-professionalisation and Re-professionalisation in Care and Welfare* (pp. 109-121). Amsterdam: Amsterdam University Press.

Marlowitz, F., Bellair, P., Liska, A., & Liu, J. (2001). Extending Social Disorganization Theory: Modeling the Relationships Between Cohesion, Disorder and Fear. *Criminology, 39,* 293-317.

Mayer, M. (2003). The Onward Sweep of Social Capital: Causes and Consequences for Understanding Cities, Communities and Urban Movements. *International Journal of Urban and Regional Research, 27*(1), 110-132.

Pain, R. (2001). Gender, Race, Age and Fear in the City. *Urban Studies, 38*(5), 899 – 913.

Pleysier, S. (2009). *'Angst voor criminaliteit' onderzocht. De brede schemerzone tussen alledaagse realiteit en irrationeel fantoom.* Proefschrift tot het verkrijgen van de graad van doctor in de Criminologische Wetenschappen.(promotor: prof.dr. G. Vervaeke; co-promotor: prof.dr. J. Goethals): Afdeling Strafrecht, Strafvordering en Criminologie – Faculteit Rechtsgeleerdheid – Katholieke Universiteit Leuven.

Portes, A. (1998). Social Capital: its Origins and Apllications in Modern Sociology. *Annual Review of Sociology, 24*(1), 1-24.

Prior, D. (2005). Civil Renewal and Community Safety: Virtuous Policy Spiral or Dynamic of Exclusion? *Social POlicy & Society, 4*(4), 357.

Putnam, R. D. (2000). *Bowling Alone: the Collapse and Revival of American Community.* New York: Simon & Schuster.

Rovers, B. (2007a). 'Ze deugen nergens voor': Het Belief effect in justitiële interventies. Lectorale Rede (9 november 2007)

Rovers, B. (2007b). Preventie van jeugdcriminaliteit en de placeborespons: toepassing van het belief-effect in het veiligheidsveld, *Marktdag van de Nederlandse Vereniging voor de Kriminologie (NVK).* Leiden.

Sampson, R. J., & Groves, B. (1989). Community Structure and Crime: Testing Social-Disorganization Theory. *American Journal of Sociology, 94*(774-802).

Sampson, R. J., Morenoff, J. D., & Gannon-Rowley, T. (2002). Assessing "Neighborhood effects": Social Processes and New Directions in Research. *Annual Review of Sociology, 28*(1), 443-478.

Sampson, R. J., & Raudenbush, W. (1999). Systematic Social Observation of Public Spaces. A New Look at Disorder in Urban Neighbourhoods. *American Journal of Sociology, 105*, 603-651.

Sampson, R. J., Raudenbush, W., & Earls, F. (1997). Neighborhoods and Violent Crime: A Multilevel Study of Collective Efficacy. *Science, 277*.

Schinkel, W. (2007). Tegen 'actief burgerschap'. *Justitiële Verkenningen, 33*(8), 70-90.

Soenen, R. (2001). *Diversiteit in verbondenheid. Paper in het kader van de Task Force Leven in de stad: thema 'sociale cohesie en etnische diversiteit'*

Witboek Stedenbeleid.

Sparks, R., Girling, E., & Loader, I. (2001). Fear and Everyday Urban Lives. *Urban Studies, 38*(5), 885-898.

Taylor, I., Evans, K., & Fraser, P. (1996). *A Tale of Two Cities*. Routledge: London.

Taylor, R. B. (1996). Neighborhood Responses to Disorder and Local Attachments: The Systemic Model of Attachment, Social Disorganization, and Neighborhood Use Value. *Sociological Forum, 11*(1), 41-74.

Taylor, R. B., & Covington, J. (1993). Community Structural Change and Fear of Crime. *Social Problems, 40*, 374-395.

The Challenge of Crime in a Free Society. A Report by the President's Commission on Law Enforcement and Administration of Justice. (1967). Washington DC: United States Goverment Printing Office.

Triplett, R., A., Sun, I., Y., & Gainey, R., R. (2005). Social Disorganization and the Ability and Willingness to Enact Control: A Preliminary Test. *Western Criminology Review, 6*(1), 89-103.

Uitermark, J. (2007). Temper de transformatiedrift met bewonersparticipatie. *Stadscahiers, 1*(1), 36-42.

Uitermark, J., & Duyvendak, J. W. (2007). Gemengde ervaringen van actieve bewoners in Rotterdam. In C. Bouw, J. W. Duyvendak & L. Veldboer (Eds.), *De mix factor*. Amsterdam: Boom.

Uitermark, J., & Duyvendak, J. W. (2008). Civilizing the City: Populism and Revanchist Urbanism in Rotterdam. *Urban Studies, 45*.

Uitermark, J., Duyvendak, J. W., & Kleinhans, R. (2007). Gentrificiation as a Governmental Strategy: Social Control and Social Cohesion in Hoogvliet, Rotterdam. *Environment and Planning, 39*(125-141).

van Caem, B. (2008). *Verborgen kracht. Burgerparticipatie op het vlak van veiligheid.* Amsterdam: Dynamics of Governance – Faculteit der Sociale Wetenschappen -Vrije Universiteit Amsterdam.

Van den Broeck, T. (2004). "Words Like Daggers"... A Community's Perspective on Social Safety Policy in Urban Areas. In K. van der Vijver & J. Terpstra (Eds.), *Urban Safety: Problems, Governance and Strategies* (pp. 107-132). Enschede: Institute for Social Safety Studies. University of Twente.

van der Vijver, K., & Terpstra, J. (2004). *Urban Safety: Problems, Governance and Strategies.* Enschede: Institute for Social Safety Studies. University of Twente.

van Ostaaijen, J., & Tops, P. (2007). Active Citizens and Local Safety. How the Active Citizens Matrix Can Help Local Government to Support Citizens in Their Efforts to Improve Safety. In P. Delwit, J.-B. Pilet, H. Reynaert & K. Steyvers (Eds.), *Towards DIY-Politics. Participatory and Direct Democracy at the Local Level in Europe* (pp. 321-359). Brugge: Vanden Broele.

Vanderveen, G. (2006). *Interpreting Fear, Crime, Risk and Unsafety.* Dissertatie (mr. N.J.H. Huls & dr. H. Elfers.) Den Haag: Boom Juridische uitgevers.

Walklate, S. (1998a). Crime and Community: Fear or Trust? *The British Journal of Sociology, 49*(4), 550-569.

Walklate, S. (1998b). Excavating the Fear of Crime: Fear, Anxiety or Trust? *Theoretical Criminology, 2*(4), 403-418.

Walklate, S. (2001). Fearful Communities? *Urban Studies, 38*(5), 929 – 939.

Young, J. (1999). *The Exclusive Society: Social Exclusion, Crime and Difference in Late Modernity.* London: Sage Publications.

Institutional distrust in Flanders. What is the role of social capital and dimensions of discontent?

Maarten Van de Velde
Lieven Pauwels

1   Introduction

Trust in modern democracies is said to be declining during the latest decades (Nye, 1997). This theme has intensely been studied by scholars in the field of psychology, sociology and especially the political sciences. The relevance of the study of trust in government is found at the basics of the democratic political system. Without trust democracy is at stake, mainly because government loses its legitimacy in the absence of trust. Without this legitimacy the government risks to lose important resources: the willingness of the public to pay taxes might disappear together with the preparedness of bright young people to apply for jobs in government. And perhaps as important: also the willingness of the public to comply with laws might diminish when government is distrusted (Nye, 1997). This would ultimately lead to the disintegration of society as a whole. Lennard motivates the importance of the study of institutional trust as follows: *'The voluntary compliance that is central to democracies relies on trust, along two dimensions: citizens must trust their legislators to have the national interest in mind and citizens must trust each other to abide by democratically established laws.'*(Lennard, 2008: 1). A lack of trust thus can have severe consequences for governmental institutions and the political system, it leads to lower civic political participation or even the actively participation in groups that attempt to undermine the democratic system (Van Cluysen,Van Craen & Ackaert, 2009).

A lack of trust hampers policy implementations and creates a favourable environment for populist and extremist political parties and movements (Neustadt, 1997). The importance of generating trust can also be confirmed by the high priority it has on the economical and political agenda (Kampen & Van de Walle, 2003). In this study attention is paid to trust in the Flemish parliament, the Flemish Government, the Belgian criminal justice system, the Belgian Federal Parliament, the Belgian Federal Government, the police, the politicians, the European Parliament and the United Nations Organisation.

Especially the criminal justice system has been found several times to be very vulnerable to the loss of trust. The criminal justice system is a complex matter. Earlier research suggests that trust in the criminal justice system leads to higher compliance with the law (Tyler, 1997, 2001). Furthermore the functioning of the criminal justice system depends on the cooperation of the public to report crime to the police, to act as a witness and to serve as a juror in a certain court case (Roberts & Hough, 2005). The perceived illegitimacy of the criminal justice system could also lead to dangerous situations in which the public takes the law in its own hands.

In the descriptive part of this study, attention is paid to the proportion of Flemish citizens that trust each of the aforementioned institutions. However, in the exploratory multivariate analyses trust is treated as a single latent trait.

In order to understand individual differences in trust, it is necessary to define determinants of trust and especially to uncover the mechanisms which play a central role in the creation of trust. In this study we set off from an integrated sociological and psychological perspective. This is done by combining social capital theory and social identity theory, a theory which gives an explanation for the causal interrelationship of anomia and ethnocentrism. We try to explain individual differences in institutional trust from: a) sociological background characteristics, indicative of the social environment, b) social capital, c) anomia (political powerlessness) and d) ethnocentrism as key psychological mechanisms.

This study is organized as follows: first, a series of exploratory regression models are run on two datasets for the region of Flanders in Belgium. By doing so a first insight into the role of characteristics derived from social capital and social identity theory is obtained. Second some confirmatory path analyses are performed to further understand how these characteristics have an influence on levels of trust. In order to test the stability of the results and to increase the external validity, the hypotheses are tested on two datasets. The first data are provided by the research service of the Flemish Government in Belgium, while the other data are forthcoming from the European social survey which was held in the same period in the region of Flanders.

2   Theoretical Framework in the present study

In this article we present an integrated sociological-psychological approach to the explanation of institutional trust. We are heavily inspired by the analytical tradition in criminology (Pauwels, Ponsaers & Svensson, 2009). This way of theorizing is a scientific realist approach of crime that is deeply rooted within the tradition of analytical sociology (for a discussion see Hedström & Swedberg, 1998; Hedström, 2005). *"Analytical sociology seeks to explain complex social processes by carefully dissecting them and then bringing into focus their most important constituent components. It is through dissection and analytical abstraction that the important cogs and wheels of social processes are made visible and intelligible"* (Edling & Hedström, 2005:1).

The problem of causation has not always been well treated in studies of institutional trust. The aetiology of trust does not understand very well the causal mechanisms and processes that are at work when individual differences and changes in trust are to be explained. That is why there are too many correlates of trust, which merely obscure a causal interpretation. The basic philosophy behind a mechanism-based explanation is that a social phenomenon is explained from precisely those mechanisms that bring about differences and changes in a condition (the explanandum or the dependent variable in a study).

In this paragraph we highlight the integrated theoretical model that is used as a framework in the explanation of trust.

Figure 1: Key determinants in the explanation of institutional trust

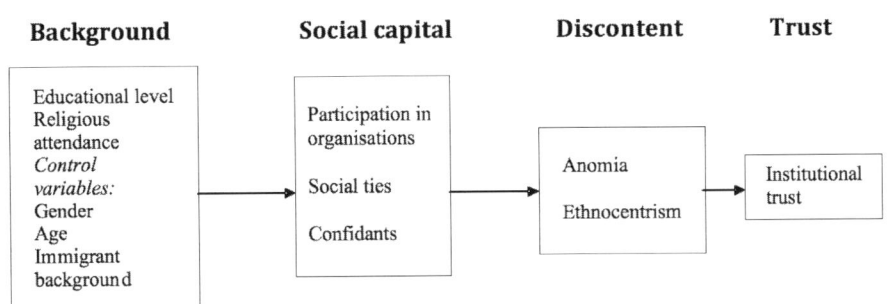

2.1   The role of background characteristics in the explanation of trust

One often-reproduced finding is that institutional trust strongly varies by background characteristics that refer to one's position in the social structure, especially educational level and religious attendance (Elchardus & Smits, 2002). Sometimes educational level is seen as an indicator of vulnerability in the meaning of lacking skills. According to Durkhemian theory the skills necessary to put trust in things are acquired through socialization processes. A higher educational level in this way is seen as a longer socialization that leads to a higher level of trust. However, the relationship between educational level and institutional trust is in a way ambiguous. While the Durkhemian approach argues that a higher educational level should lead to a higher level of trust, highly educated people are also learned to be critical and not just to accept everything what has been told. In this way a higher educated citizen tends to be more critical, an attitude which may eventually lead to a decrease in trust (Elchardus & Smits, 2002). A person's religious background is an important background characteristic that refers to the social environment one is embedded in and is regarded as an important structural cause of trust. People with strong religious beliefs, Catholic, Muslim and the liberal are more trusting than people who do not believe and are not concerned with religion. From previous studies in Flanders, we know that Catholics are more devoted and trustful than those who do not attend religious services (Elchardus, 1998). While this finding is important in itself, it cannot explain why these characteristics bring about higher levels of institutional trust. Therefore it is necessary to uncover the intervening mechanisms.

2.2   The role of social capital in the explanation of trust

Social capital has been referred to as an important mechanism in explaining the effects of one's structural background. There is however no universally accepted definition of the concept of social capital (de Hart, Knol, Maas – de Waal & Roes, 2002). In this study we follow Harell and Stolle (in press) who pose that social capital refers to cooperative relations among individuals and groups of individuals, and that these relationships are based on mutual respect, equality and norms of reciprocity. This approach actually refers to informal reciprocal relationships, such as small talk and thin trust. Other scholars have pointed to the role of participation in organisations, being a more formal aspect of social capital, referring to formal organisations such as cultural and social organisations, including voluntary organisations. Both participation in organisations and the embeddedness in the social fabric (i.e. having many persons that one can rely on) have been argued to be important mechanisms in the explanation of individual differences in trust. Social capital is also related to one's structural background, especially one's educational level. Some scholars have argued that the lowest educated are not only deprived of labour but also have smaller social networks they can rely on. Current theory argues that there is a relationship between social capital and trust. These influential scholars argue that a decline of social capital is likely to cause a loss of trust in political leadership and a loss of confidence in the institutions of government (see for example, Nye et al,1997; Norris, 1999; Pharr & Putnam, 2000). For many theorists voluntary organisations are crucial forms of social networking, and trust between citizens and their political leaders is said to be a key expression of civic virtues. In recent empirical studies we are confronted with two opposing theoretical streams uncovering the relationship between social capital and institutional trust. On the one hand there are the theorists stating that social trust and institutional trust are inextricably bound up with each other. They state that low levels of social trust are reflected in lower levels of institutional trust (Meuleman & Billiet, 2003). On the other hand, Newton (2001) proposes that the relationship between social capital and institutional trust is rather more distant and contingent than social network theories seem to assume. He argues that the relationship between social capital and institutional trust is moderate at best and therefore probably rather indirect.

2.3   The role of anomia and ethnocentrism in the explanation of trust

Many studies have already pointed to the fact that "discontent" (or a general feeling of dissatisfaction with current society) is a multidimensional psychological trait which is a key mechanism in the explanation of institutional trust. People with high levels of discontent would be a lot more distrusting. In Belgium the current study of discontent is especially situated in sociology in the form Elchardus gave it (Elchardus & Smits, 2002). He studies discontent from a holistic point of view. According to Elchardus discontent encloses feelings of anomia[12], authoritarianism and ethnocentrism but does not distinguish empirically between these concepts. The disadvantage of this holistic approach is that no insight is gained into the relative influence of each of theses dimensions on trust nor in the interrelationships of these dimensions of discontent. However these interrelationships cannot be ignored as they have formed the subject of a fierce discussion between sociological and psychological scholars. (Scheepers, Felling & Peters, 1992). During 1950s-1960s a fierce discussion raged amongst sociologists concerning the causal interrelationship between anomia, authoritarianism and ethnocentrism. In this discussion Mc Dill proposed that anomia, authoritarianism and ethnocentrism are all dimensions of what he called a 'Negative Weltanschauung' (a

---

12   Anomia was originally considered by Srole (1956) as a subjective 'feeling' responding to societal dysfunctions (Scheepers, Felling & Peters, 1992; Srole, 1956). More specifically Srole (1956) defined anomia as consisting of five sub dimensions labelled (a) political powerlessness, (b) social powerlessness, (c) generalised socioeconomic retrogression (d) normlessness and meaninglessness and (e) social isolation.(Schlüter, Davidov & Schmidt, 2007).

negative worldview) (Mc Dill, 1961: 245). This is quite similar to the discontent concept of Elchardus. Several studies empirically contradicted these findings of Mc Dill
(Struening & Richardson, 1965; Lutterman & Middleton 1970; Knapp 1971). Based on
the mixed nature of the aforementioned discussion we decided not to follow the holistic approach. Instead we focus on the interrelationship between anomia and ethnocentrism. In the same debate the theoretical backdrop of the causal relationship between
anomia and ethnocentrism is argued. In this study we state that anomia is causally
prior to ethnocentrism. This is explained by the social identity theory (Tajfel & Turner,
1979; Scheepers, Felling & Peters,1992). This theory suggests that the individual is in
a permanent need to assume for oneself a positive identity. This is done by identifying
oneself with people having perceived positive characteristics, the in-group, and contra-
identifying with people having perceived negative characteristics, the out-group. The
link between anomia and ethnocentrism can be explained as follows: "...*it may be argued that anomic people who are subject to powerlessness, meaninglessness and normlessness
and who feel socially isolated, therefore have a strong urge to re-establish a positive identity
by means of social identification, possibly accompanied by social contra-identification...*"
(Scheepers, Felling & Peters, 1992: 46). As we interpret this we could conclude that
the more an individual has anomic feelings, the more he has the urge to emphasize
the identification with the in-group and the contra-identification with the out-group,
which is equal to ethnocentrism (Sumner, 1906).

In this study however we are restricted by the measures provided by the data which
means anomia is defined only by one dimension, namely political powerlessness, and
we have no measures of authoritarianism. Despite this fact we should be able to get
an impression of the relationship between anomia, ethnocentrism and institutional
trust and approach this in a less reductionist manner as has been done in a lot of the
previous Belgian studies on the subject of institutional trust.

2.4   Conceptualizing trust

There are different ways of studying institutional trust and its sociological and psychological determinants. Scholars tend to disagree on the way trust itself is to be studied.
Sometimes, trust is studied as a general concept and sometimes trust is studied in
detail, by institution. In this study, trust is seen as a part of a broader syndrome of personality characteristics, including optimism, a belief in co-operation, and confidence
in social and political life in general. Conversely, distrusters are seen as "misanthropic
personalities", pessimists that are cynical about the possibilities for social and political
co-operation (Erikson, 1950; Allport, 1961; Cattell, 1965). On the other hand, scholars
such as Newton (2001) have argued that there is no such thing as general trust, but
rather different kinds of trust which are expressed by different types of people for different reasons. Social disaffection, in the form of low or declining social trust, is one
thing; political disaffection, in the form of low or declining political trust is another.
From that point of view it is extremely interesting to study differences in institutional
trust by institution. Also Belgian studies on institutional trust are characterized by
the returning discussion on the multidimensionality of the concept of institutional
trust. In Elchardus's view institutional trust is unidimensional (Elchardus & Smits,
2001). Others pose that this concept is bidimensional and make a differentiation between trust in the political institutions and their administrations. Institutional trust

is restricted to trust in the official political institutions and not the public administrations. From an empirical point of view it has been shown many times that trust in one governmental institution is strongly related to trust in other governmental institutions (see Elchardus & Smits 2001). Therefore we will only focus on institutional trust as a general construct in the explanatory part of this study[13].

3   Hypotheses

In the integrated sociological psychological model, individual differences in trust are explained by one's educational level and religious attendance. First, we hypothesize that religious attendance will have a positive effect on institutional trust and that a lower educational level will have a negative effect on institutional trust. Second, we assume that the effects of these structural background characteristics are strongly mediated by both institutional participation and social ties as key indicators of social capital. Third, we hypothesize that the effects of social capital are by and large mediated by anomia and ethnocentrism as key indicators of discontent. Thus we assume that the direct effects of social capital on institutional trust on its turn will be mediated by political anomia and ethnocentrism. Anomia hereby is seen as causal prior to ethnocentrism.

4   Data

The data collected for this study are drawn from two independently collected samples: the 'Social and cultural changes in the region of Flanders' (Dutch: SCV-survey, previously known as APS-survey) which is held biennial on demand of the Flemish government (Studiedienst van de Vlaamse Regering, 2003) and the ESS data (European Social Survey) (Jowell et al., 2003). We use two datasets to verify the stability of the results and to increase the external validity. In order to exclude differences caused by temporal changes in attitudes, and also because these two datasets include comparable concepts, the 2002 edition of both surveys is used. In the following section a brief overview of the sampling procedure and method of data collection in both surveys is given. There is abundant literature on both surveys, which renders detailed information, easily available (see Pickery & Carton, 2003; Jowell et al., 2003). In both surveys questions are included on educational level and religious attendance, gender, immigrant background and age, participation in organisations and social ties, anomia, ethnocentrism and trust. First we shortly introduce both samples and later we point to the differences in measurement issues.

---

13   The institutions included are all political institutions with decisive legal and political power i.e. the Flemish government, the Flemish parliament, the Belgian government, the Belgian parliament, the criminal justice system, the European parliament, the police, the United nations organisation and the politicians.

4.1   The SCV survey of the Flemish Government

This is a biennial repeated cross-sectional survey consisting of a questionnaire that has several parts: a fixed part and some periodical. The fixed part contains an amount of questions which return in each edition of the survey. These questions measure -next to social-demographic characteristics – social networks, membership in organisations and participation in cultural and leisure activities. In the periodical variable part, every second or third year questions about trust in democracy, trust in institutions, subjective safety and the like are asked[14]. The SCV-survey is conducted using the CAPI-method of data collection (computer assisted personal interviewing). The sample frame encloses the Dutch-speaking population of Belgium with the Belgian nationality aged between eighteen and eighty-five in the Flemish region and in Brussels. In total 1437 respondents were interviewed. The non-response amounted to 19%. An overrepresentation of highly-educated people and an underrepresentation of low- and uneducated respondents was observed. To remedy this, a weighing of the data has taken place[15].

4.2   The ESS survey

The European Social Survey is an international repeated cross-sectional survey covering over 20 European countries. The ESS survey (2002 round) is funded jointly by the European Commission's 5th framework programme, the European Science Foundation and academic funding bodies in each participating country, and is designed and carried out to exceptionally high standards. The ESS aims to monitor changing public attitudes and values within Europe, to investigate how they interact with Europe's changing institutions and to advance and consolidate improved methods of cross-national survey measurement in Europe and beyond. It involves strict random probability sampling, a minimum target response rate of 70% and rigorous translation protocols. The ESS-survey is conducted using face-to-face interviews that take about one hour and includes (amongst others) questions on immigration, citizenship and social and political issues. In this study we extracted the sample taken in the Flemish region in Belgium. The sample frame consists of all persons aged 15 and over resident within private households, regardless of their nationality, citizenship, language or legal status. The total sample size in the region of Flanders encloses 1234 units of analysis. In order to get a sample with a comparable social composition as the population the analyses are performed using the weighting procedure that has been prescribed[16] (Jowell et al., 2003).

---

14   Finally the variable ad hoc part is built up around a policy-relevant theme.

15   A poststratifiction has taken place. A weighing has taken place for educational level and age. The weighing factor has been calculated using the program "WEIGHT".

16   Weighting has been done with the prescribed design weight. This weighting had to be done because not all country citizens above the age of 15 had the same chance of getting selected for the sample

5 Operational measures derived from both surveys

Although the 2002 edition of both surveys included measures of the same constructs, these constructs are not measured exactly in the same way. They use identical, quite similar and quite different measures of the same underlying constructs. This can be considered both as a weakness and as a strength. A major problem is that different kinds of measurement error are present in both surveys, which we cannot control for. As measurement error is partially related to errors as a consequence of the measurement instrument (question wording, question ordering),data collection (mode effects) and others (see Billiet, 1993), one must always be careful. With regard to the present study we know that both samples have used different measures and especially different methods of data collection. Despite these differences, the ultimate goal of this study is to gain an insight into the empirical validity of the hypotheses that were stated earlier. From a theoretical point of view it is very interesting to move beyond these differences and try to gain an insight into the level of generalisation possible from both samples, especially with regard to the main hypotheses considering the direct and indirect effects of structural background, social capital, anomia and ethnocentrism in the explanation of institutional trust. A detailed description of the operational measures can be found in appendix I.

Operational measures

Background characteristics
The *educational level* has been measured by the highest obtained degree. *Religious attendance* is the frequency in which people attend religious services apart from special occasions such as funerals and weddings. Control variables included in the analyses are *gender* (male, female), *age* and *immigrant background* (at least one parent is born as a non-Belgian).

Social Capital
Social capital is measured in three ways: firstly *social ties,* this is the number of persons one can rely on. Secondly *attachment to significant others* has been measured as a summation scale of three items (α=.580) in the SCV- survey, and as a single question in the ESS- survey. Thirdly social capital is measured as *participation in organisations*. This is measured as a dichotomy in the SCV-survey and a trichotomous variable in the ESS- survey.

Anomia
Anomia has been measured by the scale of *political powerlessness*, which has been measured as a summation scale of five items (Scv: α=.796; Ess: α=.630).

Ethnocentrism
Ethnocentrism has also been measured as a summation scale, containing eight items in the SCV- survey (α=.884) and six items in the ESS-survey (α=.831)

Institutional trust

Institutional trust is a variety scale formed by the summation of the trust scores on the Belgian criminal justice system, the Flemish parliament, the Flemish government, the Belgian federal parliament and the Belgian federal government (α=.854) in the SCV survey. In the ESS-survey following institutions were included: The Belgian Federal parliament, the legal system, the police, the politicians, the European parliament and the United Nations Organization (α=.886)

6 Analytical strategy

First a series of block wise multiple regression analyses are performed to gain a first insight into the mediating effect of social capital, anomia and ethnocentrism in the relationship between structural background as educational level, religious attendance and institutional trust. The first block consists only of the structural background variables, including some important demographic controls (age, gender and immigrant background), the second block consists out of participation in organisations and social ties and the third block is formed by anomia and ethnocentrism. Second, a more critical test is performed on the relationships between all exogenous and endogenous variables. In social science research, an indirect effect occurs when the influence of an antecedent variable on the effect variable is mediated by an intervening variable. We performed a series of structural equation models to test whether the mediating variables in the analyses really can be considered as intermediate relationships (Bollen,1989; 1993).

6.1 Univariate descriptive analysis of trust in the institutions

In order to present the univariate descriptive statistics of trust in the different institutions the number of original answering categories has been reduced to three categories, namely very low to low trust, a neutral category neither low, nor high trust and a category high to very high trust. The univariate descriptives of all variables can be found in appendix II.

Table 1: Level of trust towards the governmental institutions (SCV-data)

	Very low/Low	Not low/Not high	High/ Very high	N
The criminal justice system	41%	35,8%	23.1%	1430
The Belgian Federal Government	27,9%	49,2%	22,8%	1436
The Flemish Government	27,5%	46,1%	26,4%	1440
The Belgian Federal Parliament	24,6%	51,5%	23,9%	1423
The Flemish Parliament	24,3%	50,8%	25%	1402

From table 1 it can be seen that the Flemish citizen differs in institutional trust. Most Flemish citizens seem to have very low levels of trust in the criminal justice system. All the other institutions have about the same percentage of citizens having a low or very low level of trust (about 25%), the proportion of people distrusting the criminal justice system (40%) is remarkably high.

Table 2: Level of trust towards the governmental institutions (ESS-data)

	Very low/Low	Not low/Not high	High/Very high	N
The legal system	32,8%	45,4%	21,8%	1205
The politicians	27,7%	54,5%	17,8%	1217
The European Parliament	23,2%	51,9%	24,9%	1108
The United Nations	22,1%	44,0%	33,8%	1038
The Belgian Federal Parliament	21,1%	52,4%	26,5%	1184
The police	14,0%	39,3%	46,7%	1224

From table 2 we see that the legal system receives the least trust, almost a third of the Flemish citizens (32,8%) has a low to a very low level of trust in the legal system followed by 27,7% who have a low to very low level of trust with respect to the politicians. Striking is the very high proportion (46,7%) of citizens who have high to very high levels of trust in the police.

Table 3: Results of the block wise regression analyses

Dependent variable: Institutional trust	Block 1				Block 2				Block 3			
	M1 (SCV) β (S.E.)		M1(ESS) β (S.E.)		M2(SCV) β (S.E.)		M2(ESS) β (S.E.)		M3(SCV) β (S.E.)		M3(ESS) β (S.E.)	
Controls and background												
Gender(male)	.013	(.195)	.020	(.586)	.030	(.198)	.004	(.589)	.030	(.185)	-.044	(.551)
Age	.023	(.007)	-.204***	(.018)	.030	(.007)	-.179***	(.019)	.088**	(.007)	-.120***	(.017)
Origin (immigrant background)	.060*	(.410)	.007	(.993)	.060*	(.411)	.013	(.989)	.040	(.389)	-.029	(.920)
Educational level: No certificate	.017	(.795)	-.111**	(.968)	.019	(.793)	-.095**	(.966)	.031	(.741)	-.002	(.919)
Primary education	.013	(.297)	-.150***	(.694)	.013	(.297)	-.133***	(.695)	.052	(.279)	-.055	(.654)
Secondary education (ref.cat.)												
Higher education	.080**	(.227)	-.002	(.934)	.084**	(.227)	-.002	(.927)	-.019	(.221)	-.079**	(.874)
Religious attendance	.122***	(.061)	.151***	(.216)	.111***	(.061)	.136***	(.216)	.099**	(.057)	.107***	(.200)
Social capital												
Number of confidants+					.023	(.204)	---	---	.008	(.190)	---	---
Presency of confidants ++					---	---	.010	1.021	---	---	.021	(.945)
Participation in organisations					-.035	(.121)	.064*	.689	-.058*	(.113)	.038	(.638)
Social involvement ++					---	---	.104***	.298	---	---	.051	(.277)
Attachment to significant others +					.092**	(.030)	----	---	.082**	(.028)	---	---
Discontent												
Ethnocentrism									-.123***	(.016)	-.267***	(.035)
Political anomia									-.320***	(.025)	-.241***	(.096)
Model evaluation												
R² adj.	2.1%		8.1%		2.8%		9.7%		15.3%		23.2%	
F-change	4.675***		16.095***		3.925**		7.699***		90.866***		105.173***	

***=p<.0001 **=p<0,01; *=p<0,05 ; "+"= exclusively measured in SCV, "++"= exclusively measured in ESS

6.2 Block wise regression analyses on the SCV data

In model 1 we only find small effects form background variables on institutional trust. The background characteristics only account for 2.1% of the variation in institutional trust. According to this analysis people with an immigrant background would have slightly more trust in government as well as respondents with a higher education. The more people attend religious services the more trustful they are. Gender does not seem to have a significant effect.

When controlling for social capital in model 2 the explained variation in institutional trust rises very slightly to 2.8%. We do not find great changes in the influences of background characteristics on institutional trust. In addition we find a small positive effect of 'attachment to significant others' on institutional trust.

When adding anomia and ethnocentrism as indicators of discontent in model 3 the explained variance of the model rises to 15.3%. Anomia is by far the strongest determinant of institutional trust. People with great feelings of political powerlessness trust less the governmental institutions. Also ethnocentric people have less trust in these institutions, but this effect is less strong than the latter. The effects of religiosity and attachment to significant others on trust have been further moderated by the effects of anomia and ethnocentrism. Age now has a significant but very small positive effect on trust whereas participation in organisations has a very small negative effect on trust.

6.3 Block wise regression analyses on the ESS data

Model 1 includes only background characteristics. This model explains 8.1% of the variance of trust in the government. More variables have a significant effect and the beta's are rather stronger in the ESS data than on the SCV data. We find that older people have moderately less trust in the government. People with no certificate or who only got a certificate of primary education have less trust than people who obtained a secondary education degree. The higher the level of religious attendance, the higher the level of institutional trust.

In model 2 the effects of age, educational level and religious attendance on institutional trust seem to be hardly moderated under control for social capital. As measures of social capital, participation in organisations has a very small effect and social involvement a small positive effect on institutional trust. The added value of the social capital measures to the explaining power of the model is low, the squared multiple correlation only rises with 1.7 percentage points. When adding ethnocentrism and anomia in model 3, we observe that the effects of social capital are completely moderated. The squared multiple correlation rises to 23.2%. On these data, ethnocentrism seems to have a slightly stronger effect on trust than anomia. People with high levels of anomia and /or high levels of ethnocentrism have less trust in the government. Still we find that older people have less trust in government than the younger. Social capital indicators lose their significant effects.

6.4   Testing direct and indirect effects

In this paragraph the results of a path analysis in LISREL are shown. Structural Equation Modelling is used to assess direct and indirect effects of the educational level, religious attendance, participation in organisations, social ties, anomia, ethnocentrism and trust. The control variables used in the block wise regression analyses were omitted in these models. From the block wise regression analyses we have learned that anomia and ethnocentrism have strong direct effects on institutional trust. The basic idea was to test for what exogenous variables anomia and ethnocentrism are acting as intervening mechanisms. We ran several models guided by theory but present the best fitting model based on the correlation matrix. In order to evaluate the fit of the models, the Root Mean Square Error of Approximation (RMSEA) is preferred over the Chi-square value as a measure of overall fit. Chi-square tends to be very sensitive to the size of the sample, resulting in accumulated high values of the statistic, i.e. the 'uncritical' rejection of models, in the case of large samples. RMSEA on the other hand is a measure of 'close fit', indicating that it takes into account the error of approximation in the population as well as the precision of the measure itself (Jöreskog & Sörbom, 1996). Models with an RMSEA <.05 are considered acceptable (Billiet & McClendon, 2000). Contrary to block wise regression analysis, confirmatory path analysis has far more potential to test the assumptions of equalities of effects. The results are summarized under table 4.

Table 4: summarizing results of the lisrel analyses (standardized effects)

Dependent ->	SCV-survey			ESS-survey		
Independent	Trust	Anomia	Ethnocentrism	Trust	Anomia	Ethnocentrism
Anomia	-.32 (.03)		.37 (.02)	-.25 (.02)		.32 (.03)
ethnocentrism	-.11 (.03)			-.27 (.02)		
Participation in organisations	--	-.081 (.02)	--	--	-.10 (.02)	
Social ties	--	--		.08 (.02)	-.12 (.02)	-.088 (.02)
Educational level	-.09 (0.02)	-.30	-.22 (.02)		-.40 (.02)	-.11 (.02)
Religious attendance	.13 (.02)	--	--	.075 (.02)	--	--
Model evaluation R²	14%	10%	23%	21%	22%	17%
RMSEA	.014			.018		
AGFI	0.99			0.99		
CHI²	5.70			6.17		
df	4			5		
p-value	0.22			0.29		

The LISREL analyses conducted on the SCV survey reveal that only participation in organisations and educational level have direct negative effects on anomia and that anomia, and educational level have direct effects on ethnocentrism. Social ties do not play a role in explaining individual differences in institutional trust. Interestingly religious attendance does have a direct effect on institutional trust, but this effect does not go through anomia and ethnocentrism. Educational level still has a direct effect on trust, but the direct effect is rather small compared to the indirect effects through anomia and ethnocentrism. The model withdrawn is the best fitting model and from the model fit parameters it can be seen that the fit indices are highly acceptable (RMSEA:.01; AGFI:.99)

Figure 2: direct and indirect effects on institutional trust SCV

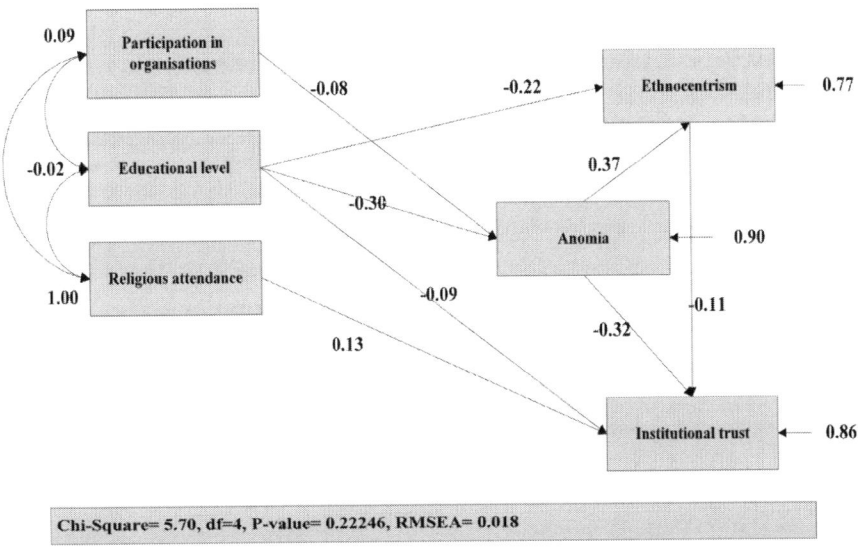

The LISREL analyses conducted on the ESS data reveal that participation in organisations, social ties and educational level have direct negative effects on anomia and that anomia, social ties and educational level have direct effects on ethnocentrism. Further the data suggest that the effect of educational level is completely indirect and anomia and ethnocentrism are the paths through which educational level effects trust. At the same time it becomes clear neither social ties nor participation in organisations are acting as paths through which educational level influences trust. Similar to the finding done on the SCV data, religious attendance does have a direct effect on institutional trust, but this effect does not go through anomia and ethnocentrism. The model withdrawn is the best fitting model and from the model fit parameters it can be seen that the fit indices are highly acceptable (RMSEA:.01; AGFI:.99).

Figure 3: direct and indirect effects on institutional trust ESS

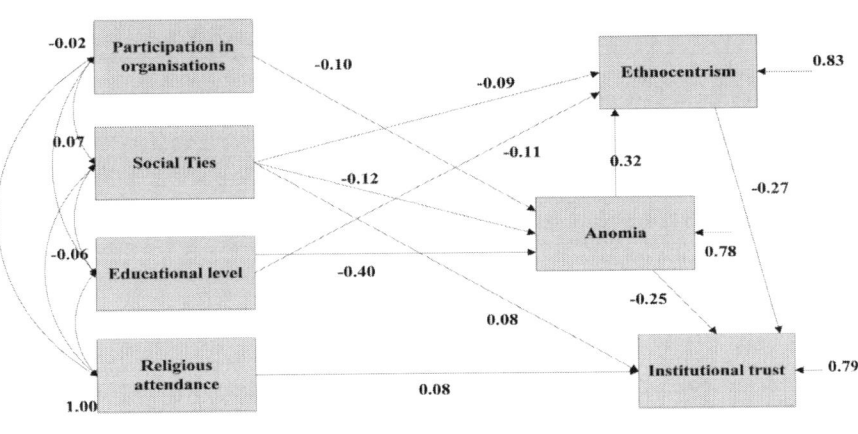

The analyses conducted above suggest that the effect of educational level is by and large an indirect effect. This is congruent with current theorizing that its effect is mediated by psychological mechanisms. Although the results from both structural equation analyses are not completely equal, one can see a similar tendency.

What is perhaps most striking in these analyses, is that our measures of social capital hardly play a role in explaining the effect of educational level, but act themselves as indirect precursors of institutional trust, especially through their effect on anomia (both social ties and participation in organisations). There is no strong support for an effect of social capital on ethnocentrism. The ESS data document a direct negative effect of social ties on ethnocentrism and the SCV data suggest a direct effect of participation in organisations on ethnocentrism. The ESS data further suggest that social capital still has a direct effect on trust. We considered two dimensions of social capital (participation in organisations and social ties) that both were measured differently. One can thus conclude that social capital matters in the explanation of institutional trust, but these effects are rather indirect and anomia is the key mechanism at work here. Further studies should however try to measure social capital more accurately as has been done here and to take into account differences between forms of social capital such as bridging and bonding.

Both data reveal that anomia and ethnocentrism are key mechanisms in explaining individual differences in trust. In these analyses anomia is seen as a mechanism that explains ethnocentrism, guided by social identity theory and not vice-verca. Models that considered ethnocentrism as causally prior to anomia had worse model fits then the models we derived from social identity theory.

7 General conclusion and discussion

In the descriptive part of this study we observed remarkable differences in the levels of trust by the public towards the different institutions analyzed in this study. In both datasets, the judicial institutions receive the lowest degrees of trust. Also politicians are less trusted. When it comes to institutions receiving high levels of trust, the police and the United Nations Organization get the highest scores. The trend of police receiving higher levels of trust than the judicial system is also found in the results of the Belgian Justice Barometer (Parmentier, Vervaeke & Goethals, 2004) and earlier editions of the SCV survey (Elchardus & Smits, 2001; Elchardus & Smits, 1998).

This study reveals that educational level and religious attendance are of substantial value in the explanation of institutional trust in both samples, regardless of age, immigrant background and gender, which do not seem to have very strong effects. In both samples the explanatory power of background variables are somewhat different, varying from very poor in the SCV-survey to moderate in the ESS-survey. This difference may be due to different sampling procedures. A more substantive interest in this study is that social capital in the definition of social ties and social participation does not have the substantive effect on institutional trust as is often suggested by social capital theory (Lin, 2001). This may mean that the scope of social capital theory is not as large as supporters of the theory suggest (Newton, 2001). On the contrary anomia and ethnocentrism are powerful predictors of individual differences of institutional trust. However it is not clear why these two variables have such important influences on institutional trust. It might be important of getting hold of anomia and ethnocentrism to maintain the well-being of the democratic political system.

It is important to recognize some major weak points in the causal analyses conducted in this study. First of all cross-sectional data are used in this study, and these are not suited to establish causal ordering between variables (Bijleveld, 2008). Panel data are much better to do so. Why that is so is explained below. Clearly, for investigating the dynamic relations of authoritarianism, anomia, ethnocentrism and trust, panel data are much more desirable. Specifically there are three reasons why panel data appear particularly adequate for such an investigation (Finkel, 1995). First, regarding the construct specifications of authoritarianism and anomia, panel data offer the opportunity to test the measurement invariance of the measurement model underlying these constructs by comparing individual responses to the indicator variables across different measurement points (Pleysier et al., 2005). Second, panel data are particularly appropriate for testing causal assumptions such as for the relations of authoritarianism and anomia as the observations are collected over two or more points in time. Third, panel data offer informative explorative insights on the dynamics of the theoretical constructs over time using different methods of longitudinal data analysis. We are fully aware of the fact that our data are too limited to empirically assess this research question. Therefore this discussion should not be held here. We can merely acknowledge the limitation of the data and stress the need for panel data to assess issues of causal ordering. Another question that cannot be answered here is the degree by which we overestimate the effect of anomia. E.g. does anomia really have an indirect effect on trust through its direct effect on ethnocentrism and does anomia predict trust, independent of authoritarianism? That is a theoretical discussion that has been going on for a long time (Scheepers, Felling & Peters, 1992) and could not be assessed due to data limitations. Both samples did not contain authoritarianism-items in the 2002 edition. Some scholars have argued that it is actually authoritarianism that leads to ethnocentrism and that anomia thus only indirectly effects ethnocentrism (Scheepers, Felling & Peters, 1992; Meloen, 1997). With regard to the explanation of trust it is a question worth for further investigations. Authoritarianism could be a mechanism that is developed as a compensation of anomia, as has been described by Levin (2002).

As with many constructs, conceptual cloudiness is a major problem that deserves attention (Pleysier & Parmentier, 2000). In these analyses we were restricted to existing measures in both samples. Many of the concepts described are in fact multidimensional and for the purpose of these analyses we were only able to use some. We conclude that much research still has to be conducted in this area, because social capital, anomia and ethnocentrism still make a powerful contribution to discussions of institutional trust from a scholarly and institutional point of view, thereby documenting the enduring qualities that these classic sociological and psychological constructs possess. Although institutional trust may be seen as a latent trait, i.e. the fact that one distrusts political institutions, it may be worth (from a criminological point of view) studying the criminal justice system more in detail. This might be done by studying actors, and other dimensions of the criminal justice system (such as procedures, perceived fairness). The findings provided in this study can therefore be seen as a general point of departure that may stimulate such further research.

8   Bibliography

Allport, G. W. (1961) Pattern and Growth in Personality. New York: Holt, Rinehart and Winston.

Bijleveld, C.C.J.H. & Commandeur,J.J.F. (2008). *Multivariate analyse: Een inleiding vooor criminologen en andere sociale wetenschappers.*

Billiet, J., (1993). *Ondanks beperkt zicht. Studies over waarden, ontzuiling, en politieke veranderingen in Vlaanderen.* Brussel: VUBPress.

Billiet, J.B. & McClendon, J. (2000). Modeling acquiescence in measurement models for two balanced sets of items. *Structural Equation Modelling* 7(4): 608-628.

Bollen, K.A. & J.S. Long (eds). 1993. Testing Structural Equation Models. Newbury Park, CA: Sage.

Bollen, K. A. 1989. Structural Equations with Latent Variables. Wiley Series in Probability and Mathematical Statistics. New York: Wiley.

Cattell, R. B. (1965) The Scientific Analysis of Personality. Baltimore: Penguin Books.

De Hart, J., Knol, F., Maas-De Waal, C. & Roes, T.,(eds.) (2002). *Zekere banden: Sociale cohesie, leefbaarheid en veiligheid.* Rijswijk: Sociaal en Cultureel Planbureau.

Edling, C. & Hedström, P. 2005. *Analytical Sociology In Tocqueville´S Democracy In America. Working Paper No. 3.* Department Of Sociology. Stockholm University.

Elchardus, M. & Smits, W.(1998). Vertrouwen. Het vertouwen van de Vlaming in politiek, overheid en instellingen in tijden van affaires. In Ministerie van de Vlaamse Gemeenschap (Ed.), *De Vlaamse overheid en burgeronderzoek 1998.* Brussel: Ministerie van de Vlaamse Gemeenschap.

Elchardus, M. (1998). Was u vandaag al slachtoffer of werd u al beschuldigd? In M. Elchardus (Ed.), *Wantrouwen en onbehagen* (pp. 222). Brussel: VUBPress.

Elchardus, M. & Smits, W. (2001). Een wantrouwig landje. Maatschappelijk vertrouwen in Vlaanderen. In Ministerie van de Vlaamse Gemeenschap (Ed.), *Vlaanderen Gepeild! De Vlaamse Overheid en Burgeronderzoek 2001.*Brussel: Ministerie van de Vlaamse Gemeenschap.

Elchardus, M. & Smits, W. (2002). *Anatomie en oorzaken van het wantrouwen.* Brussel: VUBPress.

Erikson, E. H. (1950). Childhood and society. New York, NY: Norton.

Finkel, S.E.,(1995) *Causal Analysis with Panel Data,* Sage University Paper Series on Quantitative Applications in the Social Sciences. Beverly Hills: Sage Publications.

Harrel, H.. & Stolle, D., (in Press). "Reconciling Diversity and Community? Defining Social Cohesion in Developed Democracies." In Hooghe, M. (ed.). *Theoretical Perspectives on Social Cohesion and Social Capital.* Brussel: Royal Flemish Academy of Belgium for Science and the Arts.

Hedström, P., 2005. *Dissecting The Social. On The Principles Of Analytical Sociology*. Cambridge: Cambridge University Press.

Hedström, P. & Swedberg, R., (Eds.). 1998. *Social Mechanisms: An Analytical Approach To Social Theory*. Cambridge: Cambridge University Press.

Jowell, R. T. & The Central Co-ordinating Team. (2003). European Social Survey 2002/2003: Technical Report. London: Centre for comparative social surveys.

Kampen, J. K. & Van de Walle, S. (2003). Het vertrouwen van de Vlaming in de overheid opgesplitst naar dimensies en indicatoren. In Ministerie van de vlaamse Gemeenschap (Ed.), *Vlaanderen Gepeild! 2003* (pp. 370). Brussel: Ministerie van de Vlaamse Gemeenschap.

Knapp, R.J. (1976). Authoritarianism, alienation, and related variables, a correlational and factor-analytic study. *Psychological Bulletin, 83,* 194-212

Lennard, P. T. (2008). Trust your compatriots, but count your change: the roles of trust, mistrust and distrust in democracy. *Political Studies, 56,* 312-332.

Levin, J. (2002). *The violence of hate: Confronting racism, anti-Semitism, and other forms of bigotry*. Boston: Allyn and Bacon.

Lin, N. (2001). *Social capital: A theory of social structure and action*. Cambrdidge: University Press

Lutterman, K.G & Middleton, R. (1970). Authoritarianism, anomia and prejudice. *Social Forces, 48,* 485-492.

Mc Closky, H. & Schaar, J.H. (1965). Psychological dimensions of anomy. *American Sociological Review, 30,* 16-40

Mc Dill, L. E. (1961). Authoritarianism, Prejudice, and Socioeconomic Status: An Attempt at Clarification. *Social Forces, 39,* 239-245

Meloen, J. (1997). De geschiedenis van de autoritaire persoonlijkheid als voedingsbodem van rechts-extremisme. Consequenties voor de bestrijding van politiek racisme, in H. De Witte (ed.), *Bestrijding van racisme en rechts-extremisme. Wetenschappelijke bijdragen aan het maatschappelijk debat.* (pp. 119-132). Leuven: Acco.

Meuleman, B. & Billiet, J. (2003). De evolutie van de perceptie van etnische dreiging tussen 1991 en 2004 en de relatie met institutioneel vertrouwen. In M. v. d. V. Gemeenschap (Ed.), *Vlaanderen Gepeild 2003* (pp. 370). Brussel: Ministerie van de Vlaamse Gemeenschap.

Neustadt, R. E. (1997). The politics of Mistrust. In J. S. J. Nye, Zelikow, D.P.& King, D.C. (Ed.), *Why people don't trust government* (pp. 340). Cambridge Massachusetts: Harvard University Press.

Newton, K. (2001). Trust, social capital, civil society and democracy. *International Political Sciences Review, 22*(2), 201-214.

Norris, P. (1999). *Critical Citizens. Global Support for Democratic Government*. Oxford: Oxford University Press.

Nye, J. S. J. (1997). Introduction: The Decline of Confidence in Government. In J. S. J. Nye, Zelikow, D.P. & King, D.C. (Ed.), *Why people don't trust government* (pp. 340). Cambridge Massachusetts: Harvard University Press.

Parmentier, S., Vervaeke, G. & Goethals, J. (2004). *Justitie doorgelicht: de resultaten van de eerste Belgische justitiebarometer.*Gent: Academia Press.

Pauwels, L., Ponsaers, P. & Svensson, R.,(2009). "Analytical Criminology : A style of theorizing and analysing the micro-macro context of acts of crime", in: *Contemporary Issues in the Empirical Study of Crime*, Cools, M., De Kimpe, S., De Ruyver, B., Easton, M., Pauwels, L., Ponsaers, P., Vander Beken, T., Vander Laenen, F., Vande Walle, G. & Vermeulen, G. (eds.), Maklu, Governance of Security Research Papers Series I, Antwerpen, p. 129-140

Pharr, S. J. & Putnam, R.D. (2000). *Disaffected Democracies: What's Troubling the Trilateral Countries?* Princeton: Princeton University Press.

Pickery, J. & Carton, A. (2003). Hoe representatief zijn telefonische surveys in Vlaanderen? APS Not@S, 4, 27. Retrieved from http://aps.vlaanderen.be/statistiek/publicaties/pdf/nota/2005-03_telefoon.pdf

Pleysier, S. & Parmentier, S. (2000). Het vertrouwen in de instelling: enkele beschouwingen. *Panopticon, 22,* 63-79

Pleysier, S., Pauwels, L., Vervaeke, G. & Goethals, J. (2005). Temporal invariance in repeated cross-sectional 'fear of crime' research. *International Review of Victimology, 12,* 273-292.

Roberts, J. V. & Hough, M.J. (2005). *Understanding public attitudes to criminal justice.* Maidenhead: Open University Press.

Scheepers, P., Felling A. & Peters,J. (1992). Ethnocentrism: Update of a Classic Theme and an Empirical Test. *Politics and the Individual, 2,* 43-60.

Schlüter, E., Davidov E. & Schmidt, P. (2007). Applying Autoregressive Cross-Lagged and Latent Growth Models to a Three-Wave Panel Study. In K. Monfort, Oud, J., & Satorra, A. (eds) (Ed.), *Longitudinal Models in the Behavioral and Related Sciences.* Mahwah: Lawrence Erbaum Ass.

Srole, L. (1956). Social Integration and Certain Corollaries: an Exploratory Study. *American Sociological Review, 21,* 709-716.

Struening, E.L. & Richardson, A.H. (1965). A factor analytical exploration of the alienation, anomia and authoritarian domain. American Sociological Review, 30, 768-777.

Sumner, W. G. (1906). *Folkways: a study of the sociological importance of usages, manners, customs, mores and morals.* Boston: Ginn and Co.

Studiedienst van de Vlaamse Regering (2003). *Survey naar sociaal-culturele verschuivingen in Vlaanderen 2003.* Brussel: Ministerie van de Vlaamse Gemeenschap.

Tajfel, H. & Turner, J. (1979). An integrative theory of intergroup conflict. In: Austin W.G. & Worchel, S. (Eds.). *The social psychology of intergroup relations*. Monterey: Brooks/Cole

Tyler, T.R. (1997). Procedural fairness and compliance with the law. *Swiss Journal of Economics and Statistics*, 133, 219-240

Tyler, T.R. (2001). Trust and law abiding behavior: Building better relationships between the police, the courts, and the minority community. *Boston University Law Review*, 81, 361-406.

Vancluysen, K., Van Craen & M., Ackaert, J. (2009). *Gekleurde steden: Autochtonen en allochtonen over samenleven*. Brugge: Vanden Broele.

Appendix I: measurement of constructs

SCV-survey

Background characteristics	
Gender	0= Female 1= Male
Age	Continuous variable
Highest certificate	0= None 1= Primary education 2= Secondary education 3= Higher education
Origin of the parents	0= Both parents were born with Belgian nationality 1= At least one parent was born with a non-Belgian nationality
Religious attendance: ('Once in a while people attend religious servi- ces in special occasions as weddings, funerals etc.. If we do NOT take these special occasions into account, how often do you attend religious services?)	1= Never 2= Very seldom 3= Only on religious holidays as Christmas and Easter. 4= Monthly 5= Several times a month 6= Weekly 7= Several times a week

Social capital

Participation in organisations*	-1= Member of no organisation 0= Member of one organisation 1= Member of at least two organisations
Number of confidants**	-1= More than one standard deviation below mean 0= Between one standard deviation below mean and one standard deviation above mean 1= Above one standard deviation above mean

* Membership of the following organisations has been taken into account: a youth movement, a youth association, a youth club, an environmental or a nature association, a fanclub, an association which helps the disabled, elderly or poor, the national health service, a drama or arts club, a hobbyclub, a women's movement, a social-cultural association, a sports club, a political association or political party, a religious association, a community organisation or a carnival club, a labor union, an association of merchants, a municipal advise board or a school board, an association for families, a society attached to a local pub (darts, soccer, ...), the red cross, the Flemish cross, the voluntary fire brigade, general aid-services, a self-help group.

**'Once in a while most of the people discuss important personal matters with others, for example when they have had a fight with someone very dear to themselves, when they have troubles at work or problems of the same kind. Who are the people you discuss these personal matters with?' We counted the number of different persons that one can rely on, and made a distinction between respondents with less social ties than one standard deviation below the mean (-1), respondents with social ties equal or between one standard deviation below the mean and one standard deviation above (0), and respondents with social ties that exceeded one standard deviation above the mean (1).

Attachment to significant others*

Alpha=.580	Factor loading
To what extent do you feel attached to your circle of friends?	.513
To what extent do you feel attached to the most of your relatives?	.634
To what extent do you feel attached to the most of your neighbours?	.551

Political powerlessness*

Alpha=.796	Factor loading
To vote is of no use	.691
The promises made during the elections almost come to nothing	.708
Poltical parties are only interested in my vote, not in my opinion	.762
So many people vote, my vote is of no importance	.520
Polticians do not listen to normal citizens	.626
It is very difficult to change an unfair law	.505

*All five items could be answered on a five point scale : 'I totally disagree (1), I disagree (2), I agree nor I disagree (3), I agree (4), I totally agree (5)'

Ethnocentrism**

Alpha=.884	Factor loading
Immigrants contribute to the welfare of Belgium (R)	.708
You cannot trust immigrants	.688
Immigrants take advantage of the welfare system	.786
Muslims form a threat for our culture/customs	.728
A diversity in cultures forms an enrichment (R)	.696
When the amount of jobs diminishes immigrant should be sent back	.745
Foreigners who want to settle here should be welcomed (R)	.609
A stranger who legally resides in Belgium for more than 5 years should get the right to vote for the local council (R)	.618
Strangers who want to settle here have to adapt to our customs	.511

**All propositions could be answered on a five point scale: 'I totally disagree (1), I disagree (2), I agree nor I disagree (3), I agree (4), I totally agree (5)'

Institutional trust***

Alpha =. 854	Factor loading
Degree of trust in Belgian criminal justice system	.442
Degree of trust in Flemish Government	.786
Degree of trust in Flemish Parliament	.814
Degree of trust in Belgian Parliament	.846
Degree of trust in Belgian Federal Government	.858

*** Could you tell for each of these institution if the amount of trust you have in them is : very low (1), low (2), not high, not low (3), High (4), very high (5)?

ESS-survey

Background variables	
Gender	0=Female 1=Male
Age	Continuous
Highest certificate	0= none 1= Primary education 2= Secondary education 3= Higher education
Immigrant background	0= Both parents were born with Belgian nationality 1= At least one parent was born with a non-Belgian nationality
Religious attendance	1= Never 2= Less often 3= Only on special religious holidays 4= At least once a month 5= Once a week 6= More than once a week 7= Every day

Social capital

Participation in organisations*	0= Member of no organisations 1= Member of at least one organisation
Is there someone whom you can discuss intimate and personal matters with?	0= No 1= Yes
In comparison to other people with your age, how often do you take part in social activities?	1= a lot less than most 2= Less than most 3= About the same 4= More than most 5= A lot more than most

*Membership of the following organisations has been taken into account: a sports club, a club for out-door activities, an organisation for cultural and hobby activities, a trade union, a business, professional or farmers' organisation, a consumer or automobile organisation, an organisation for humanitarian aid, human rights, minorities or immigrants, an organisation for environmental protection, peace or animal rights, a religious or church organisation, a political party, an organisation for science, education or teachers and parents, a social club, club for the young, the retired/elderly women, or friendly societies and any other voluntary organisation such as the ones that have just been mentioned.

Political powerlessness

Alpha=.630	Factor loading
How often does politics seem so complicated that you cannot really understand what is going on?*	.545
Would you say politicians are just interested in getting people's votes rather than in people's opinions?**	.598
How difficult or easy do you find it to make your mind up about political issues? ***	.586
Do you think that politicians in general care what people like you think?****	.477
'Do you think that you could take an active role in a group involved with political issues?'*****	.391

* Answer categories go from : Frequently (1), Regularly (2), occasionally (3), Seldom (4), Never (5)
* Answer categories vary from : Definitely (1). Probably (2), Not sure either way (3), Probably not (4), Definitely not (5)
*** Answer catagories go from : Very easy (1), Easy (2), Neither difficult nor easy (3), Difficult (4), Very difficult (5)
**** Answering possibilities were: Most politicians care what people like me think (1), Many care (2), Some care (3), Very few care (4), Hardly any politicians care what people like me think (5).
***** Nearly all politicians are interested in people's opinions (1), Most politicians are interested in people's opinions (2), Some politicians are just interested in people's votes, others are not (3), Most politicians are just interested in votes (4), Nearly all politicians are just interested in votes.

Ethnocentrism

Alpha=.831	Factor loading
Would you say that people who come to live here generally take jobs away from workers in Belgium, or generally help to create new jobs?*	.660
Most people who come to live here work and pay taxes. They also use health and welfare services. On balance, do you think people who come here take out more take out more than they put in or put in more than they take out?**	.667
Would you say it is generally bad or good for Belgium's economy that people come to live here from other countries?***	.766
Would you say that Belgium's cultural life is generally undermined or enriched by people coming to live here from other countries?****	.589
Is Belgium made a worse or a better place to live by people coming to live here from other countries?*****	.775
Are Belgium's crime problems made worse or better by people coming to live here from other countries?******	.610

* Scores go from : 0 create new jobs to 10 Take jobs away)
** Scores go from 0 Generally put in more to 10 generally take out more
*** Scores go from 0 Good for the economy to 10 Bad for the economy
**** Scores go from 0 Cultural life enriched to 10 Cultural life undermined
***** Scores go from 0 Better place to live to 10 Worse place to live
****** Scores go from 0 Crime problems made better to 10 crime problems made worse

Insitutional trust*

Alpha =.886	Factor loading
Trust in the Belgian Parliament	.786
Trust in the Belgian Legal system	.717
Trust in the Belgian Police	.648
Trust in Belgian Politicians?	.828
Trust in the European Parliament	.820
Trust in the United Nations	.724

* Scores go from 0 No trust to 10 complete trust

Appendix II: Descriptives of all variables included in the analysis.

SCV		Frequency	Percent
Gender	Female	759	51,4
	Male	718	48,6
	Total	1477	100
Educational level	No certificate	22	1,7
	Certificate primary education	212	16,3
	Certificate secondary education	703	54,2
	Certificate higher education	360	27,8
	Total	1297	100
Immigrant background	Both parents Belgian	1385	94
	At least one parent non-Belgian	89	6,0
	Total	1474	100
Religious attendance	Never	647	43,9
	Very seldom	341	23,1
	Only on religious holidays	169	11,5
	Monthly	91	6,2
	Several times a month	59	4,0
	Weekly	155	10,5
	Several times a week	13	0,9
	Total	1475	100
Number of confidants	Low	206	14,2
	Moderate	1081	74,8
	High	159	11,0
	Total	1446	100
Participation in organisations	Low	622	43,7
	Moderate	410	28,8
	High	391	27,5
	Total	1423	100

Parametric measures

	Min.	Max	Mean	Median	Std. Dev.	N
Social Bond	3	21	15,30	15	3,37	1477
Political anomia	8	30	20,69	21	4,15	1477
Ethnocentrism	9	45	29,39	29	6,85	1477
Trust in government	5	25	14,52	15	3,40	1477

ESS		Frequency	Percent
Gender	Female	569	46,1
	Male	664	53,9
	Total	1233	100
Educational level	No certificate	176	14,5
	Certificate primary education	373	30,7
	Certificate secondary education	510	42
	Certificate higher education	155	12,8
	Total	1214	100
Immigrant background	Both parents Belgian	1112	90,4
	At least one parent non-Belgian	118	9,6
	Total	1230	100
Religious attendance	Never	581	47,1
	Less often	254	20,6
	Only on special holidays	165	13,4
	At least once a month	94	7,6
	Once a week	123	10
	More than once a week	14	1,1
	Every day	2	0,2
	Total	1233	100
Presence of confidants	Yes	1111	90,3
	No	119	9,7
	Total	1230	100
Participation in organisations	None	341	27,6
	Member of at least one organisation	893	72,4
	Total	1234	100
Social involvement	A lot less than most	192	15,6
	Less than most	359	29,2
	About the same	443	36,0
	More than most	191	15,5
	A lot more than most	44	3,6
	Total	1229	100

Parametric measures

	Min.	Max.	Mean	Median	Std. Dev.	N
Age	15	95	45,08	44	18,30	1229
Anomia	7	25	17,63	18	3,34	1234
Ethnocentrism	5	60	33,92	33	8,55	1234
Institutional trust	0	60	30,16	32	10,52	1234

Conceptualising the role of police culture in change strategies

Marleen Easton and Dominique Van Ryckeghem

1   Introduction

The history of police reform in Belgium reveals that police culture has never been a real concern to our policy makers. The focus has always been on the structural reorganisation of the policing landscape (Enhus & Ponsaers, 2005). Similarly, literature about the Belgian police culture(s) is almost non-existent. Academics who tried to shed a light on the various dimensions of the former police force have never ventured to consider police culture as a research subject.[1] Nonetheless many of them have endorsed the importance of this culture. The police departments often made limited analyses, but the findings were seldom or never published[2].

Moreover, the culture of police departments has been effectively researched at international level. To conceptualise the role of police culture in change strategies, a review of this literature is necessary and this should take a number of elements into account. First, a study of a culture (the culture of a group of people, an organisation or a nation) is in itself beset by theoretical and ideological schisms. Second, a survey of a culture invariably means the researcher him/herself subscribes to a specific cultural trend. A grasp of these trends and their research methods is nearly indispensable in order for us to gain a full understanding of police culture and, almost important, for us to be able to interpret the findings.

This contribution starts with a theoretical review of the problems encountered when studying culture in general and the trends it inspires. Secondly, we focus on police culture and describe the key findings of the available studies, while showing how they reflect certain cultural trends. Thirdly, we move away from the unidisciplinary approach and we seek to adopt an interdisciplinary stance for each of the cultural study problems we explained in the first part of this paper. Therefore, the study of police culture is rather complex: various approaches and research methods have to be factored in. Conversely, the simplicity lies in the fact that opting for a specific approach and type of research allows this approach to be incorporated, its limitations to be recognised and the research findings to be assessed in the light of their true value. Therefore, understanding culture through research paves the way for the incorporation of culture in change strategies, which is an important point of interest in current reforms of police systems across the world.

---

1   Els Enhus, Christian Eliaerts, Cyrille Fijnaut, Paul Ponsaers, Lode Van Outrive, Tom Van den Broeck, e.a.

2   E.g.: In 1993 the Belgian Gendarmerie carried out an opinion survey amongst its members to discover the level of motivation and the attitudes of state police employees (N=1800).

2   The study and research of culture

'Culture is a blank space, a highly respected, empty pigeonhole. Economists call it 'tastes' and leave it severely alone. Most philosophers ignore it – to their own loss. Marxists treat it obliquely as ideology or superstructure. Psychologists avoid it, by concentrating on child subjects. Historians bend it any way they like. Most believe it matters, especially travel agents." (Mary Douglas, 1987)

The study of culture, whether it concentrated on a nation, an organisation or a profession, has always involved a variety of theoretical and ideological premises. The study of culture always forces the researcher to adopt a specific viewpoint on culture. To fully understand and interpret the results of those studies it is important to fully grasp the different theoretical and ideological premises and the research methods that are applied. In the following, we will therefore focus on the study and investigation of culture and apply the insights in a second part on the study and investigation of police culture.

A study of 'culture' means examining an issue that has been surveyed elaborately by sociologists, historians, anthropologists, philosophers and politicians. The studies have neither resulted in an identical approach to culture nor in a harmonised definition or a single methodological approach. Various overviews of cultures also reflect the complicated nature of the concept. Stuart Hall (1980) claims the cultural concept is the site of converging interests rather than a logical or clear conceptual idea. Nevertheless, in what follows we try to get grip on this complexity by giving an outline of theoretical and empirical literature on culture. The aim is to gain a better understanding of cultural theories and beliefs that also underpin the study of police culture and the way research on this topic is conceptualized.

2.1   The content-based approach and defining culture

Within this section we pay attention to the theoretical approaches to culture on the one hand and cultural definitions and analysis on the other hand.

2.1.1   Approaches to culture: free will versus determinism

Cultural theory is based on two dimensions: the *subjectivist* and the *mechanistic* interpretation of human activities.

According to the subjectivist interpretation, human activities are primarily motivated by human qualities: feeling, perception or susceptibility. People 'follow' a specific cultural system more for voluntary reasons than for strictly mechanistic ones. Human experience is a central feature as a source of meaning: people make history. Culture determines human activities.

According to the mechanistic interpretation human behaviour is first of all a reaction or a reply to environmental factors. Experience is not authentic but predetermined. The reasons for following a cultural system are therefore not voluntary but more a question of unconscious action.

These approaches have provided a framework for specific cultural definitions and research methods.

2.1.2   *Cultural definitions and cultural analysis*

The *subjectivist* interpretation defines culture in terms of values or ideas (symbols). The *mechanistic* approach focuses on structures within which these values and ideas are formed. Depending on the authors, the structure is understood in terms of the historical social structure (described as class structure by many Marxist writers) or the specific conceptual structure (thought structures including 'the world').

In the case of the *subjectivist* approach (E.g. Parsons, 1952) *values* take shape around issues in the social system, such as equality versus inequality and respect for authority versus critical behaviour. In this perspective, the key aim of cultural research is to obtain knowledge about these values as well as the process which allowed the values to become part of the social system. These analyses are carried out through (quantitative) structured, uniform interviews in order to gain insights into the values of the respondents.

Other authors (such as Geertz, 1973; Edgerton, Langness, 1979; Van Hoewijk, 1988; Denzin, 1989) reject the idea that culture is made up of values, claiming it involves *symbolic universes* which the participants build onto this culture. Culture is a process of developing meanings. Culture is defined, depending on the specific trend, in terms of people's intellectual background (culture comprises ideas, is a system of knowledge), of reciprocal human activities (culture is a social-cultural system) and of an almost formal procedure (culture is a regulatory system). The content is assessed in the light of quantitative techniques or interviews, observations and conversation analysis.

The *mechanistic* interpretation covers various ideological views. We single out the materialist view, the structural approach of culture and the structuralistic approach seeking to reconcile the former two.

Materialists believe that culture also involves values and meanings. The difference with subjectivists is that ideas and opinions are not so much formed as they are rooted in a social structure. The source of meaning is not human experience itself, but the material (historical) circumstances determining the experience or, in materialistic terms, 'consciousness'. Research in this context is primarily historical. Later approaches (Gramsci, 1980; Williams, 1963, 1965; Thompson, 1968; Hall, 1980) have tempered this way of thinking in terms of structure and culture as separate poles: there is dialectic between being social and consciousness. Various groups and classes derive values and meanings from their given historical circumstances and relationships[3]. Conversely, lived traditions and practices in which insights are expressed in turn play a role in determining history.

In the structural approach to culture, with Lévi Strauss (1992) as its main exponent, people are primarily regarded as bearers of structures. Experience is not a source of authenticity but a result of *unconscious thought structures and categories*. They create practices that give meaning. These unconscious structures are studied in the light of a qualitative assessment of the social discourse that is best compared with the analysis of a language's grammar.

Finally, the structuralists claim that culture represents *all practices giving meaning*, created by categories and frameworks of thoughts, which in turn are *rooted in the so-*

---

3   The values are determined by the economic conditions of existence. Consequently, the values of a higher class are different from those of a lower class.

cial (materiel, historical) structure (for example, Laclau, 1977; Althusser, 1976). Various intelligence gathering techniques, all qualitative, are used in order to identify this empirical basis.

2.2   Cultural focus

The cultural focus specifies the kinds of questions that arise during an analysis of culture. This refers to questions about the origins and the implications of a specific culture or about its nature and characteristics.

A specific cultural focus is first of all chosen in view of the required research outcomes/questions. Three levels are singled out (Louis, 1985):

- The natural level comprises research questions considering the origin of a specific culture ('where does the culture come from?'), the outcomes and their effects ('where do they lead to?') and the actual manifestations of this culture ('how does it appear?').
- The efficiency level is connected to the 'management' of culture. This approach is characterised by questions such as what can be done with or for a specific culture. The priority is the potential power of a culture and to what extent this power can be managed and controlled or guided.The reflective level covers all the attempts to understand the nature and characteristics of a culture as well as the debate concerning the related philosophical assumptions, structures and targets of a culture. This level offers the broadest approach: the research questions for the previous levels may also form part of this.

2.3   The limitations of culture

The limitations of culture refer to both the scope of the environment under study (Is there reference to the culture of a country, region, city, organisation?, etc.) as well as the group of people whose culture is identified (does this involve the culture of police officers, 'management cops' or 'street cops'? etc.). Given that police officers form part of a very specific organisation, the effective connection between organisation and culture has to be researched. The literature offers two perspectives for understanding this connection: first culture as a variable in the organisation and secondly culture as a 'root metaphor'.

2.3.1   Organisation and culture: culture as a variable in the organisation

Various studies of organisations (for example, Schein, 1967, 1985) define culture as shared basic assumptions. In this perspective, groups and organisations have a culture only if they are fairly stable. Groups and organisations develop as a result of an adaptation to the environment and internal integration issues (concerning power, rewards, punishments and the like). This idea of culture as the 'cement' of the organisation forms the theoretical basis for developing cultural typologies.

A number of authors (such as Deal and Kennedy, 1982; Peters and Waterman, 1982) took these ideas one step further by looking for the key to commercial success in cultural characteristics or specific 'types'. The resulting 'one best way strategy' is the belief that successful organisational cultures are universally applicable and therefore malleable. This was qualified by Bollinger and Hofstede (1980): a successful organisational culture also has to be compatible with a country's culture. Not every culture is uniformly amenable to a specific organisational culture, nor is this compatibility a guarantee: in the case of one of the companies of Peter and Waterman, for example, there was no essential conflict between its organisational culture and the national culture and this was described as highly successful in terms of cultural characteristics. Nevertheless, the company still went bankrupt.

What these approaches have in common is the way they define culture in terms of equal assumptions: culture as a cement for the whole. It is less important whether there assumptions are referred to at the level of the country, organisation or suborganisation, culture is a variable *within* this level.

2.3.2 *Organisation and culture: culture as 'root metaphor'*

Organisations are in themselves cultural phenomena (Smircich, 1983; Frissen and Van Westerlaak, 1990). Culture is the determining factor and therefore the 'modeller' of the organisation, which has a number of implications. First, the limitation of culture is problematic. Since culture does not exist as a variable, there is no actual 'level' for it to be analysed on. Culture is everywhere but difficult to grasp. Secondly, looking at organisations as cultural phenomena implies that all the cultural differences outside the organisation (such as the ethnic group, the professional category and training) will also have an impact within the organisation and its subcultures. Thirdly, seen from an 'organisations as cultures' perspective, it is unimportant to pay attention to the question of effectiveness. Organisations exist and have or generate one or multiple meanings for the people who constitute the organisation. Consequently, the cultural concept in this theoretical stream offers no indication about how organisations function effectively.

2.4 Conclusion

This extremely brief overview shows exactly how complicated studying and investigating culture is and the extent to which it involves a variety of theoretical and ideological premises. The belief that human beings are the creators of culture and society is contrasted with the fact that a human being is not a completely autonomous being but is largely determined by the structures within which he/she exists. There are also various viewpoints about the content of culture: does this involve values, ideas, symbols, does culture consist of patterns of behaviour or a set of rules, or is it more a question of shared assumptions? These questions and various starting points underpin all cultural studies, irrespective of whether a culture is studied in terms of a general cultural context or an organisation culture.

Both the content of culture and the way its relationship with the social structure is understood affect the way culture is researched. A quantitative methodology is contrasted with qualitative methods and historical research with purely contemporary research.

Just as important for cultural research are the reasons for investigating culture. One option may be the desire to gain more knowledge and a better understanding of culture, as well as the aim of deploying culture in a change strategy. In the latter case there is a need to guard against simplification and normative judgements about the limitations of culture, its content or its 'power'.

This paper is about police culture. The question is how far the problem areas, mentioned above, are encountered in the study and investigation of police culture. Our examination is based on international studies as well as (the admittedly limited) research in Belgium.

3   The study and investigation of police culture

Published in the 1950s, 1960s and 1970s, the first studies on police culture originated in the Anglo-Saxon world (such as Skolnick, 1966, Westley, 1970, Wilson, 1968, Van Maanen 1974, 1975, 1979, Cain, 1973). The initial theorists were extremely reduction-ist and we might say that the discipline of the researcher sets the tone for the way peo-ple understand police culture: sociologists focus on socialisation processes; psycholo-gists consider the culture of the individual police officer at field level; organisation theorists mainly discuss the way the members of an organisation behave within the context of the organisation. It is also important for the various disciplines to provide a variety of insights. The following studies are arranged in chronological order but the list is by no means exhaustive.

3.1   The pioneers: 'working personality' and 'police behaviour'

Back in 1950 Westley (1970)[4] undertook the first sociological analysis of the American police organisation. He paid attention to the uniqueness of the police profession and the implications for the identity of each policeman or policewoman. One of the im-plications is the informal code of behaviour of officers and the careful attention for secrecy: it is always better to say too little than too much. New recruits are socialised into this culture via rituals: *"Their occupational culture is transmitted and maintained through the various initiation rites that mould the rookie policeman, a process culminating in the rookie's recognition of the rules and values of the group as defining his own self-esteem"* (Westley, 1970: 105)

Skolnick (1966), taking his cue from Westley, regards the police culture as a set of values, attitudes, rules and practices that have an impact on the way police officers carry out their duties. This culture develops in relation to the specific characteristics of policing, such as the dangerous nature of the work in the community, the authority of the supervisors and the pressure to boost the productivity and effectiveness of the organisation. This unique combination of factors generates a cognitive response and a behavioural one. A *'working personality'* is developed with the following characteris-tics: suspicion, social isolation, internal solidarity, a deep sense of mistrust between the police and the community, the exercise of authority (which exacerbates the misun-derstandings, an ambiguous relationship with moral principles (power and manliness are essential qualities for combating evil), the closed nature of the force, the conserva-tism and informal pressure for effectiveness (Skolnick in Parienté, 1994). Skolnick also points to an interaction between the psychological and the social characteristics of

---

4   His thesis was completed in 1950 but not published until 1970.

the environment and the profession: the former will form part of the *'working personality'*. Comparable studies which focused on the British and American police systems, Banton (1964) and Clark (1965) draw roughly the same conclusions.

Wilson (1968) takes this idea one step further and researches 'police behaviour' in eight different communities[5] in Great Britain. Police officers appear to tailor their working style to the environment they are operating in. Wilson describes three types: the 'watchman style', 'legalistic style' and 'service style'. He attached a lot of importance to the values and standards that develop both within the police organisation and at field level referring to the idea that the policeman/woman is a member of a 'trade'. Police officers are reported to absorb the applicable cultural rules via 'leadership' rather than formal training. Experienced beat officers involved with 'real' policing consequently make a comparatively large impact on police culture.

The focus on the socialisation process leads to the next stage in the study of culture. The research covers not only recruitment and training but also socialisation at field level. An investigation is made in particular of the extent to which these processes affect the development of police officers' personalities and action style.

3.2 Transfer of the 'working personality' and 'police behaviour': formal and informal socialisation

Interest is growing in the formal training of newcomers, based on the assumption that the values and standards of recruits and the police culture are interconnected (McNamara, 1967; Harris, 1973; Van Maanen, 1973, 1978a; Hopper, 1977). Attention is paid to the content of culture and the way its values and standards are transferred. Gray (1975) shows how the American police organisation cultivates and safeguards the uniformity of its culture. In the light of individual screening, for example, the police force recruits individuals who can demonstrate an affinity with the police organisation and its culture. Consequently, police training is regarded as being of little use for determining the behaviour of police officers. Weiner (1974) even goes as far as to suggest the police force's role in society neutralises the police training. Socialisation at field level is often more important in gaining skills than the training itself.

Authors increasingly feel that the police culture itself primarily develops in the workplace. Informal socialisation is gaining in importance as a research subject.

3.3 Nurturing the police culture

Police culture covers the values and standards that affect behaviour patterns and work practices, as they are applied by the officers (Van Maanen, 1974, 1975, 1979, 1984; Cain, 1973; Brown, 1981; Punch, 1983; Reiner, 1985).

According to Sandler and Mintz (1974) police culture is nurtured by the specific image, the 'self image' of police officers, that is largely built upon the basis of the image of the *'crime-fighter'*, the fight against crime.

---

5   "I try to describe the behavior of patrolmen discharging their routine law-enforcing and order-maintaining functions, to explain how that behavior is determined by the organizational and legal constraints under which patrolmen work, to discover the extent to which it varies among police departments, and to determine insofar as the evidence permits what accounts for these differences and especially how local politics contributes to them." (Wilson 1968:10).

Speaking about police culture, Punch says *"...An unwritten and almost unconsciously learned occupational code exists which articulates and informs actual police work."* (Punch, 1978: 517).

Brown (1981) reproduces the image of the 'crime-fighter'. Police officers seem to pay little attention to rules and procedures and operate from an 'action perspective'. A key component of this perspective is the loyalty and bond of solidarity among police officers. The police culture thus defined is a source of friendship, protecting the officers against outside criticism.

Other studies nonetheless show that crime-fighting accounts for only a small part of policing (Manning, 1978; Punch, 1979). The duties of police officers are more often located in a grey area between policing and social work[6]. However, Punch[7] claims they would rather not know this. They are not keen on associating requests for help with policing, so prefer to avoid them, with the result that the police service is focused on acting in response to clear problematic situations, giving rise to a reactive culture, according to Punch.

This analysis has in turn given rise to the assumption that police culture is determined less by the nature of the work than by the prevailing ideas about policing.

3.4   Implicit assumptions of police culture

Chatterton (1975) studied the police force's world view and the attitude of police officers towards their job and the way they perform their duties. Manning (1977) also paid attention to these implicit assumptions. A recurring factor here is the way police officers use stereotypes to organise the world and, more importantly, to simplify: *'policing labelling practices'* (Van Maanen, 1978b).

The authors argue that this view of the world and work does not have an immediate impact on police activities on the ground, as these activities still also depend on the specific situation: *"In a sense, the freedom of the patrolman to enforce or not to enforce has the effect of creating rules. His rules are not the administrative rules, which derive substantially from the criminal code or municipal regulations, but are those 'rules of thumb' that mediate between the departmental regulations, legal codes, and the actual events he witnesses on the street."* (Manning, 1977: 162). *'Good police work'* is defined according to each situation.

Recognised and problematized for a longer period of time, the discretionary responsibility issue (Bittner, 1967, 1970; Goldstein, 1964, 1977; Muir, 1977)[8] is in this context related to police culture. Police officers are acknowledged to enjoy a degree of scope and autonomy at field level to interpret rules and procedures and also to enforce them depending on the situation. Culture is said to play a key role precisely in this type of interpretation and enforcement.

---

6   This is why they are described as the 'secret social service'.
7   He was the first to carry out research into police culture in continental Europe. In the early 1970s he researched the activities of the Amsterdam police.
8   Goldstein identifies the discretionary responsibility in the 'gaps' created by the dynamic interplay between a) the rules of the broader community and the police departments and b) the informal culture of the police service as a world full of rules.

3.5   Organisation theories and police culture

Research into the implicit assumptions of police culture and police activities at field level, the recognition that police officers largely construct 'their' culture themselves as well as the general opinions of organisation theory cause multiple cultures to be identified within a single police organisation. Reuss-Ianni and Ianni (1983) proved that police culture is not a monolith. In light of conflicting perspectives in procedures and practices, they make a distinction between two cultures: that of the 'street cops' and that of the 'management cops'.

Applying conclusions drawn from organisation theories results in police culture being regarded as a component of the control system within the police organisation (Brown, 1981; Lipsky, 1980). Apart from the formal, military-bureaucratic control based on a strict line of command, police officers are also subject to an informal control in light of the prevailing culture. And this is often the most important. According to Brown (1981:83) this is because: "*Hierarchical controls are directed toward (...) the more mundane aspects of job-related behaviour. The group controls, on the other hand, are directed to those behaviors of immediate concern to the performance of the police task*".

Finally, consideration is also given to the interaction between the police culture and environment of the organisation. This is known in organisation theories as the 'contingency approach'. According to Alderson (1979), who was part of the police force for many years, there is a clear relationship between the culture of the community and its police force. He sums this up in the winged expression: '*To police us you should live with us*'.

3.6   Police styles, implicit assumptions, organisation theories and informal police culture

In the 1980s and 1990s, authors generally built upon the earlier conclusions. The research is often more fundamental, the findings more detailed.

Reiner (1985) and Broderick (1987) continue to research categories of police styles. Reiner (1985) makes a distinction between 'New Centurion' (the genuine crime fighters), 'Uniform Carrier' (the cynics), 'Professional' (sympathetic officers) and 'Bobby' (those who regard policing as a profession). Broderick (1987) makes a distinction between 'Enforcer', 'Realist' 'Idealist' and 'Optimist'. Reiner (1985) also describes the characteristics of police culture (having a mission, cynicism, machismo, racism, political conservatism and pragmatism), referring to seven types of stereotyping of citizens as a functional requirement for police officers at field level: ('*good class villain, police property, rubbish, challengers, disarmers, do-gooders, politicians*'). Police officers are said to use these stereotypes to legitimise their activities on the ground. Van der Torre (1992) also reproduces the idea of a police style, making a distinction between 'restorers of order, 'social workers', pragmatic loyalists' and 'pessimists'[9], on the basis of his research in the Netherlands. He suggested that *what* the various types of officers did was not much different, in contrast to the differences in *how* they did their work.

---

9   Tom Van den Broeck (2001:40-48) recapitulates the work of Wilson (1968), Reiner (1985), Broderick (1987) and Van der Torre (1992), believing there are similarities in the way the four authors rank police style.

In the case of implicit assumptions, more and more authors contend that police officers are more than passive objectives absorbing the police culture[10] (Fielding, 1988; Shearing and Ericson, 1991; Reiner, 1992:109; Shearing, 1994; Chan, 1996). Fielding regards police officers as the intermediaries of the structural and cultural effects of the profession. Their individual experience interacts with the requirements of the work and the pressure of the job. Shearing and Ericson claim that police officers actively construct their culture so as to arrange reality into some kind of order. Culture is transferred less via a process of socialising and internalising rules than via a combination of stories and aphorisms indicating to the officers how they should regard the work and how they should behave within this context. Culture is cultural knowledge, taking the form of *'police stories'*: they form the plans and instructions police officers use as a basis for seeking information, organising it into categories, taking action and legitimising this action. Chan also defines culture in terms of cultural knowledge, based on the work of Bourdieu (1990, 1992). Her theory acknowledges the interpretative and active role police officers play in relating their skills to the social and political context of policing.

Organisation theories continue to have an impact on cultural research (Roberg and Kuykendall, 1990). Research on corruption within the Dutch police force prompted Punch (1985) to conclude that the environment had a major impact on the interaction between police culture and those involved in the culture. Others point out that environmental conditions represent a key factor in the development of police culture (Alpert and Dunham, 1988; Wasserman and Moore, 1988; Greene and Decker, 1989).

Greene et al. (1992) fleshed out the distinction between 'street cops' and 'management cops'. They made a qualitative investigation of how police culture is connected to the interrelationships between core staff, between the core staff and supervisors and between both staff categories and the population. Manning in turn makes a distinction (1993) between three subcultures: the culture of the command, middle management and core staff. The individual variable he uses as a basis for making this distinction cover various items, such as personality, generation and career development. Added to these are structural variations in terms of rank, task and specialisation (Reiner, 1992; Chan, 1996; Parienté, 1994; Punch, 1999).

Goldsmith (1990) sees police culture as an informal regulatory framework which is also functional. The informal rules regulate the behaviour of police officers, enabling them to survive in a profession regarded as dangerous, unpredictable and alienating. As well as being a variable to explain and predict policy behaviour, culture is regarded by Goldsmith as a way of finding solutions for misbehaviour and even influencing discretionary responsibility. In order to be able to control and regulate the police force, police culture also has to be 'confirmed'. The police officers themselves have to regard this as useful. This raises the tricky question of how police culture is involved in formulating rules for giving structure to the exercise of discretionary responsibility.

---

10   In the case of the "transfer' of culture, Fielding (1984, 1988) and Southgate (1988) researched the formal training of newcomers entering the police force. Monjardet and Gorgeon (1992, 1993a, 1993b) carried out a longitudinal study in France on the socialisation of police officers during their training.

3.7   A specific organisation theory : management

In the late 1980s management jargon began to replace the military terminology more and more. The NRC Handelsblad[11] reported: *"The police is called (here) a 'powerful brand' that 'has to be communicated'. The police is spoken of as a 'business', citizens are called 'customers' and policing consists of 'products' that need to be 'clear products'"*. Conclusions drawn in the private sectors are transposed into the police organisation. The manager specifies the 'required' values with which each police officer has to comply (Cachet et al., 1993; Greene et al., 1994; Mastrofski, 1998) in order to achieve a result-based, customer-oriented and quality 'product'. The police officer is called upon to devise a strategy to close the assumed gap between existing and required values, such as, recruitment, training, discipline, leadership and opportunities for promotion.

In the Netherlands, research in support of the police organisation in the early 1990s was based on management theories (*for example,* Fokkinga[12] et al., 1992; Wilders, 1994; Van Keulen and Brink,1993; Hagenaars and op ten Noort, 1993). Van Keulen and Brink apply a 'culture diagnosis model' to the Dutch police service. They examine the market strategy, the innovative capacity, the power, management style, information, communication, the social climate and the differences of opinion within the police force. They define how the cultural characteristics should appear according to the new style. Hagenaars and op ten Noort research the cultural and innovative capacity of the regional police in Gelderland-Zuid[13]. They raise the question of how the management of the existing culture can be steered in the required direction.

3.8   French viewpoint

In the 1990s, French researchers also decided to get involved in the study of police culture. Parienté (1994) and Monjardet (1992, 1993, 1994) did not agree with the Anglo-Saxon definition of culture in terms of shared values. They concluded that police culture is not as uniform as it is often assumed to be. According to Monjardet, individual officers take up positions in relation to the 'law' (defer or interpret) and the 'others' (openness or otherwise). That police officers are subject to the rules of the law is also indicated by way of legalism. The fact that they also interpret the law is a reflection of their view of the law as a limitation, framework or a contract. The positions individual officers take up in relation to the 'others'/'*les non-policiers*' refers to the degree of transparency/secrecy, their position in relation to other 'players' on the field of security and their openness to partnerships. These two axes ('law' and the 'others') or this co-ordinate system should determine the position of police officers in relation to all other values, which in fact allow for a certain degree of variation.

---

11   'Officer seeks customer in Happyland', NRC Handelsblad of 15 October 1997, p 3.
12   The Fokkinga research team used the standard Hofstede questionnaire (1980) to research the basic values of the police. The key question in this cultural research is whether the organisation culture of the police region under consideration is consistent with its internal integration function and external adaptation function and whether cultural differences between the police forces may create problems for the region in the future.
13   They use the Sanders &and Nuijen (1987) model which, subsequent to in-depth research, drew the conclusion that six cultural dimensions may be singled out (result-based versus process-based, people-oriented versus work-oriented, professional versus organisation-based, open versus closed, tight versus loose, pragmatic versus normative).

This approach puts the importance of values in the police culture into perspective, making it dependent on the connection with the environment, the regulatory framework in which the police operate and the personal interpretations of the police officers

3.9   Belgian research on police culture

Police culture has been a neglected field of study in Belgium. All in all, there has been only one study (explicitly related to police culture) commissioned by the Ministry of Home Affairs. As a result of the impending merger at local level, Antwerp University's 'Institute for Management and Administration' researched the cultural differences between the state police and the municipal police (Ceuppens and Mertens, 2001).

As is often the case in management theories, the researchers take organisation theory conclusions as a starting point, in which the psychologist Schein (1980, 1992) was a key influence. They assess the organisation culture with a standardised questionnaire whose reliability had already been established in business circles. An assessment was made in particular of the competition amongst employees, competition with other organisations and the willingness within the organisation to take risks. The upshot is that the study focuses more on the management style(s) of municipal police and the state police than on cultural differences. Both police services are said to be hallmarked by an oppositional style[14]. This finding was presented to the supervisors of both organisations. The following quotation needs no comment: *"The supervisors who were interviewed acknowledged the aforementioned recapitulation. The belief is that this explanation at organisation culture level is completely true but there is still a factor that transcends this organisation culture: police culture. It was terribly difficult to get the respondents to elaborate upon this theory, everyone senses the difference in culture but cannot put it into words."* (Ceuppens and Mertens, 2001:13).

Meanwhile, research is conducted in Belgium which generated insights that are closely related to police culture. One example is an empirical research which examined the extent to which Community (Oriented) Policing (COP) takes shape (or not) in multicultural neighbourhoods and through contact with ethnic minorities, and how both parties feel about this (Easton et.al., 2009)[15]. This type of research ties in closely with what is often described as police sociology, which gives a picture of how policing actually takes shape in the social processes between police officers and citizens and between the officers themselves (van der Torre, 1999). The assumptions of this study rely on Shearing and Ericson's (1991) claim that police officers actively construct their culture so as to arrange reality into some kind of order.

One of the main conclusions is that the building blocks of experience from which policemen construct perceptions of the community are fragmentary and the perception is selectively created. The research results, gathered in six neighbourhoods through observations and interviews, indicate that the police assess "regular (tough)

---

14   The researchers use the focus questionnaire to compare the Organisation Culture Profiles of the municipal police and the state police. For further details about these instruments, see Van Muijen et al. 1996.

15   This research (01/12/05-31/05/08) was financed by Belgian Science Policy (www.belspo.be) and a contribution is published in this volume under the title 'The view of the police on community policing in Belgian multicultural neighbourhoods' by Marleen Easton & Paul Ponsaers.

customers" and classify them in their own categories, for which they create strong perceptions. The boundaries of these categories sometimes coincide with ethnicity, but just as much with other characteristics, such as age, lack of parental control, gender aspects, marginality, immigration status, etc. These categories refer to the 'policing labelling practices' Van Maanen (1978b) developed in his earlier work. In relation to the work of Skolnick the relationship between perception, internal reasoning and approach in the field seemed to be quite complex and influenced by different factors such as cultural and social capital of the policemen on the beat & situational factors. In this respect this research reflects different trends in the study of police culture mentioned above.

3.10   Conclusion

The preceding overview shows how most of the studies described take *the subjectivist interpretation of human activities* as their starting point. Therein, police officers are deemed to 'determine' their own organisation culture. Conversely, Gray (1975) suggests the cultural socialisation begins much earlier. His investigation into an American police department showed that solely individuals that already display an affinity with the culture of the police organisation, on the basis of their own physical backgrounds, are recruited. Chan (1996) attaches importance to the (partly unconscious) thought structures of police officers, which, in turn, are of key importance for the action they take.

There is a lot more variety in the approaches to defining culture. Some authors consider the *content* of police culture: the values, the personality, the police style, the implicit assumptions, the informal culture, etc. [16] Others focus on *its development*: the police profession or rather the ideas or perceptions that are available about the profession should determine the culture[17]. *The transfer* of the culture is another factor: an examination is made of the formal socialisation process, such as recruitment and training, as well as informal processes arising at field level[18]. Others interpret police culture on the basis of *organisation theories*: of key importance in this respect are the control within an organisation, the line of command, plus the existence of subcultures and the relationship with the environment[19]. In the final analysis, 'managers' mainly regard culture as an instrument to be manipulated in order to achieve a specific type of behaviour or push through a specific policy.

This latter group researches culture on the basis of a purely *rational cultural focus*. This is obvious in the statements made about the power of culture and the complete faith in its controllability and manageability[20]. The research questions of their predecessors mainly originate in the *natural cultural focus*: how does the culture appear in practical terms (cynical, macho, conservative, solidarity-based,...), what law enforcement approach does it give rise to (action perspective, reactive,...) and how does it emerge (job requirements, perceptions,...).

---

16   See for example, Westley, Skolnick, Wilson.
17   See for example, Van Maanen, Cain, Punch, Reiner, Manning.
18   See for example, Van Maanen, McNamara, Harris, Hopper.
19   See for example, Reuss-Ianni & Ianni, Brown, Lipsky, Alderson.
20   See for example, Fokkinga, Wilders, Van Keulen & Brink, Hagenaars & op ten Noort, Ceuppens & Mertens.

Lastly, culture is primarily researched as a *variable in the organisation*. The impact the environment has on culture formation is increasingly acknowledged: some regard this as a precondition, for others it is almost essential[21]. The limitations of culture in terms of organisation culture are also seconded by a belief in culture as the 'cement' of an organisation and the basis for determining typologies[22]. Solely the French authors refuse to regard culture in these terms and assign more importance to personal interpretation[23].

In sum: both culture in general and police culture in particular have been researched in different ways or more effectively within various academic disciplines. First and foremost, this points towards the complicated nature of culture and the difficulty of investigating it. In this respect, all the approaches considered are more complementary than exclusive. However, as far as they are independent and not related to other conclusions their understanding of culture continues to be comparatively limited and unsatisfactory, hence an interdisciplinary approach is recommended (see also Paoline, 2003).

4   Culture revisited and how culture "works"

In this last part, we hope to make a case for the future study of police culture. For this, we opt for an interdisciplinary approach and consider the problem areas in cultural studies based on this perspective. This results in an analytical formula that clearly shows the complementary nature of the diverse theoretical visions or trends. From a practical point of view, the formula constitutes a possible basis for the development of a change strategy in which culture is given a place among other variables. In particular, the relationship with the environment within which the change must occur, and the policy, is explicitly set out.

4.1   The intrinsic approach and definition of culture "revisited"

Today, it seems outdated to hold on to the pure form of the cultural approaches. One cannot understand "culture" without connecting this to subjective meanings, nor without taking into account its limits, typical of the social structure. One cannot interpret social behaviour without assuming that someone is following codes invented by someone else, nor can one ignore the fact that humans are the ones who create a changing environment for each cultural code.

To summarise, culture consists of cultural and structural "elements".

Cultural "elements" include values and attitudes (for police culture, *e.g.* manliness, internal solidarity, bravery, authoritarianism), but also ideas, assumptions or unconscious thought processes (for police culture, *e.g.* the use of stereotypes), beliefs, symbols, rituals (for police culture, *e.g.* the "police stories", badges).

---

21 See for example, Wilson, Alderson, Punch, Alpert & Dunham, Wasserman & Moore, Greene & Decker.
22 See for example, Broderick, Van der Torre.
23 See Parienté and Monjardet.

Structural elements in a police environment include the line of command or hierarchy. They can be manifested in specific cultural practices (*e.g.* the manner of greeting).

These cultural and structural elements overlap and "articulate" with one another, resulting in something which we call (work) culture in its entirety. These may refer to the organisation or the police environment as such, but also to the society of which they are a part.

In order to make an inventory of all these "elements", it is best to use both quantitative and qualitative methods. Interviewing people systematically and uniformly should provide an insight into their values and the way in which these have been acquired. Ideas and underlying assumptions emerge mainly during more in-depth, open conversations and observations. It goes without saying that an idea of the historical context will always help in interpreting and possibly clarifying values, ideas and human behaviour, also in a broader context.

4.2   The cultural focus "revisited"

The research questions which guide the study within the natural and practical focus of culture (see above), are not about the culture itself, but about how culture "works": how it comes into being, what the effects it generates (natural focus) and what kind of "power" it has. It is not the content of culture or the acquisition of knowledge in relation to culture that is central, but its origin, its effects or "what people can do with it".

However, if one does want to examine the content of culture, a reflexive focus is essential. This implies as broad an approach and definition of culture as possible, as was argued above (how culture "works", can only be discussed when developing a possible change strategy).

4.3   The boundaries of culture "revisited"

Here, culture as a variable is opposed to culture as a metaphor. In this field, culture as a metaphor would imply a study of the police organisation as a concrete manifestation and historical outcome of Western society and culture. But this is not the kind of study we have in mind. That is why we cannot help but see culture as a variable, while taking into consideration the existing points of criticism and the limitations of this option.

In view of further research, culture is foremost a variable which needs to be "filled" with values, norms, symbols, ideas, behaviour, etc. However, it is important, precisely within this kind of research, to have a reflexive focus and to avoid lapsing into clichés regarding the value or power of a culture.

It is also important to recognise the interaction of the cultural variable with other variables. The following, last part discusses how this is possible.

4.4   How culture "works"

Culture does not exist in itself. It always exists in interaction with other variables. The most obvious of these variables, often occurring in the development of theories around culture and police culture, is the environment. Policy is a variable as well: during the possible implementation of (here: police-related) policy measures, one speaks of "cultural resistance" or "cultural (in)appropriateness". As a result, a certain interaction or mutual influence between culture and policy is likely.

The question moreover arises as to how all these variables are related to the output in the field, i.e. the actual policing. The following formula draws attention to precisely those relationships. It can, therefore, serve as a guideline to draw up a change strategy, to the extent that a policymaker wants to take cultural diversity into consideration. The definition is analytical: it shows three ways in which culture is linked to variables such as policing, policy and the environment (Klitgaard, 1992; Van Ryckeghem, 1995; 1996a; 1996b).

4.4.1 Culture and policing in an analytical definition

(a) Policing   = P (Policy, Environment, Culture, etc.) [24]
(b) Culture   = C (Policing, Policy, Environment, etc.) [25]
(c) Utility$_C$ = U$_C$ (Policing, Policy, Environment, etc.) [26]

(a) The combined action of policy, environment and the existing culture leads to a certain outcome in policing, e.g. a particularly reactive or preventive approach, a high or low degree of clarification depending on the criminal phenomena, etc.

(b) But culture is not a static, and therefore unchangeable, variable. It is itself subject to a changing environment, a specific policy and the policing outcome that accompanies this.

In practice of course, the two equations merge into one another: they continuously influence one another. Often, however, only one level (and often also a very limited number of variables) is taken into consideration when drawing up policies. One expects to find a direct relationship and outcome based on the policy measures enacted.

(c) A third level is the utility function: cultural subjects only subscribe to a certain kind of policing and policy if they also derive an individual or social added value from this. This "human" aspect can best be explained in terms of reward systems, legal benefits, etc. The manner in which one rewards or is rewarded is largely determined by culture. Suppose the decision is taken to integrate different police services with each another. If such a reorganisation would take place in an environment where the police services distrust one another or are in some way prejudiced against each other, then it seems advisable to combine the reorganisation or the policy measure with additional benefits. Naturally, these benefits must also be regarded by the players as adding real value in comparison to the earlier situation. This softens the original policy measure.

---

24   The first equation is a production function for policing. A production function is an input-output relationship. It shows the relationship between the dependent variable (the output: "policing") and the factors "deployed" for that (the input: "policy", "culture", "environment", etc.). In other words, policing is a function of policy choices (e.g. the organisation of the force, the priorities with regard to police tasks), environmental factors (e.g. rural or urban), cultural elements and others.

25   The second equation is a production function for culture. Culture is, in its turn, a dependent variable and a function of the (outcome of) policing, the policy, the environment and other factors.

26   The third equation is a benefit or utility function. Policy measures, environmental factors and the outcome of policing are assessed by the players based on their social benefit.

In making policy choices, it is opportune to consider culture at the three levels. More knowledge of the variables and their mutual interaction in the equations can open up opportunities for rethinking policy strategies in light of culture.

4.4.2 *Police cultural studies and the analytical definition*

The studies on police culture, which we have discussed in part 2, are mainly situated at the first level of the definition. Policemen create the culture and culture help in determining the outcome of policing. Organisational theorists acknowledge the link with environmental factors, but see the interaction with culture as primarily unidirectional. The extent to which the existing culture also influences the environment, and how this combined action determines, among others, the policing, is less discussed.

4.4.3 Policy and/or change strategies and the analytical definition

The formula can be helpful in devising policy strategies because it is a guideline for listing the diverse variables and because it takes into consideration the way in which they influence each other and are influenced by one another. We illustrate this formula with an example before and after the significant reform of the Belgian law enforcement system in 2001[27]. As an example of the past we refer to the introduction of the "Basic Police Care with Quality" (Basispolitiezorg met Kwaliteit or BPZ-K) project in the Belgian state police in 1992. The recent example refers to the introduction of "The vision on an 'excellent police function'" in the integrated police force in 2007.

In 1992, the government asked the police services to pay more attention to the safety needs of the citizen. The Belgian state police responded to this with a specific project: Basic Police Care with Quality (BPZ-K). The inspiring forces behind the project thought they would be able to bring about a real cultural change by developing specific initiatives in brigades and districts and introducing methods for a problem-oriented analysis (compared to the incidental approach). This would lead to an increased readiness to help each other and to more cooperation with the population. Concrete initiatives included asking the help of postmen for providing information, or developing contacts with inhabitants with a view to more (direct) cooperation[28] (Van den Broeck and Easton, 1997). It was hoped that this form of project-based working would develop into a process-based work method throughout the organisation and the metaphor of a spreading drop of oil (the oil spot phenomenon) was used to describe this. But at the end of the 1990s, the BPZ-K movement, died down, which was partly due to the social circumstances (police reorganisations), (Easton, 2001).

If we look at the BPZ-K project based on the formula for culture and policing, the basic question that arises is whether the existing culture and the environment were amenable to the adopted policy measure. We deem it possible that the policymakers did not truly ask themselves this question. The reactions from the *environment* with regard to the notable initiatives mentioned above were, in any case, negative: the in-

---

27 With the law on the integrated police force (07/12/1998) structured on a local (local police) and a federal (federal police) level, the state police, municipal police and Criminal Investigation Department, who used to be separate agencies, were integrated.

28 Despite other initiatives, these have become the most well-known, partly due to the media attention and the fact that they appeal to the imagination.

habitants, the management of the Post, the press, everyone spoke out against "the use of postmen as the ears and eyes of the police".

In addition, there was a great deal of resistance at other levels of the state police organisation (*culture*): staff members did not feel involved, felt that there were too many changes within a short time, etc. Subsequently, the mutual interaction of policy, culture, environment (the independent variables) has not resulted in a real change in the police output (the dependent variable), except in a rather limited and temporary way.

Nevertheless, it must also be said that the BPZ-K movement was responsible for small but still significant changes in the existing culture (the dependent variable). The relationship between the state police and citizen was, for the first time, problematized by the organisation itself, the isolated position of the state police with respect to the population was brought under discussion, and new methods were introduced for dealing with problems. In addition, the brigade personnel could, for the first time, initiate projects themselves.

Finally, at a third level, we discuss the benefit of the policy, the environment and the outcome of policing for the actors themselves. In our opinion, the social benefits lie mainly in the shared idealism, the cooperation with like-minded people, the "exciting" belief in the change of an old and powerful organisation and working for the benefit of the population, and in making an individual contribution to that change. Since these "benefits" remained limited to a relatively small group within the state police, the movement did not even carry enough weight to actually push through the policy. This also partly explains why the oil spot has stopped spreading at some point.

Since 2007 "The vision on an 'excellent police function'" is the guideline in the reform of the Belgian integrated police. The main goal of the vision is to integrate the 'community oriented police function' (mentioned in the Law on the integrated police function and implemented by circular in 2003), 'intelligence led policing' and 'optimal management' (as a combination of management models or – theories made especially applicable to the police organization) into a government policy in the field of 'societal security' (Bruggeman, et.al., 2007).

This ambitious vision generated the pitfall that a community oriented police function could be obtained by implementing intelligence led policing (a working method) and principles of 'optimal management' (a management model). No effort has been made to take into account the bottom-up input from the professionals on the street (the existing culture), which have to implement the policy of community oriented policing. Nonetheless, the empowerment of these professionals is a crucial factor of success in changing the overall organization. This refers explicitly to the utility function mentioned above (Easton, 2008).

To some degree this policy document generates the same concerns as the BPZ-K-project in 1992. The existing culture of the policemen/women on the street was not the point of departure to generate the vision on an 'excellent police function'. Besides, a great deal of resistance within the organisation arose due to the fact that the COP-reform had been installed since 2003 and people felt a new challenge was presented much too soon. As a consequence the utility of the policy was questioned (Easton & Dormaels, 2008). Nevertheless, the policy document seems to be kept alive amidst leaders in the organisation. The future will show whether the new policy will succeed and whether it will take the organisation one step further in the change process.

For recent international examples dealing with policy and change strategies in relation to police culture see Armacost, 2004; Brunetto, 2005; Chan, 2007, Davies, 2008.

5   Conclusion

In this article, the key perspectives in cultural studies have formed the framework for a discussion of studies on police culture. It is primarily the *subjectivist* interpretation of human behaviour that seems to predominate here. People make their own history and culture, policemen their own police or organisational culture.

In this interpretation, most researchers have defined the culture based primarily on their own discipline: psychologists, sociologists, anthropologists speak respectively over working personality, values and norms, ideas, symbols and rituals. At the organisational level, this is mainly about culture as the "cement" of this organisation. But this vision is challenged by the exposure of subcultures on the one hand (Anglo-Saxon authors), and the role of personal interpretation on the other (the French authors).

The management theories, which had a very strong presence in the 1980s as well as in the 1990s, have narrowed the concept of culture further down. Their first concern was not culture, but rather the "modernisation" of organisations and the development of change strategies. From this perspective, culture appeared to be simultaneously important and malleable, but intangible as well.

It is certain that it will never be possible to define culture completely. However, the most complete option for cultural studies continues to be a combination of the insights from various disciplines. This interdisciplinary approach, which is consistent with a reflexive cultural focus, must also be reflected in the research methods: the use of a single method means shedding light on certain cultural aspects and ignoring others.

It is also certain that there is no success formula for cultural change. But one can reduce the chance of "failure" by taking into consideration various factors and their mutual interactions, but in such a way that a change strategy remains workable and manageable.

Our police organisations are beyond question manifestations of contemporary society. Therefore, it is unrealistic to assume everything can be controlled or managed; realism and modesty, together with a pinch of luck are likely to remain key ingredients of our formula.

6   Bibliography

Alderson, J., *Policing Freedom: a commentary on the dilemmas of policing in western democracies*, Plymouth, Latimer Trend en Company Ltd, 1979.

Althusser, L., Ideologie en ideologiese staatsapparaten, *Recht en Kritiek*, 1976, 2, 116-135, (oorspronkelijke uitgave: 'Idéologie et appareils idéologiques d' Etat', La Pensée, 151, 1970, herdrukt in: Positions, Parijs, Editions Sociales, 1976, 67-125).

Alpert, G., Dunham, R., *Policing urban America*, Prospect Heights, Illinois, Waveland Press, 1992.

Armacost, B., Organizational culture and police misconduct. *George Wahington Law Review*, 72 (3): 453-546, 2004.

Banton, M., *The Policeman in the Community*, New York, Basic Books, 1964.

Bittner, E., The police on skid row: a study of peace keeping, *American Sociological Review*, 1967, 32, 5, 699-715.

Bittner, E., *The Functions of the Police in Modern Society*, Washington, DC, U.S. Government Printing Office, 1970.

Bourdieu, P., *In Other Words: Essay Towards a Reflexive Sociology*, Cambridge, Polity Press, 1990.

Bourdieu P., Wacquant, L., *An Invitation to Reflexive Sociology*, Cambridge, Polity Press, 1992.

Bordua D. Ed., *The Police: Six Sociological Essays*, London, Wiley, 1967.

Broderick, J., *Police in a time of change*, Prospect Heights, Waveland Press, 1987.

Brown, M., *Working the street, Police Discretion and the Dilemmas of Reform*, New York, Russell Sage, 1981.

Bruggeman, W., Van Branteghem, J-M, Van Nuffel, D. (Ed.). *Naar een excellente politiezorg*, Brussel, Politeia, 2007.

Brunetto, Y., Farr-Wharton, R. The role of management post-NPM in the implementation of new policies affecting police officers' practices. *Policing- an international journal of police strategies & management*, 28 (2): 221-241 2005

Cain, M., *Society and the Policeman's Role*, London, Routledge and Kegan Paul, 1973.

Ceuppens, E., Mertens, E., Organisatiecultuur: de zachte kant van de politiehervorming, in *Handboek Politiediensten*, onderdeel: organisatieverandering, Diegem, Kluwer, 2001.

Chan, J., Changing police culture, *British Journal of Criminology*, 1996, 1, 109-134.

Chan, J., Making sense of police reforms, *Theoretical Criminology*, Vol. 11(3): 323–345; 1362–4806, 2007.

Clark, J., Isolation of the Police: a Comparison of the British and American Situations, *Journal of Criminal Law, Criminology and Police Science*, 1965, 56, 307-319.

Crank, J., *Understanding police culture*, Cincinatti, Anderson Publishing, 1998.

Davies, A., Thomas, R., Dixon of dock green got shot! Policing identity work and organizational change. *Public Administration*. 86 (3): 627-642 2008.

Deal, T., Kennedy, A., *Corporate Cultures. The Rites and Rituals of Corporate Life*, Rading, Addison-Wesley, 1982.

Denzin, N., *Interpretive Interactionism*, Newbury Park, Sage (Applied Social Research Methods Series, 16), 1989.

Easton, M., *De demilitarisering van de rijkswacht*, Criminologische Studies, 7, Brussel, Vubpress, 351p, 2001.

Easton, M., Ponsaers, P., Demarée, Ch., Vandevoorde, N., Enhus, E., Elffers, H., Hutsebaut, F., Gunther Moor, L. (eds.), Multiple Community Policing: Hoezo?, Gent: Academia Press, 2009.

Easton M., Van den Broeck, T., Community Policing bij de rijkswacht: return to sender? Ok-kaarten en de inzet van postbodes nader beschouwd, in *Politeia*, 1997, 2, 16-19, deel I.

Easton, M., Ponsaers, P., Demarée, Ch., Vandevoorde, N., Enhus, E., Elffers, H., Hutsebaut, F., Gunther Moor, L. (eds.), Multiple Community Policing: Hoezo?, Gent: Academia Press, 2009.

Easton, M. Excellente Politiezorg: gewikt en gewogen..., in: Collier, A., Devroe, E. & Hendrickx, E. (Eds.), '*Politie en Cultuur*', Cahiers Politiestudies, nr. 9., 111-125, Brussel: Politeia, 2008.

Enhus, E., Ponsaers, P., "Onmacht tot cultuurverandering – Politiehervorming in België", in : Tijdschrift voor Criminologie, Boom Juridische Uitgevers, Den Haag, 2005, (47) 4, pp. 345-354 (ISSN 0165 – 182x).

Easton, M., Dormaels, A., "Burgemeesters en korpschefs aan het woord inzake de politiehervorming" in Van Cauwenberghe, K. (Eds.), *De Hervormingen bij politie en justitie: in gespreide dagorde*, Mechelen, Wolters Kluwer, 2008, 49-62.

Edgerton, R., Langness, L., *Methods and Styles in the Study of Culture*, Los Angeles, Chandler and Sharp Publishers, 1979.

Ericson, R., Rules For Police Deviance, in Shearing, C., *Organizational Police Deviance: Its Structure and Control*. Toronto, Butterworths, 1981.

Ericson, R., *Reproducing Order: A Study of Police Patrol*, Toronto, University of Toronto Press, 1982.

Fielding, N., Police socialisation and police competence, *British Journal of Sociology*, 1984, 35, 4, 568-90.

Fielding, N., *Joining Forces: Police Training, Socialization, and Occupational Competence*, Londen, Routledge, 1988.

Fokkinga, L., Jaspers, K., Jongen, P., Vollenberg, M., Bijlsma-Frankema, K., '*En dan de cultuur nog... een cultuursociologisch onderzoek bij de Politieregio Gooi en Vechtstreek*', Universiteit van Amsterdam, Faculteit der Politieke en Sociaal-Culturele Wetenschappen, juli 1992.

Frissen P., Van Westerlaak, J., *Organisatiecultuur. Van toverwoord tot bruikbaar begrip*, Schoonhoven, Bedrijfskundige Signalementen, 1, 1990.

Geertz,C., *The Interpretation of Culture*, New York, Basic Books, 1973.

Goldstein, H., Police Discretion: the Ideal vs. The Real, *Public Administration Review*, 1964, 23, 140-148.

Goldstein, H., *Policing a Free Society*, Cambridge, Mass, Bollinger, 1977.

Goldsmith, A., Taking Police Culture Seriously: Police Discretion and the Limits of Law, *Policing and Society*, 1990.

Gramsci, A., *Grondbegrippen van de politiek. Hegemonie, staat en maatschappij*. Nijmegen, Sun, 1980.

Gray, T., Selecting for a Police Subculture, in Skolnick J. and Gray T. Ed. *Police in America*, Boston, Educational Associates, 1975.

Greene, J., Decker, S., Police and Community Perceptions of the Community Role in Policing: The Philadelphia Experience, *The Howard Journal of Criminal Justice*, 1989, 22, 8, 105-123.

Hagenaars P., Op ten Noort, H., Regio komt bij van Cultuurschok: Onderzoek naar cultuur en innovatievermogen regiopolitie Gelderland-Zuid, *Het Tijdschrift voor de Politie*, 1993, 9, 263-266.

Hall, S., Cultural Studies: Two Paradigms, *Media, Culture and Society*, 1980, 2, 57-72.

Harris, R., *The Police Academy: An Inside View*, New York, John Wiley, 1973.

Hofstede, G., *Culture's consequences. International differences in work-related values*, London, Sage, 1980.

Hopper, M., Becoming a policeman: socialization of cadets in a police academy, *Urban Life*, 1977, 6, 2, 149-170.

Klitgaard, R., *Taking culture into Account. From 'Let's' to 'How'*. Paper for the International Conference on Culture and Development in Africa, Washington DC, World Bank, April, 1992.

Laclau, E., *Politics and Ideology in Marxist Theory*, London, Verso, 1977.

Levi-Strauss, C., *Tristes Tropiques*, London, Penguins Books, 1992.

Lipsky, M., *Street-Level Bureaucracy*, New York, Russell Sage, 1980.

Manning, P., The Police: Mandates, Strategies and Appearances, in Douglas, J., Ed., *Crime and Justice in American Society*, Indianapolis, Bobbs-Merrill, 1971.

Manning, P., *Police Work: The Social Organization of Policing*, Cambridge, Massachusetts, Massachusetts Institute of Technology Press, 1977.

Manning, P., The Police: mandate, strategies and appearances, in Manning, P., Van Maanen, J., Ed., *Policing: a view from the street*, new York, Random House, 1978, 7-31.

Manning, P., Van Maanen J. Ed., *Policing: A View from the Street*, Santa Monica, California, Goodyear, 1978.

Manning, P., The Social Control of Police Work: Observations on the Culture of Policing, in Holdaway, S., Ed., *British Police*, London, Edward Arnold, Ltd., 1979.

Manning, P., Occupational Culture, in: Bailey, W.G., Ed., *The Encyclopedia of Police Science*, New York and London, Garland, 1989.

McNamara, J., Uncertainties in Police Work: The Relevance of Police Recruits Backgrounds and Training, in *The Police: Six Sociological Essays*, Bordua, D. J., Ed. New York, Wiley, 1967.

Monjardet, D., Gorgeon, C., '*1167 recrues: description de la 121e promotion des élèves-gardiens de la paix de la police nationale*', Paris, CNRS et IHESI, 1992.

Monjardet, D., Gorgeon, C., *La Socialisation Professionelle des Policiers, tome 1: la formation initiale*, Paris, CNRS et IHESI, 1993.

Monjardet, D., La Culture Professionnelle des Policiers, *Revue Française de Sociologie*, 1994,35, 393-411.

Muir, W., *Police: Streetcorner Politicians*, Chicago, University of Chicago Press, 1977.

Paoline, E., Taking stock: Toward a richer understanding of police culture. *Journal of Criminal Justice*, 31 (3): 199-214, 2003.

Parienté, P., Les valeurs professionelles: une ressource pour l'encadrement de l'activité policière?, *Les Cahiers de la Sécurité Intérieure*, 1994, 16.

Parsons, T., *The Social System*, New York and London, Tavistock and Routledge and Kegan Paul, 1952.

Peters T., Waterman, R., *In Search of Excellence*, New York, Harper and Row, 1982.

Punch, M., *Policing the Inner City*, London, Macmillan, 1979a.

Punch, M., The Secret social service, in Holdaway, S., Ed, *The British police*, London, Arnold, 1979.

Punch, M., Officers and Men: Occupational Culture, Inter-Rank Antagonism, and the Investigation of Corruption, in Punch, M., Ed., *Control in the Police Organization*. Cambridge, MA, MIT Press, 1983.

Punch, M., et al, Politiecultuur, in: Fijnaut C. (Ed.), *Politie. Studies over haar werking en organisatie*, Alphen a/d Rijn, Ced. Samsom, 1999, p263-281.

Reiner, R., *The politics of the police*, Wheatsheaf Books, 1985.

Reuss-Ianni E., Ianni, F., *Two cultures of policing*, New Brunswick, Transaction Books, 1983.

Reuss-Ianni, E., Ianni, F., Street Cops and Management Cops: The Two Cultures of Policing, in Punch, M.,Ed., *Control in the Police Organization*, Cambridge, MA, MITPress, 1983.

Roberg, R., Kuykendall, J., *Police Organization and Management: Behavior, Theory and Processes*, California Brooks, Cole Publishing Company, 1990.

Sanders, G., Neuijen, B., *Bedrijfscultuur : diagnose en beïnvloeding*, Assen, Van Gorcum, 1987.

Sandler, G., Mintz, E., Police organizations: their changing internal and external relationships, in *Journal of Police Science and Administration*, 1974,2, 4.

Shearing, C., Ericson, R., Culture as Figurative Action, *British Journal of Sociology*, 1991,42, 4, 481-506.

Shearing, C., *Transforming the Culture of Policing*, paper Centre of Criminology University of Toronto en Community Peace Foundation, University of Western Cape, 1994.

Schein, E., *Organizational Psychology*, Englewood Cliffs, Prentice Hall, 1980 (First edition: 1967).

Schein, E., *Organizational culture and leadership*, San Francisco, Jossey-Bass, 1992 (first edition: 1985).

Smircich, L., Concepts of Culture and Organizational Analysis, *Administrative Science Quarterly*, 1983, 28, 3, 339-385.

Skolnick, J., *Justice without Trial: Law Enforcement in Democratic Society*, New York, Wiley, 1966.

Skolnick, J., Woodworth, J., Bureaucracy, information and social control: a study of a morals detail., in Bordua, D., Ed., *The Police: six sociological essays*, New York, Wiley, 1967.

Southgate, P., *New Directions in Police Training*, 1988.

Thompson, E., The making of the English Working Class. Harmondsworth, 1968.

Van den Broeck, T., Copland: over politiecultuur, in *Handboek Politiediensten*, onderdeel: organisatieverandering, Diegem, Kluwer, 2001.

Van den Broeck, T., Easton, M., Community Policing bij de rijkswacht: return to sender? Ok-kaarten en de inzet van postbodes nader beschouwd, in *Politeia*, 1997, 3, 16-19, deel II.

Van der Torre, E., De Nuchtere Werkelijkheid van het Basispolitiewerk, *Tijdschrift voor de Politie*, 1992, 54, 4, 163-167.

Van der Torre, E., Culturele Diversiteit: een verschillende beleving van het alledaagse basispolitiewerk, *Tijdschrift voor de Politie*, 1992, 54, 5, 200-204.

Van der Torre, E. (1999). *Politiewerk: politiestijlen, community policing, professionalisme*. Alphen aan de Rijn: Samson.

Van Hoewijk, R., De Betekenis van Organisatiecultuur: een Literatuuroverzicht, *M en O, Tijdschrift voor Organisatiekunde en Sociaal Beleid*, 1988, 42, 1, 4-46.

Van Maanen, J., Observations on the Making of Policemen, *Human Organization*, 1973, 32, 407-18.

Van Maanen, J., Working the Street: A Developmental View of Police Behavior, in *The Potential for Reform of Criminal Justice*, Jacob, H., Ed., Beverly Hills, California, Sage Publications, 1974.

Van Maanen, J., Police Socialization: A Longitudinal Examination of Job Attitudes in an Urban Police Department, *Administrative Quarterly*, 1975, 20, 207-28.

Van Maanen, J., People Processing: Strategies of Organizational Socialization, *Organizational Dynamics*, 1978a, 7, 1, 18-36.

Van Maanen, J., The Asshole, in Manning, P., Van Maanen, J., Ed., *Policing: A View from the Streets*, New York, Random House, 1978b.

Van Maanen, J., Schein, E., Toward a Theory of Organizational Socialization, in Staw, B., Ed., *Research in Organizational Behavior*, New York, JAI Press, 1979, 209-269.

Van Maanen, J., Making Rank: becoming an American police sergeant, *Urban Life*, 1984, 13, 2-3, 155-176.

Van Muijen, J., Koopman, P., De Witte, K., *Focus op organisatiecultuur*, Schoonhoven, Academic Service, 1996.

Van Keulen-Schellart, G., Bink, H., *Herordenen of Vernieuwen: Politiecultuur in de fusie*, Politiestudiecentrum LSOP, Warnsveld, 1993.

Van Ryckeghem, D., Information Technology in Kenya: A Dynamic Approach, *Telematics and Informatics*, 12,1, 57-65, 1995.

Van Ryckeghem, D., *Cultuur en informatietechnologie in een derde wereld-context. Een onderzoek naar de invloed van een niet-Westerse cultuur bij concrete implementatie- en gebruiksprocessen van computertechnologie, op basis van Kenyaanse en Belgische case-studies*. Niet gepubliceerde doctoraatsverhandeling. VUB, 1996a.

Van Ryckeghem, D., Computers and Culture: cases from Kenya, in Roche, E.M., Blaine, M.J., Ed., *Information Technology, Development and Policy*, Aldershot, Avebury, 1996b, 153-170.

Wasserman, R.,Moore, M., Values in Policing, *Perspectives on Policing*, 8, Washington, DC, National Institute of Justice, November, 1988.

Weiner, N., The effect of education on police attitudes, *Journal of Criminal Justice*, 1974, 2, 317-328.

Westley, W., *Violence and the Police: A Sociological Study of Law, Custom and Morality*, Cambridge, Massachusetts, MIT Press, 1970.

Wilders, O., Een Huwelijk maakt nog geen relatie. De politiereorganisatie vanuit een cultuurbenadering, *Het Tijdschrift voor de Politie*, 1994, 3, 12-15.

Williams, R., *Culture and Society, 1780-1950*, Harmondsworth, 1963.

Williams, R., *The Long Revolution*, Harmondsworth, 1965.

Wilson, J., *Varieties of Police Behavior, the Management of Law and Order in Eight Communities*, Cambridge, Harvard University Press, 1968.

Wilson, J., Kelling, G., Broken windows: the police and neighbourhood safety, in *The Athlantic Monthly*, 1982, 29-38.

The view of the police on community policing in Belgian multicultural neighbourhoods

Marleen Easton
Paul Ponsaers

1   Context and research focus

As part of Belgium's police reforms, which were consolidated by the Law of 7 December 1998, the government opted to introduce Community (Oriented) Policing (COP) as the official policing model. It was meant to be the cultural reorganisation of the Belgian police based on five pillars: service orientation, partnership, empowerment, problem-solving and accountability[1]. It was only to be expected that the implementation of this model would be far from straightforward and would give rise to all kinds of issues and difficulties. Despite the fact that the COP model is historically rooted in a number of ethnically-coloured conflicts, the most pressing question was that of how this model might be applied in multicultural neighbourhoods with extremely complex and diverse social contexts in Belgium. Earlier studies show that both police officers and members of ethnic minorities are of the opinion that relations between them are frequently problematic and marked by distancing and mistrust (further elaborated on in 2.). These relationships have become strained as the result of problematic mutual perceptions, ascribed meanings, visions and expectations. Factors such as these, combined with structural neighbourhood factors (discrimination and a heterogeneous social and ethnic mix), are a severe hindrance to the implementation of COP, but further obstacles are encountered in the intrinsic ambiguities wrapped up in the COP model itself as regards the meaning to be attached to "the community". It is for example not at all clear what "the" community means.

In order to, on the one hand, investigate the "transition" to COP in multicultural neighbourhoods, and on the other, to support it in the future, it is essential to gain an understanding of day-to-day practices, of how interactions between the police and ethnic minorities take shape and break down. The present study aims to come to grips with these factors and processes by *examining the extent to which COP takes shape (or not) in multicultural neighbourhoods and through contact with ethnic minorities, and how both parties feel about this.* This type of police study is related to what is often described as police sociology, which gives a picture of how policing actually takes shape in the social process between police officers and citizens and between the officers themselves (van der Torre, 1998).

The present study uses the theoretical perspective of social constructionism. In studying social problems, social constructionism steps back and looks at *who* calls 'something' a (social) problem and *how* they define (and explain) this problem (Burr,

---

[1]   Steered by Ministerial Circular CP1 27/05/03.

2003; Clarke, 2006). In other words, it puts the emphasis on meanings and ascribed meanings. This is because meanings, visions and perceptions make "sense" of a subject and lend structure to its reality. They build up a social reality which is continually (re)produced through social interaction. There is also the question of to what extent an (internal) representation or discourse, being a reflection of how the actors construct "their own" reality, actually manifests itself in practice and whether or not it affects social (inter)action. The central question in the present study is that of how (and why) police officers build up their view of the world and people ("the community" included) and categorise and label on the basis of these constructions (Van Maanen, 1978). To what extent (and why) do they reason in terms of "limits" of "feasibility" or workability and how meaning attribution affects social action (Swidler, 1986) (in this case, policing) and thus the implementation of a policy (in this case, community oriented policing) (Boussard, Loriol & Caroly, 2006).

This starting point is also reflected in the following three core objectives of the present study.

1. Examine whether or not and how the Belgian COP model takes shape in interactions with ethnic minorities in the framework of handling routine tasks and incidents [reality].
2. Examine how ethnic minorities perceive/experience interactions with the police in the framework of their handling of routine tasks and incidents, what their expectations are in this area and examine how congruent these expectations are with key points in the Belgian COP model [perception of immigrants].
3. Examine the extent to which police officers working in the field see the Belgian COP model as applicable in their interactions with ethnic minorities when handling routine tasks and incidents, the problems and possibilities they perceive and their expectations in this area [perception of the police].

The focus in this article lies on the analysis of the perceptions, views and behaviour of street cops on the beat and during interventions.

An understanding of this matter provides fixing points from which to support the application of a Belgian interpretation of community (oriented) policing and provides added value for policymakers – and other authorities – wishing to help the police develop in line with this philosophy (see 5. support for decision-making).

This contribution is based on research commissioned and financed by the Belgian Science Policy Office as part of the "Research programme in support of 'Society and Future'"[2].

---

2   It was carried out by two researchers (Chaim Demarée & Natascha Vandevoorde) at the University College of Ghent and supported by a team consisting of Lodewijk Gunther Moor (expert & former director linked to the Dutch Society, Security and Police Foundation) and five promoters (Prof. Els Enhus, Free University Brussels, Prof. Frank Hutsebaut, Catholique University Louvain, Prof. Henk Elffers, Free University Amsterdam, Prof. Paul Ponsaers, University Ghent & Prof. Marleen Easton, University College Ghent).

2   Relationship to (inter)national research

To position the current study in relation to (inter)national research, roughly two main research lines can be referred to. The first relevant line of research concerns (reciprocal) perceptions and images between police and ethnic minorities. The second line refers to research on the implementation of COP (in multicultural neighbourhoods).

Foreign and (limited) Belgian research focusing on the relationship between the police and ethnic minorities, indicate a tense relationship and complicated interactions (Bowling, Phillips & Shah, 2003; Casman, Gailly, Gavray, Kellens & Lamaître, 1992; Haen Marshall, 1997). These studies often point out that the cause of a difficult relationship originates in conflicting cultural backgrounds and a mutual construction of negative images, which often lead to self-fulfilling prophecies. Concerning ethnic minorities, this is frequently connected to (perceived) racism (Casman et al, 1992; Brunson & Miller, 2006; Goodey, 2006) and to social vulnerability and exclusion (Sun, Payne & Wu, 2008). Studies suggested this may lead to the development of a 'vindictive subculture', which is also directed towards the police (as 'delegates' of the Belgian government) (Hebberecht, 1995; Bekkour, 2001). On the other hand, negative images the police embrace of ethnic minorities are often linked to a belief in the validity of crime figures and to the frequency of negative confrontations, which lead to stereotyping processes and to an inadequate knowledge of ethnic communities (Casman e.a., 1992; Feys, 2002; Hebberecht, 1995; van San & Leerkens, 2001; Walgrave, 2002). The bulk of the research is, however, usually fragmented, in a sense that it focuses on particular elements of this relationship (e.g., institutional racism, see Lea, 2004) or treats it merely as a secondary subject, e.g. not making it the focus of research in any way (see Van San & Leerkens, 2001; Vercaigne, Mistiaen, Walgrave & Kesteloot, 2000).

In Belgium, the research focus is generally characterised by successive narrowing. First of all, mainly ethnic minorities, entering Belgium during the first and second wave of migration, have been researched. The third -post industrial- wave of migration has been the focus of much less attention, although a recent catch-up trend can be discerned (Feys, 2002). Secondly, within the first and second wave of migration, it is often Moroccan and Turkish immigrants who are the subjects of research. Thirdly, within these groups the research focus narrows even more by focusing on youngsters and especially young men (Feys, 2002). Consequently, there is a lack of research on *other* segments of the immigrant population. Research on the so-called 'new immigrants' (third migration wave) and other ethnic minorities is much less prevalent. Similar research patterns can be noticed in other European countries characterized by equivalent immigration history (Bowling, Phillips & Shah, 2003). As a result the global picture remains quite patchy.

Concerning the second line of research, (Belgian) studies about community policing usually examine implementation of a COP model in police force(s) and focus on problems and obstacles encountered (by interviewing senior and rank-and-file officers, doing observations and analyzing documents) (Vandevenne, 2006; Vandevoorde, Vaerewyck, Enhus & Ponsaers, 2003). This kind of research does not explicitly focus on multicultural neighborhoods, although research results can provide insight with regards to our current study. However, this means research focusing on the (Belgian) implementation of COP is quite limited. In other (especially Anglo-Saxon) countries (where COP originated), a research tradition focusing on the implementation of this model does exist (see for example Eck & Rosenbaum, 1994; Mastrofski, 1998; Sparrow, 1988). Those studies, however, are not always relevant or useful to the Belgian police context and the way COP was conceptualized in our country.

In sum, *empirical* studies to date have barely touched the relationship between the police and ethnic minorities, in Belgium or abroad. Moreover, limited research has taken community (oriented) policing as its starting point, which underpins the originality of this study. On the subject of the implementation of COP in relation to multicultural neighbourhoods, no Belgian research has been done whatsoever, meaning that this study attempts to explore unknown research 'territory'. The research design is located at the crossing of the two research lines mentioned above and aims to integrate under-researched elements in previous research. For example, the third wave of migration is taken into account. Moreover, expectations, possibilities and problems concerning COP are not only examined through a police perspective, but from the perspective of ethnic minorities also. This research reflects an interest in what COP means to those working in 'the field' (police) and to those who 'receive' it (the population, e.g. ethnic minorities).

In the following section, the methodology of the current study is illustrated in detail.

3   Methodology

Given that empirical research on community (oriented) policing in multicultural neighbourhoods is rare, in Belgium and elsewhere, the option for qualitative research methods has been taken (semi-structured interviews with accompanying "on the spot" observations and conclusive round table talks). Where the research objectives are concerned, there are three arguments in favour of this methodology. The first argument rests on the suitability of these methods in order to gain insight into the de facto manifestation of these interactions in reality. The second argument for this choice is that little is known about the exact research context and these methods actually make it possible to explore context in terms of influential factors. The third argument lies in the ability of these methods to give insight into the interplay of meanings and perceptions attached to social phenomena by the actors (both the police and ethnic minorities) on the one hand, and insight into policing practices on the other (Swanborn, 2004; Wester & Hak, 2003; Verschuren & Doorewaard, 2005).

After a preparatory phase (explorative literature study), qualitative studies of five multicultural neighbourhoods, lasting six months, were carried out in five Belgian police zones. Besides an even spread over Belgian soil and cooperation from the top ranks of the force; the main intrinsic criteria for selecting the multicultural neighbourhoods in the police zones were the degree of urbanisation and the diversity of the ethnic minorities' residential histories (see diagram 1). The diversity of the ethnic minorities' residential histories is linked to the three migration waves mentioned earlier. In these terms diversity is an important point of departure in this study.

Diagram 1: The selection of five Local Police Forces (LPF).

History Migration	Brussels	Flanders	Walloon Provinces
Short History	LPF1 Urbanization: Type 2 Cooperation Chief of Police?	LPF2 Urbanization: Type 1 Cooperation Chief of Police?	LPF3 Urbanization: Type 1 Cooperation Chief of Police?
Long History	LPF 1 Urbanization: Type 2 Cooperation Chief of Police?	LPF4 Urbanization: Type 2 Cooperation Chief of Police?	LPF5 Urbanization: Type 2 Cooperation Chief of Police?

A context analysis of each police zone facilitated the selection of (representatives from) the immigrant community (37 in total) for the semi-structured interviews and with this, an entrance was made to the research field. Semi-structured interviews (24 in total) with police officers from the selected police zone were held prior to the "on the spot" observation period (182 days in total). The fieldwork in the five multicultural neighbourhoods was brought to a conclusion with two round table talks (one in each part of the country) in which the research results and policy recommendations were put before hands-on experts from immigrant and police milieus for refinement. For a schematic overview of the methodology see Diagram 2 below.

Diagram 2: Overview of the methodology.

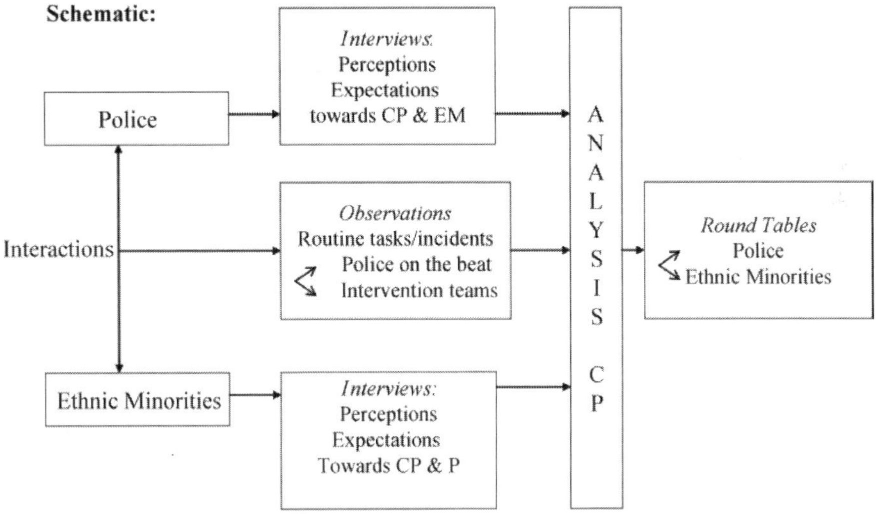

CP= Community Policing; EM= Ethnic Minorities ; P = Police

During the fieldwork in the second Flemish police zone the illness of one of the researchers prevented the completion of the planned observations and interviews on the spot and this case study has not been fully executed. The results from this case study have been critically assessed in the light of the overall research results and no anomalies have been detected. Nevertheless, it is fair to say that this study relies on four and a half case studies instead of five.

In what follows the reseach results reflect the analysis of the perceptions, views and behaviour of street cops on the beat and on interventions in the four and a half case studies mentioned above. The focus lies on the conclusion of this analysis and not on illustrations of the actual wording of our interviews with police officers and the variant encounters between the police and minority groups in the selected multicultural neighbourhoods.

4   Research Results

In essence this research is about perceptions, or in other words, the views which citizens and the police have of each other and the way in which these come about, in relation to Community (Oriented) Policing (COP). Advocates of COP argue in favour of a police force with a presence in the community. In other words, in this way of thinking, the police, on the one hand, construct their perception of the community on the basis of what happens across society as a whole, and the community, on the other, constructs its perception of the police on the basis of what happens in the police force as a whole (see figure 1). Or to put it another way: they are the mirrors through which people observe reality, or better yet: the degree to which these mirrors are not distorted by an overly reduced or fragmentary perception. Of course the COP aspiration provides a sort of idealised picture which might never be approached or realised in an optimal way, i.e., a non-distorted perception. Nonetheless, is it a strong picture, because it tells us that, in this situation, mutual perceptions rest on adequate and balanced detailed information from both realities.

Figure 1: Visual representation of COP
Optimal relationship of police / community

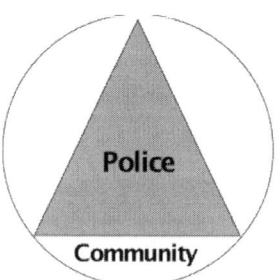

4.1   "Over and underpolicing" of groups, not neighbourhoods

The research distinctly reveals that the police regularly come into contact with only parts of the community, certainly in the so-called "problematic neighbourhoods".
These groups of individuals in the neighbourhood community, the so-called "regular customers", are subject to "overpolicing" (also referred to in the literature by the term "police property") (Reiner, 1994). On the other hand, a neighbourhood community contains groups with which the police hardly, if ever, come into contact, or, to put it

another way, who they don't actually know or know well. Here the term "underpolicing" is used de facto (see figure 2).

Figure 2: Visual representation of COP – Poor relationship of police / community

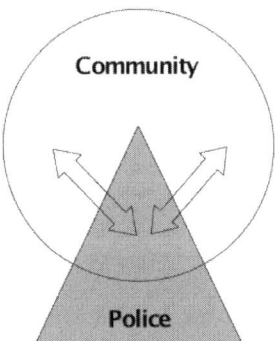

The study focuses on neighbourhoods with a heterogeneous demographic composition. This was one of the starting points when selecting the observation sites. From this study, it is obvious that both sorts of groups (subject to "overpolicing" and "underpolicing") are present in precisely the same geographical location. In other words, it is not a geographical characteristic in these neighbourhoods, but a group characteristic. Or, to put it more clearly: it is not the police perception of so-called "problematic neighbourhoods", but the perception of so-called "problematic groups" in heterogeneously composed neighbourhoods.

The dividing line cannot simply be drawn in terms of ethnicity: the interplay of different factors have to be faced and these will be covered in logical order below. The research tells us that the line of separation between both groups does not run clearly between immigrant groups on the one hand, and non-immigrant groups on the other. In these problematic neighbourhoods, there are ethnic minorities and individuals who are subject to "overpolicing" while other ethnic minorities and individuals are more "underpoliced". In the same way, there are also ethnic minorities in the same neighbourhoods who are "overpoliced", while other non-immigrants are faced with "underpolicing".

It is obvious that the neighbourhood community as advocated in the vision of COP does not exist for police officers on the beat[3]. They experience the conceptual vagueness of a notion like "community" on a daily basis. The perception of policemen tends to suggest the existence of an extremely fragmented social patchwork of origins, behavioural patterns, preferences, statuses, "cultures" and ages. In brief, they experience multiple neighbourhood communities.

Conclusion: the police come into contact with only a (small) part of the community. Or to put it another way: the building blocks of experience from which policemen construct perceptions of the community is fragmentary, and, not only that, but the perception is selectively created.

---

3   The term "police officer" here refers to the profession of policeman in the sociological meaning of the word and not to the rank of "officer" in the force.

4.2   "Overpolicing" of regular (tough) customers

This study reveals that "overpolicing" applies to only a small minority of groups and individuals in neighbourhood communities. It is with these groups that the police experience a feeling of proximity (which does not imply "mental" proximity: on the contrary). At times the police encounter these "regular customers" as victims and at others as offenders. The boundary between both roles appears extremely vague. These are "regular customers" who "involve a lot of work" for the police, who repeatedly call on police services and interventions ("trifling" or "cat-in-the-tree" interventions), or who make the police feel obliged to watch out for them on their own initiative. On some occasions, the police intervention is of a repressive kind; on others, it is of the servicing or caring kind. What is also striking is that these groups have little or no care facilities of their own and that their ability to cope independently is extremely limited. In other words, these are groups who do not have solid social networks to rely on. The result is that they only have the police to call on, for this is the only authority with which they are familiar when problems arise. Work is always inconvenient and the police quickly come to see these "regular customers" as "tough customers". It is therefore only logical that the police construct strong, distinct perceptions of these groups.

Significant groups of these "regular (tough) customers" can be found in economically weak sectors of the neighbourhood community, who have ended up in a rather marginal situation with little in the way of social capital, low schooling, with their own lifestyles, limited verbal skills and knowledge of the official language, little home stability, of a particular age group, perhaps with a criminal past, etc. and, now and then, a particular ethnicity[4]. Policemen who regularly come into contact with these groups ascribe them with specific attributes, such as certain irritating attitudes and specific characteristics (including outward characteristics). Before arriving on the scene again, policemen have perceptions and patterns of expectations about the situation they will encounter and about the behaviour they are likely to face.

The police assess these "regular (tough) customers" and classify them in their own categories, for which they create strong perceptions. The boundaries of these categories sometimes coincide with ethnicity, but just as much with other characteristics, such as age, lack of parental control, gender aspects, marginality, immigration status, etc. Therefore, police categories are certainly not constructed solely on the basis of ethnicity, although at some moments, ethnicity can actually be of importance in the construction of some specific categories (e.g., Moroccan youths).

Some of these "police property" groups are given nicknames, e.g., "drunks" (schelen), "petty crooks" (groseilles), as well as "jerks" (kaks), "tramps" (carapilsiens) and "amoebas" (eencelligen), etc. The police create their own meaningful categories on the

---

4   It is also notable that precisely this set of risk characteristics is very similar to that observed among other actors in the administration of criminal justice. Other research (Beyens, 2000) also shows that magistrates and judges use these extra-legal perceptions as a basis for their decisions in their daily practices. They use a similar set of ascribed characteristics to allow them to judge the chances of social integration or the risk of non-integration. Among magistrates, this often involves similar, pure perception, which does not rest on systematic observation, but on more anecdotal hands-on knowledge. We should stress here that this does not in any way amount to forms of so-called "discrimination", but a judgement on the chances of integration which has little substantiation. However, this observation is not unimportant. In this sense, the perceptions of the police do not essentially differ from those of other actors from the echelons of the criminal justice system, and the perceptions that do exist are confirmed in the minds of the servants of criminal justice.

basis of hands-on experience. Policemen are expressing complex social phenomena with these nicknames.[5]. This is functional, because the terms are meaningful within the police subculture and allow for internal communication. Outside the police organisation, however, these terms lose their meaning or relevance to a great extent. For policemen, developing a vocabulary of this type is a means of quickly identifying specific groups in a detailed and refined way. This practice is highly locally embedded and diverse, but only inter-subjective if the full richness of this local language is understood. The police identify "fault lines" and differentiate between groups with these nicknames. The observations on this matter again clearly reveal that policemen do not see the (multicultural) neighbourhood and its residents as a single community, nor do they treat it as a single community.

As mentioned earlier, the *"overpolicing"* of certain groups in the community creates a situation in which the police construct perceptions which are extremely specific and selective. This does not mean that these perceptions have come from thin air or that they are not based on "hands-on" experience. The police have a fairly adequate, intricate and realistic picture of these so-called "property" groups. This is because it rests on knowledge beyond the common knowledge in society. It is precisely through repeated and daily confrontation and intervention, through the frequent grinding down of perceptions, that perceptions become more tangible and categories are created. The police are perhaps the only actors in society to have such regular dealings with extremely problematic groups in neighbourhood communities.

If these interactions are not meaningful, it is not because policemen interact with only a part of the neighbourhood community. In the field, policemen come into contact with these problematic groups willingly or otherwise and have a very broad arsenal of practices, varying from a continuum of social-preventive to purely repressive means, to deal with this (something elaborated on later in more detail). As far as they are concerned, one thing is clear: they will have to make choices and take action. Some of these practices can be seen as useful and effective by the hierarchy, others not. The point is that these practices are largely unknown to the hierarchy and are barely, if at all, discussed in the force (or beyond), and therefore, scarcely valorised.

Police cynicism is growing in the minds of policemen as a result, and frustration and professional isolation are on the rise. Policemen on the beat and investigating officers try to clarify where the problems lie, but the force does not pay enough attention to what they have to say (or so they feel). As a result, the internal police subculture mentioned above arises, in effect a police vocabulary, which is difficult, if not impossible, to communicate within the force or to the outside world. This phenomenon is reinforced when policemen on the beat are faced with problematic realities, which can in fact only be solved by other actors (emergency services, social sectors, youth protection, etc.). Often, however, policemen on the beat can only observe that the desired solution does not materialise. The result is an increase in feelings of powerlessness and/or frustration. Policemen, and certainly policemen on the beat, who bring these problems home after work know these problems only too well, but feel powerless to remedy their causes. They are left with the overwhelming feeling of "fighting a losing battle".

---

5   We will make no statement as to whether these categorisations should be seen as discriminating. Any form of categorisation can be discriminating. In the first instance, we note here that the categorisations do not follow ethnic lines and in this sense are not designed to discriminate on the basis of ethnicity.

Many aspects of police organisation contribute to this growing feeling of power-lessness. Understaffing, bigger beats, the "hurried idea" of intervention teams, the paperwork involved in running the district and the fact that countless officers, in the metropolitan fabric in particular, are not sufficiently familiar with the local context, undoubtedly play an added role. A great many policemen feel left out in the cold be-cause they operate in a context in which the broader care professions do little, in which they perceive the judicial system as "too lax" with their attempts at corrective action, and because they are faced with a population that is at times very anti-authoritarian. In some cases, this leads to *a policy of pseudo-tolerance*.

There is also a growing feeling of not being listened to or understood. Much more so than in the past, it is recommended that the available hands-on knowledge of the police be systematised and put to use. Indeed, dismissing this hands-on knowledge as pure stereotyping would be all too easy. Policemen on the beat have a right to speak when it comes to groups they know well and from close quarters. Should police cyni-cism increase without any form of dialogue, the gulf between the police and the com-munity will grow and this is precisely what should be countered in a COP vision, i.e., the feeling of "us" and "them".

4.3 "Underpolicing" and neglect

On the other hand, some groups in neighbourhood communities are little-known to the police. One of the most striking observations of this study is that, where opinions on desired police conduct are concerned, there are no significant differences between problematic and less problematic groups in the neighbourhood, between well-known or little-known communities, or between immigrants and non-immigrants residents. Just about everyone wants a pleasant and helpful police force. Everyone thinks that the police should be present, reachable, available, familiar with the problems (however, small) and consulted.

This contrasts starkly with the view held by less problematic groups in neighbour-hood communities on the necessary policing of ... "others". These others are in need of law and order, i.e., repressive attention from the police, or so this study shows. This observation ties in with Lerner's theory on The belief in a Just World (Lerner, 1980). It seems that residents in neighbourhoods and districts always see "others" as the cause of the problem. No one counts himself among the others. Problems occur in certain districts of a city, in certain streets of a district and in certain houses of a street. Residents seek constant confirmation that they are good and that others are not. The same mechanisms are at play between non-immigrants and immigrants resi-dents. There are remarkable similarities between immigrants and non-immigrants residents' perceptions of policing.

(1) Some of the less well known groups (faced with "underpolicing") tend to be un-problematic. "Underpolicing" means that the police do not come into sufficient contact with the majority of less problematic groups, which are able to cope inde-pendently in the neighbourhood community. Many of the positive neighbourhood and district dynamics pass the policeman on the beat and the investigating officer by. Policemen do not see it as their job to get to know these dynamics in more de-tail. As a result, large areas of community life are unknown to policemen and they

are in danger of making an exception to the rule. They are not familiar enough with the positive forces emanating from large groups in the neighbourhood community thereby strong perceptions of small problematic groups are insufficiently corrected and are in danger of becoming too generalised. The antidote for police cynicism is present in problematic neighbourhoods and districts, but it is not sufficiently used, so that again, the gulf between the police and the neighbourhood community grows.

(2) Besides these more unproblematic groups, there are also neighbourhood communities, certainly in the neighbourhoods investigated in this study, which are confronted with countless societal problems and can certainly therefore be considered as socially problematic from this standpoint. The police cannot really assess or can only guess at these groups' activities and the problems or conflicts that exist. This category of groups is in some sense less well defined and much vaguer. For that matter, these groups are not infrequently cast in the role of victim. These groups rarely turn to the police when problems arise and often resort to their own mechanisms to find solutions and cope for themselves. The perception is growing in these neighbourhood communities that they are the subject of police neglect and under-protection. The police do have some perceptions of these groups, but they tend to be vague and not particularly explicit. It should be stressed once again that these groups are made up of both immigrants and non-immigrants. Let's take a closer look at this.

(2a) The first unknown group is that of travellers, mobile groups of people. These are groups who stay in multicultural neighbourhoods for only a short time (e.g., people passing through, asylum seekers, people without papers, newcomers who return home, gypsies and students). One of the characteristics typical of the multicultural neighbourhood is a high turnover of population. It is because of this high turnover that policemen are not able to gain a clear picture of these groups or assess their expectation patterns, let alone develop a relationship with them. In practice, what infrequent contact there is with these groups is usually confined to district activities, such as administrative procedures when they first register as residents. In other words, the police have neither the opportunity nor the time (nor even the capacity) to get to know them.

(2b) The police have a rather stereotypical perception of other unknown groups in multicultural neighbourhoods, which differs little or not at all from the dominant perceptions in society. Lack of knowledge and tangibility or a feeling of social distance on the part of the police seems to stimulate the use of clichéd perceptions, which can quickly become associated with certain criminal or troublemaking profiles. Thus, for example, policemen on the beat are relatively quick to associate Eastern Europeans with "heavy weapons" (due to an alleged "history of war") and Gypsies are always "thieves" and "tinkers". However, these perceptions are not particularly based on experience and policemen on the beat copy them almost seamlessly from broader social perceptions. The observation that perceptions of known groups ("overpolicing") and unknown groups ("underpolicing") are intertwined tends to convince policemen that the latter are equally grounded in reality and experience.

(2c) There are also groups that have lived in the multicultural neighbourhood for some time, often for several generations. These are so-called sedentary groups, who have developed a preference for "managing their own affairs" and orientate towards social networks in their "own" (ethnic) group. Among those worthy of mentioning here are the Belgian Turks and Pakistanis. In practice, policemen observe that these groups rarely turn to the police force for help because they would rather settle their own conflicts than involve the police. Few complaints are received from these population groups and they are not particularly willing to report themselves as the victims of crime. Policemen on the beat find it very difficult to obtain information (e.g., concerning a request from the tribunal for further investigation). Despite physical proximity and even the impression of good relations and an attitude of respect from members of this community, the police often have a sense of distance and lack of tangibility. These are thresholds which police officers sense, and which prevent them from assessing what is "really" going on in these communities, the perception they have of the police, exactly what the community expects and the activities in which they are engaged. These observations may well relate to the cultural preferences of certain ethnic sections of the community. Nonetheless, they are also relevant in the framework of a COP policy. Indeed, they illustrate significant problems in the area of partnership, working to solve problems and effective external orientation.

Two points can be made, however, in relation to the observations above.

(1) The police have identified these obstacles in order to gain a better knowledge of large sections of the population. However, only very rarely does this relate explicitly to COP. Policemen tend to attach the most importance to cultural aspects, if in their eyes, these are of relevance to interactions in the field. In particular, policemen stress the things that come across as derogatory to police officers (misunderstandings, misinterpretations, language problems or attitudes). Although the police representation often stresses these "cultural differences" (depending on the zone), this study shows that they occur less frequently than the (internal) representation would lead us to believe, and that more subtle cultural preferences are of significance, i.e., the notion of "closed communities" and the frequently-stated preference for "managing one's own affairs".

(2) A second matter, which relates to repeated reports from officers of the distancing they experience with certain ethnic minorities, is that this appears diagonally opposed to the tenor of the talks which were held with countless members of these communities wherein demand for closer contact (albeit in a neutral context) is one of the most salient expectations from the police. These are the groups who want to see tougher and firmer policing of troublemaking, which they often ascribe to other parties.

4.4   Two sides of the same coin

This study makes it apparent that overpolicing and underpolicing are two sides of the same coin. The police can devote extra attention to certain groups of "regular customers" who they know to be repeated offenders or troublemakers in a neighbourhood or district. But it is precisely this extra attention to fellow residents which can lead to other residents in the neighbourhood being neglected. The police appear unconcerned about the latter residents – they are out in the cold as it were. In other words, this gives rise to the perception that extra attention paid to one leads to indifference about the other.

Examining the diversity of communities in one and the same multicultural neighbourhood, more groups face "underpolicing" today than "overpolicing". In line with foreign research (Crowther, 2004), some groups of citizens in multicultural neighbourhoods are crying out, as it were, for more attention. This is a form of "underpolicing", which is moving towards an evasive, reactive style in respect of these neighbourhoods, but which certainly cannot be ascribed to police perceptions of the neighbourhood and its residents.

These two idealised types of police perception of the population can often be placed on a continuum, i.e., with well-known communities at one end and lesser-known communities at the other.

In the case of known "regular customers", the research shows that the police make choices between several representations and practices or modes of action. There is a palette of representations and practices from which the policeman can choose and use, depending on the situation. These can be social, organisational or even individual in nature. Throughout this study it became clear that, in practice, more choices are open (within the arsenal of available practices) when it comes to the better known groups. To a certain point, modes of action and practices are found to be in line.

It is this palette which is missing for groups in which tangibility is low and clichéd representations set the tone, where insufficient knowledge of the context exists and the stereotypical impressions formed by colleagues are seldom corrected. At such points, policemen fall back on a one-dimensional discourse of consensus, which is not infrequently repressive in nature. Where this involves lesser-known groups, it appears that police perceptions all too quickly tend towards the pessimistic, the distrustful and sometimes even the dominantly negative. Again, this mechanism does not seem to relate to "immigrants" or "non-immigrants". This perception is expressed clearly in the internal representation, in the form of statements which leave little to the imagination. What is also notable is that these discourses do not necessarily translate into practice, in the field or during interaction with the population groups in question. The conclusion must therefore be that there is not necessarily, in this case, a direct link between the representation on the one hand, and conduct in the workplace on the other.

4.5   Aversion to the COP discourse, yet still acting in accordance with it

The detailed categorisation of specific groups and the subculture that ensues from this relates to the way in which policemen try to resist attempts to drastically change police culture through policy. While close contact, cooperation and a service-minded set up are the main policy task objectives, it seems that the priorities of policemen lie elsewhere. This is because policemen at the grass roots level, as well as those in the middle ranks, are not particularly sympathetic towards this way of doing things, which they feel is imposed from above. This research shows that investigating officers in particular, and to a lesser extent, policemen on the beat, excel in disregarding, belittling and even ridiculing the COP philosophy.

The approach is seen as neither applicable nor expedient due to the (multicultural and metropolitan) context. Their officers see a COP policy as being (too) "soft". In terms of knowledge, many inspectors never progress further than blindly copying out a few COP principles. Only a minority recognise its usefulness and manage to apply the overall concept of the philosophy to their own role or the police force as a whole. Policemen find COP "too abstract" and, as they say themselves, they find it difficult to see its relevance. The ultimate argument is often that COP "won't work here". The impression is that COP policy implementation is not driven from the bottom up, but is a "managerial" top-down approach and foreign to "real" policing [6].

Nevertheless, this is a strange observation given that the findings of this research suggest that the approach taken on the work floor does in fact contain many COP elements. Depending on population group or context, policemen (can) use all kinds of approaches and strategies in line with the detailed knowledge they have of specific societal problems, individuals, groups, neighbourhoods and even social networks. It is notable that some policemen on the beat appear to be extremely proficient in this area. Practices being used by these policemen, allowed them to make the problems they face workable and controllable, albeit the proverbial bandage on a wooden leg. In some sense, these are "COP-real practices" because they are faced with these phenomena and no other authorities are deeply concerned with them. To keep this reality controllable or workable to some extent, the police develop their own specific ways of dealing with things, characterised by mediation, consultation, conciliation (i.e., focusing on de-escalation), as well as monitoring, and if necessary, taking a "hard" approach when there is no better alternative.

Not only does this research suggest that the police approach already contains COP elements, but that, because of the varied approaches and strategies that make the categorisation mechanism a central theme, there can even be talk of *multiple* COP practices.

Given the suggestion of a COP approach (albeit to a small extent) and that there are variations in approaches, based on communities, police resistance to COP policy remains truly astonishing, and this resistance is not adequately explained by a lack of knowledge of COP. Policemen may well see COP as part of the so-called "external" world, which *in their eyes* has little or no connection with substantial "actual policing realities in the field". COP has no place in the world in the way that policemen perceive and structure it themselves. The philosophy comes across as a *corpus alienum* and appears incompatible with the unique way in which policemen perceive, (re)construe and, as they say themselves, are forced to deal with the context of work. COP is perceived as something "foreign", something imposed by an "external world", and this creates strong resistance.

---

6   This is a remarkable observation in itself, given that COP is all about trying to prevent a top-down approach.

4.6   Relationship between internal representation and approach

What are the factors that cause so-called overpolicing or underpolicing? What is the relationship between perception, internal reasoning and approach in the field? The inductive factors identified here sketch out the more complex reality. Here it concerns general context factors such as types of formal control to do with racism, migration history and degree of ethnic density to an abstract level.

(1) The first factor relates to the tendency of policemen on the beat to *set priorities* themselves, more or less independent of implemented policy. These policemen prioritise much more on the basis of perceived and experienced realities in the workplace (in terms of criminality, troublemaking and neighbourhood problems). Policemen on the beat construct their own mental "crime profiles", which they relate to specific communities and to which they tailor their own approach. These profiles largely rest on anecdotal hands-on experience and not on systematic analysis. Nor do they bear much relation to policy control. According to Boussard et al. (2006) this tendency reflects that which policemen on the beat consider as "real policing" or "real customers".
     Some well-defined minorities, such as Gypsies, suffer directly through this. This is a group of the population from which policemen (want to) experience mental and physical distancing, and to whom they will even systematically attach a crime profile. It is a group whose negative image always comes through during interactions, regardless of the police officer(s) involved or the specific situation (with the exception of neighbourhood policing).

(2) The second factor follows on from the first. Once again, this involves conflicting priorities for the policemen, as regards what is considered *"real" and "unreal"* policing, and so it can also run contrary to envisaged policy. In this respect, the many curt interactions noted with marginalised non-immigrant population groups can be quoted. In one sense, this behaviour is understandable, since the results of the study suggest that these groups are seen as tough customers, and the police are driven beyond the frustration threshold because of the powerlessness they feel in remedying the causes of the recurring problems. In situations like these, all that usually remains to be done is an attempt to de-escalate.

(3) As third factor, *situational* aspects often influence the tone of an interaction and the approach taken by policemen. It is in the concrete situational context that underlying perceptions or preferences are expressed. This tends to be the case if the situation, and more particularly, the behaviour of the citizen overlap in many areas with the so-called attributes that policemen give to the person or group in question through their categorising system. In other words, when, in the eyes of policemen, the "preconceptions" open to them are confirmed. This finding is extremely consistent and is strongly related to the way in which policemen define their own role and how they interpret the behaviour of the citizen. It is important to stress here the fact that officers are more interested in the respect they expect from citizens due to their position of authority, than the respect they are required to show these citizens.

(4) During interactions with the so-called "property groups", policemen can choose from several aspects of reasoning and approach, depending on the situation. For

example, they can depart from the dominant social reasoning and, thanks to their varied experience, opt for specific strategies or a custom-made approach. It may be tough and controlling, or conciliatory or mediating. It would appear that there is quite a lot of choice. However, these choices appear to be missing in interactions with groups for which only *partial (hands-on) knowledge* is available. At such moments, policemen seem to lose grip of the situation and fall back on a "consensus approach", which leans closely towards underlying but clichéd perceptions and exaggerated criminality profiles.

(5) The fifth factor is the *"cultural and social capital"* of policemen on the beat. This covers all the underlying attitudes, political preferences, knowledge, social skills, as well as experience, knowledge of various social representations and trends, social networks, environments frequented in their private life, etc. In particular, this factor clarifies the precarious relations between policemen and young North Africans. In general, these youths suffer from a (considerably) negative image. This is reflected in the defeatist reasoning noted among forces in a number of police zones. Relations between the police and these groups cannot be described as anything other than problematic.

Perceptions of North African youths rest on explicit mental crime and troublemaking profiles. We were, however, unable to note any so-called "consensus approach", as opposed to the observations with regard to Gypsy families. On the contrary, during interactions with Moroccan youths, for example, we noted circumspection in the main. In practice (despite the negative image), we even noted a palette of varying "soft" and "hard" approaches or approach strategies, which varied greatly from officer to officer. However, some policemen follow a black & white reasoning. These policemen prefer tougher overtures, regardless of the situation. This type of approach could be observed mainly among policemen who place themselves on the right of the political spectrum and are quicker than their colleagues to criticise multicultural society. They appear less ready to take a mediating line. Other policemen showed a more subtle picture, and one that departed from the at times "hard" reasoning of the force. They were also the ones who took the trouble to put their colleagues' negative statements about policing multicultural neighbourhoods into context. They appear to opt for other forms of action, notably often with success for that matter, and corrected their colleagues' statements or doings, but never in the citizen's presence.

5   Support for decision-making

In relation to the above research results, three clusters of recommendations can be proposed to optimise the implementation of the Belgian COP model in multicultural neighbourhoods.

5.1 Validating knowledge & approaches in the field of known problematic communities

It is recommended that the available hands-on policing knowledge of policemen in the field concerning *known problematic communities*, and the approach taken in relationship to this, be validated.

A significant step towards this would be to systematise (if possible) and record this hands-on knowledge. Not to keep a check on police officers, but to increase the internal learning effect. It would appear that this type of hands-on knowledge does not fit in with the registration and performance measurement systems currently used in the police force. In particular (but not exclusively), the writing up of a police report is easily traceable in the current system, and in this way, it is disproportionately validated as a part of police work. Indeed, it is a long way off the type of hands-on knowledge brought to light in this research and used to affect the overwhelming majority of police work in (multicultural) neighbourhoods. In other words, more thought must be put into how this hands-on knowledge can be recorded, so that its value can be assessed.

One essential step in the exchange of hands-on knowledge may be found through (horizontal) inter-vision moments in which officers exchange their knowledge, experience, tactics and techniques among themselves (and relevant partners from the integrated safety sector), but above all are able to discuss the dilemmas they experience in the field. This can help demarcate the boundaries between policing and other actors in the safety domain, in relation to mediation, for example. This is also an opportunity for police leaders to recognise their people's freedom to implement policy in the field and to give these people room to consider a responsible way of responding to this freedom in the field. This process can generate an understanding of the degree to which it may or may not be possible to formulate separate choices relating the discretionary powers accorded police officers in the (multicultural) field. In addition to intervision moments, coaching can be used to focus greater attention on the validation of hands-on police knowledge on the work floor. The final goal would be to contribute towards forms of "smart" policing.

Additionally, the learning effect of both recommendations can be converted into a learning process, assuming that hands-on knowledge finds a place in the method of assessing and remunerating the work of policemen. Rewarding creativity in dealing with complex problems in (multicultural) neighbourhoods can be an "incentive" of great importance in revaluing neighbourhood policing, in relation to officers and investigating officers alike. An important note in the margin here is that the method of remuneration should not be such that officers are taken away from the neighbourhoods. An upgrade in the police service often implies more office work, and this should not the case when it comes to developing COP.

Finally, it should be said that these recommendations stand or fall in relation to the structural and statutory context in which they are implemented. This is because a great many standards and statutes have been introduced since the police reforms of 1998 and these do not always allow for the necessary flexibility. What is required is a rethinking of these elements to ensure the provision of quality services for the residents of (multicultural) neighbourhoods.

5.2 Facilitating and stimulating knowledge on lesser known (problematic) communities

It is recommended that the available hands-on policing knowledge of policemen in the field concerning *lesser known problematic communities*, and the approach taken in relationship to this, be enlarged, facilitated and positively stimulated. It is apparent from the present study that this is the most underestimated challenge to modern policing in multicultural neighbourhoods. This is because there are communities of which the police have little or no knowledge, and this is a real void in the framework of COP, given that each of these communities in a democratic constitutional state is entitled to expect an equal service provision.

To address this void, it is important that policemen expand their contacts in (multicultural) neighbourhoods by developing more and systematic contacts, also at the micro-level, with a view to building up trust with members of communities who are not immediately apparent or known. This implies an investment in contacts, whereby the immediate result (let alone effect) in terms of current police performance measurement systems will not be known right away. Thought must be given as to how this work can be valorised with a view to a better social embedding of the police in (multicultural) neighbourhoods.

In building up these contacts, it is recommended that a peep be taken over the "wall" to investigate methodologies from other sectors (such as outreach street work) to provide better services for communities. In turn, this evolution can be positively stimulated by working out "incentives" to facilitate police creativity in this area and, in this way too, contribute to the development of "smart" cops.

Finally, a function-oriented recruitment of policemen in this context can be a useful path for the future. Whereas recruitment is uniform at present, meaning that every recruit can reach similar jobs over time, a form of selectivity when recruiting, and above all during selection, would provide an opportunity to respond more effectively to the personnel requirements relating to COP. Indeed, it is utopian to assume that uniform recruitment according to an amorphous profile will automatically contribute to a larger selection of "smart" cops. Additionally, the various training courses in the police landscape can be screened in terms of COP requirements in order to better organise the objectives set on a "points" basis[7]. The aim is to stimulate a force-wide community-oriented approach and bridge the pitfall of ascribing this to one or more functions within the police organisation (e.g., neighbourhood policing).

5.3 Two-way communication between communities and the police

This study reveals that it is not just the police, but the communities in multicultural neighbourhoods that suffer from distorted, incomplete, fragmentary perceptions of the (local) police. This is something that takes shape in practice through an inadequate flow of knowledge and information, incorrect or unrealistic patterns of expectation and highly generalised impressions of (overly) apparent police practices, while other less apparent practices are barely heard of. For their part, policemen voice difficulties,

---

7   PONSAERS, P., Thursday, 16 October 2008 – Egmont Palace, Brussels. International Seminar: Police education and training in Belgium: on the way to Bologna? Organiser: Centre for Police Studies. Lecture: "Conclusions of the seminar".

sometimes even resistance, towards assessing the expectations and wishes of communities, or pride themselves on knowing them "well" in any case. It is therefore recommended that the available *hands-on knowledge of the police in all problematic neighbourhood communities* be further validated.

It is crucial here that the exchange of information between the police and the communities be stimulated to allow for two-directional traffic. All too often, communities are seen merely as "suppliers" of useful information (from an instrumental viewpoint) and less as a party requesting information on exactly what the police do and where the emphasis on safety policy lies. Nonetheless, this implies a challenge in terms of COP, given that the provision of information is an important task in providing a service to communities. The present study also shows that communities usually expect a more proactive attitude from the police. Both parties stand to benefit from a mutual assessment of problems, desires, expectations, rights and duties. Explaining police practices in certain (parts of) neighbourhoods (and their limitations) can help communities to understand the role and function of the police in their society. Informed people can be, for example, more understanding of potential trouble generated in their neighbourhood through certain police action. This is an opportunity for the police to become more embedded in communities with which they generally have little contact, a goal which is clearly congruent with the ideas of COP. Finally, this is the meaning of one of the principles of COP in Belgium, the so-called principle of "empowerment".

When communicating with communities, it is also important to search for the most suitable way of giving concrete shape to the exchange of information. Face-to-face contact is without a doubt an extremely powerful way of meeting this requirement, but at the same time, it implies heavy investment in order to apply some sort of system. It is becoming more important, however, that the police not only work on a so-called "approachability", but (pro)actively succeed in addressing citizens in their face-to-face contacts.

5.4   Reconsidering community policing

Finally, there is a need to reflect on the body of thought enshrined in the concept of COP. The present study shows that, during the talks and interviews, policemen were barely able to reproduce the philosophy or sketch out its implications for the force as a whole. Nonetheless, observations in the field have shown that several police practices do lean towards the COP philosophy. In other words, although the five principles (external orientation, partnership, problem-solving action, accountability and empowerment) seldom roll off the tongue, some current practices do illustrate that COP is possible in the (multicultural) arena.

It would be reckless to claim that this is sufficient to assure the success of the democratic process of renewing the Belgian police landscape, which was set in motion with the integrated police act of 1998. This study patently shows the challenges that are still waiting when it comes to effectively implementing COP. If a next step should be taken, it is essential that current police policy take the police practices illustrated in the present study into account.

Therefore the present study can be used as a source of information in the further development of COP in Belgium. The time has come to realise that extra policy documents and guidelines on this subject, such as the "excellent police function" document, barely get through to the policemen in the field. The time has come to treat policemen in the field as "practical professionals" and take their hands-on knowledge as a starting point for all further management of the change process. This will give competent police involvement (internal empowerment) a real chance and create leverage for competent neighbourhood resident involvement (external empowerment) (in this case, ethnic minorities in multicultural neighbourhoods). With this recommendation, the present study aims to contribute towards the further democratisation of the Belgian police system.

The police receive a disproportionately high amount of enquiries from some communities and, to a large extent, this affects the time available for and how other communities are treated. This is the day-to-day reality in which policemen in the field operate and in which they are being asked to work in a community-oriented way. The five principles of this philosophy require that police policy pay greater attention to this reality and the available hands-on knowledge. In this way, policemen can be appropriately involved in the performance of their duties in a (multicultural) society.

6   Bibliography

BEKKOUR, M. (2001). *Interview met professor Lode Walgrave over criminaliteit en de multiculturele samenleving.* Elektronische kopie dd. 16 maart 2005, URL: http://www. kifkif.be/modules.php?op=modload&name=News&file=article&sid=629 (Kif Kif – de interculturele site van Vlaanderen).

BOUSSARD, V., LORIOL, M., & CAROLY, S. (2006). Catégorisation des usagers et rhétorique professionnelle. Le cas des policiers sur la voie publique. *Sociologie du travail,* 48(2), 209-225.

BOWLING, B., PHILLPS, C. & SHAH, A. (2003). Policing ethnic minority communities. In: Newburn, T. (ed.), *Handbook of Policing.* Collumpton: Willan Publishing, p. 528-555.

BURR, V. (2003). *Social Constructionism* (tweede editie). Londen: Routledge.

BRUNSON, R.K. & MILLER, Y. (2006). Young black men and urban policing in the United States. *British Journal of Criminology.* Vol. 46, pg. 613–640.

CASMAN, M., GAILLY, P., GAVRAY, C., KELLENS, G. & LEMAITRE, A. (1992). *Police et immigrés.* Brugge: Vanden Broele.

CLARKE, J. (2006). Social Constructionism. In: MCLAUGHLIN, E. & MUNCIE, J. (eds.), *The Sage dictionary of Criminology.* Londen: Sage.

CROWTHER, C. (2004). Over-policing and under-policing social exclusion. In: BURKE, R. H. (ed.), *Hard cop, soft cop. Dilemmas and debates in contemporary policing.* Cullompton: Willan Publishing.

ECK, J. & ROSENBAUM, D. (1994). The new police order: effectiveness, equity and efficiency in community policing. In: ROSENBAUM, D. (ed.), *The challenge of community policing: testing the promises.* Thousand Oaks: Sage, p. 3-23.

FEYS, J. (2002). Politie, een afspiegeling van de samenleving? De relatie tussen politie en etnische minderheden. In: DUHAUT, G., PONSAERS, P., PYL, G. & VAN DE SOMPEL, R. (red.), Voor verder onderzoek... Essays over de politie en haar rol in onze samenleving. Brussel: Politeia, p. 771-793.

GOODEY, J. (2006). Ethnic profiling, criminal (in) justice and minority populations (editorial). *Critical Criminology*. Vol. 14, pg. 207-212.

HAEN MARSHALL, I. (1997). Minorities and crime in Europe and the United States: more similar than different! In: HAEN MARSHALL, I. (ed.), *Minorities, migrants and crime*. Diversity and similarity across Europe and the United States. Thousand Oaks: Sage Publications, p. 224-241.

HEBBERECHT, P. (1995). De politiediensten en etnische minderheden: de geschiedenis van een gestoorde verhouding. *Cultuur en Migratie, Themanummer: De politie, Ali's vriend... Over de verhouding tussen politiediensten en allochtonen in België, het binnenhalen en het binnenhouden...*, 13 (1), p. 11-19.

LEA, J. (2004). From Brixton to Bradford: official discourse on race and urban violence in the United Kingdom. In: Gilligan, P. & Pratt, J. Crime, Truth and Justice: Official Inquiry, Discourse, Knowledge. Willan Publishing, p183-240.

LERNER, M. J. (1980). *The belief in a just world. A fundamental delusion*. New York: Plenum Press.

MASTROFSKI, S. (1998). Community policing and police organization structure. In: BRODEUR, J. (ed.), *How to recognize good policing: problems and issues*. Thousand Oaks: Sage, p. 161-189.

REINER, R. (2000). *The politics of the police* (derde editie). Oxford: University Press.

SPARROW, M. (1988). Implementing community policing. *Perspectives on policing*, 9, 1-11.

SUN, I.Y., PAYNE, B.K., WU, Y. (2008). The impact of situational factors, officer characteristics, and neighbourhood context on police behavior: A multilevel analysis. *Journal of Criminal Justice*. Vol. 36, pg. 22-32.

SWANBORN, P.G. (2004). Kwalitatief onderzoek en exploratie. *Kwalon*, 26 (2), 7-13.

SWIDLER, A. (1986). Culture in action: Symbols and strategies. *American sociological review, 51 (april)*, 273-286.

VAN MAANEN, J. (1978). The asshole. In P. K. MANNING & J. VAN MAANEN (eds.), *Policing: A View from the Street* (pp. 302-328). New York: Random House.

VAN SAN, M. & LEERKES, A. (2001). *Criminaliteit en criminalisering. Allochtone jongeren in België*. Amsterdam: University Press.

VANDEVENNE, Y. (2006). Gesprek met Yannic Vandevenne, verantwoordelijke van het Programma Community Policing van de Directie van de Relaties met de Lokale Politie (CGL), inzake de tussentijdse evaluatie van de implementering van community in testzones, Gent dd. 23 mei 2006.

VANDEVOORDE, N., VAEREWYCK, W., ENHUS, E. & PONSAERS, P. (2003). *Politie in de steigers: bouwen aan gemeenschapsgerichte politiefuncties in een lokale context.* Brussel: Politeia.

VERCAIGNE, C., MISTIAEN, P., WALGRAVE, L. & KESTELOOT, C. (2000). *Verstedelijking, sociale uitsluiting van jongeren en straatcriminaliteit.* Leuven: onderzoeksrapport in opdracht van de Dienst voor Wetenschap, Technologie en Cultuur.

VERSCHUREN, P. & DOOREWAARD, H. (2005). *Het ontwerpen van een onderzoek.* Utrecht: Lemma.

WESTER, F. & HAK, T. (2003). De methodologie van kwalitatief onderzoek. *Kwalon (speciale editie over de praktijk van kwalitatief onderzoek), 22(1),* 7-17.

WALGRAVE, L. (2002). Taboe of vooroordeel. Verantwoord onderzoek van allochtone jongeren en criminaliteit. *Alert,* 28 (1), p. 84-92.

Population density, disadvantage, disorder and crime. Testing competing neighbourhood level theories in two urban settings.

Caroline Mellgren
Lieven Pauwels
Marie Torstensson Levander

1   Introduction and aim of the study

The relationship between population density and neighbourhood crime levels is a classic one found in criminological research. Urbanization has often been seen as detrimental to the preservation of social ties among neighbourhood residents. Early scholars such as Park (1929) were concerned by this trend and very pessimistic in their views. They considered the maintenance of social ties to be more difficult for residents in areas characterized by high population density because of the anonymity that is produced by high levels of population density in cities, thus setting the stage for crime and disorder.[1] In this paper we ask the question to what extent population density, defined as the ratio of residents per unit of area, can be similarly seen as a structural background condition that sets the stage for ecological concentrations of crime and disorder in two different settings. Many scholars have addressed the question of why population density is related to crime and disorder from a theoretical point of view (Shaw and Mc Kay, 1942, Sampson and Groves, 1989). We know from numerous studies that neighbourhood levels of crime tend to correlate with structural characteristics such as population density and disadvantage (poverty) at multiple levels of aggregation (Ouimet, 2000; Wikström, 1991). However, scholars seem to disagree in some respects: (1) regarding the effect of population density independent of structural disadvantage, (2) regarding the actual mechanisms at work in the relationship between population density and crime – amongst other things questions have been raised as to the role played by social trust in this relationship (Sampson, Raudenbush and Earls, 1997; Wikström and Dolmén, 2001) and, (3) there has also been some debate as to the independent role played by neighbourhood levels of disorder in the causation of neighbourhood levels of crime (Bursik and Grasmick, 1993; Sampson and Raudenbush, 1998).

While the literature already contains a large body of empirical evidence relating to these questions, only very few scholars have addressed these issues from a comparative perspective. Studies are often restricted to a single setting. Sampson and Wikström

---

1   Other scholars (e.g. Brantingham and Brantingham, 1980) have also referred to non-residential land use as an important structural characteristic that is present in cities, but this is not the focus of the present study. We ask the reader to bear this restriction in mind and we return to this point in our concluding discussion.

(2008) have previously argued that more cross-national research is needed in order to understand both how social structure and social mechanisms affect crime levels and how general the findings of ecological studies are across countries. Indeed, only very rarely have studies shown how the distribution of crime can be similarly explained in different countries (Sampson and Wikström, 2008). Sampson and Wikström employed identical measures of the key constructs and showed that similar processes were at work in explaining neighbourhood differences in Chicago and Stockholm.

The present study draws on such previous attempts to produce insights into the comparability of the relationship between population density and crime (Sampson & Wikström, 2008; Eisner & Wikström, 1999) but our effort is more modest since we do not have identical measures of our theoretical constructs. Thus this paper is not a comparative study in the strict sense of the term but rather an examination of a much debated criminological issue. Although we acknowledge that this may be seen as a shortcoming, other attempts to compare different samples without perfect comparability in the measures employed have nonetheless contributed to the knowledge of how similar mechanisms shape crime and disorder in different settings (e.g. Svensson and Pauwels, 2008). Theories often have a middle-range status and thus allow for some level of generalisation. If a general theoretical model holds, the empirical relationships should be found in multiple settings and independently of the specific measures used as indicators of the relevant constructs.[2] Thus, this article contributes to the existing literature by examining the extent to which two theory-driven models of the relationship between population density and crime can be equally applied to data collected in two cities, Antwerp in Belgium and Malmö in Sweden. More specifically, we examine what happens to the relationship between population density and crime when neighbourhood disadvantage, social trust and disorder are introduced into the equations. In particular, we will focus on evaluating the strength of the relationships between population density, disadvantage, social trust, disorder and crime, and in doing so we will be highlighting the "comparability" of results by pointing to similarities in the findings from the two cities.

This study is an aggregate-level study at the neighbourhood level. The study of crime at aggregate levels is necessary in order to improve our theorizing and understanding of the consequences of area structures for the unfolding of criminal events. It also contributes to the development of cross-level theories on the occurrence of crime (e.g. how area characteristics contribute to the selection of targets).

2 Population density, disadvantage, disorder and crime from an ecological perspective

In this section we present the reader with a brief overview of theoretical rationales as to why the central concepts included in this study are related to crime. Our approach has its theoretical backdrop in the Chicago School (Park, 1929). Scholars following this tradition have consistently shown that ecological variations in crime are strongly related to neighbourhood structural conditions in the form of population density, neigh-

---

2   We are aware of the fact that measurement issues may flaw any results, but that is also true for comparative studies employing highly similar measures and for social science studies in general.

bourhood poverty and residential mobility (Shaw and McKay, 1969 [1942]). The role of population density in explaining ecological differences in crime was first discussed by scholars such as Quételet, who found large differences in crime levels between urban and rural areas, and the issue has attracted much theoretical attention since the early efforts made in this area by the first exponents of the Chicago School. Scholars were at that time concerned with the negative consequences of rapid urbanisation, and the consequent enormous increase in the number of inhabitants per area. Such scholars were highly pessimistic in their views. The city was seen as iniquitous, whereas the countryside was thought of as being ideal for the maintenance of order and the avoidance of problems such as crime. Thus, these scholars hypothesized that increased population density would lead to an increase in crime (Kasarda & Janowitz, 1974). Population density was blamed because it was suspected to lead to anonymity among neighbourhood residents, while in less dense and rural areas people still knew each other and could avoid problems of crime and disorder. Thus social ties (sometimes referred to as social cohesion), being the opposite of anonymity, were seen as the explanatory mechanism in the population density – crime relationship.

The role of disadvantage has been especially stressed in both classical and contemporary versions of social disorganisation theory (for an overview, see Rovers, 1997; Dolmén, 2002; Pauwels, 2007). According to social disorganisation theory, neighbourhood conditions, particularly poverty (further referred to in this article as disadvantage), residential mobility, and concentrations of immigrants would lead to a decrease in social control and ultimately to what has been referred to as a (permanent) state of neighbourhood social disorganization (Shaw and McKay, 1969), thus allowing crime to penetrate neighbourhoods. Scholars following social disorganisation theory have drawn special attention to the fact that the effects of population density are confounded with indicators of poverty, which they have labelled structural disadvantage, or the ecological concentration of poor people. In short, it has been argued that population density is merely one of several important indicators of urbanisation, and that urbanisation may result in the ecological segregation of poor people or disadvantage. The effect of population density is thus supposed to be indirect rather than direct. Yet, the question remains: why does disadvantage cause crime and disorder?

From a theoretical point of view, social disorganization theory has evolved into a "systemic" model of community organization, stressing the importance of the presence of social ties and social organization rather than disorganization (Granovetter, 1973; Kasarda and Janovitz, 1974). The contemporary approach stresses the importance of studying and measuring the actual social mechanisms at work, i.e. dimensions of social organisation. Basically, according to this line of thought neighbourhoods with low population density would be characterised by low levels of disadvantage, which would produce fruitful structural conditions for the development of local social ties, which in turn foster informal control.

Bursik and Grasmick (1993) introduced to the field of criminology the idea that weak neighbourhood ties are important inhibiting forces with regard to the neighbourhood distribution of crime. From an empirical point of view, such contemporary variants of the "social disorganization perspective" have been successfully employed in the explanation of delinquency rates (Ouimet, 2000), crime occurrence rates, victimization rates (Smith and Jarjoura, 1988) and neighbourhood levels of fear of crime (Taylor and Covington, 1993; Taylor, 1997).

Contemporary theoretical developments have largely been based on social capital theory (Coleman 1988), and particularly on Sampson and colleagues' refinement of social capital as a theory of collective efficacy (Sampson, Raudenbush and Earls, 1997; Sampson, Morenoff and Earls, 1999). Collective efficacy is clearly linked to social cohesion, since the concept is defined as social cohesion among neighbours combined with their willingness to intervene on behalf of the common good (Sampson, Raudenbush and Earls, 1997). This definition clarifies the association between 'social trust' on the one hand and 'informal social control' on the other. The existence of social trust constitutes an essential condition for the fostering of informal social control in neighbourhoods, and thus for the willingness to intervene for the common good. Generally speaking, the concept is based on shared norms and values within the community in question (Sampson, Raudenbush and Earls, 1997; Sampson, 2002), and is conceptually not so different from Thrasher's (1929) view of community organization, as outlined in his study of Chicago youth gangs. One particular merit of contemporary collective efficacy theory lies in the way in which it clearly defines the mechanisms of informal control that lead to low neighbourhood crime rates. Thus, social trust is regarded by contemporary scholars as an important mechanism of informal social control in explaining the relationship between population density and crime.

Both early and contemporary scholars who have worked from a social disorganisation perspective have considered crime and disorder as outcomes of a common cause. Sampson and Raudenbush (1999) found empirical evidence for the fact that crime and disorder are probably produced by the same underlying forces and should thus be viewed as two different aspects of the same phenomenon. But this issue in particular still divides scholars. Broken Windows theory (Wilson and Kelling, 1982; Kelling and Coles, 1996; Skogan, 1990) views disorder as a direct cause of crime and argues that disorder has an effect on crime, independent of population density, disadvantage and social trust. The major difference between these two interpretations of the crime-disorder link is that whereas "broken-windows" sees disorder as a direct cause of crime, the collective efficacy perspective views weakened social control as the cause of both crime and disorder.

3   Testing two competing models

On the basis of the above discussion we have derived and constructed two competing models that explain the relationship between population density and crime. We choose to refer to these competing models with the terms the "collective efficacy" derived model of crime and disorder (as presented in Figure 1) and the "Broken Windows" derived model of crime and disorder (as described in Figure 2). The strong version of the "collective efficacy" derived model states that population density is indirectly related to crime and disorder only through its effects on disadvantage and that disadvantage is only related to crime and disorder because of its effects on social trust (Sampson, 1985; Sampson and Groves, 1989). In a weakened version of the collective efficacy derived model some effects of neighbourhood disadvantage on crime and disorder are still allowed (Bursik and Grasmik, 1993). However, a key element in both versions is that no direct effect should be found between disorder and crime.

Figure 1: The "collective efficacy" derived model of crime and disorder

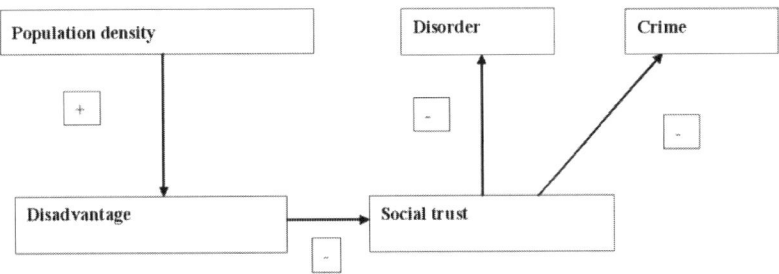

The *"Broken Windows" derived model of crime and disorder* states that disorder has an independent effect on crime, regardless of population density, disadvantage and social trust.

Figure 2: The "Broken Windows" derived model of crime and disorder

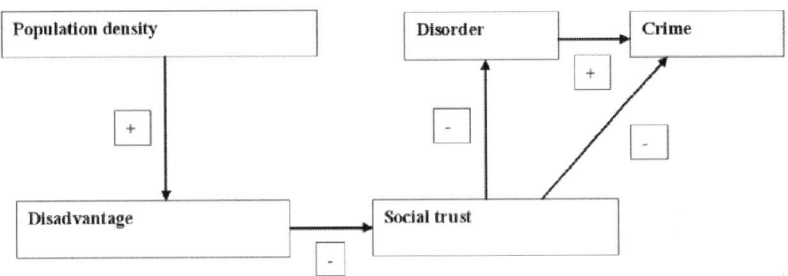

4   Data employed in the study

4.1   Administrative data at the census tract level

One major problem faced by neighbourhood studies of crime and disorder is the definition of the concept of neighbourhoods (Leighton, 1988). In this study we use geographical neighbourhoods as defined by city planning authorities and which are homogenous on key variables such as housing tenure. The areal units can be best described as neighbourhood clusters and they constitute the lowest aggregate level at which aggregate-level analyses can be conducted using the available official data. Many contemporary authors (e.g. Oberwittler and Wikström, 2009; St. Jean, 2007) have argued, in relation to the definition of neighbourhoods as units of analysis, that census tracts are too large to capture variations in social trust, social control, and crime and disorder. These authors favour smaller units of analysis, such as street segments or blocks consisting of a few hundred people. The argument is that smaller units are cohesive and exert informal control, and may be bordered by blocks that are completely different in this respect. However, in this study it has been impossible to employ such small units of analysis. The neighbourhood clusters contain between 205 and 6486 inhabitants in Antwerp and between 622 and 9370 in Malmö, figures which are similar to those representing the number of people per unit in previous studies on neighbourhood level conditions (e.g. Sampson and Wikström 2008). A total of 210 neighbourhood clusters are employed in the analyses in Antwerp. These represent 70 percent of all Antwerp neighbourhoods. In Malmö, only neighbourhood clusters in which at least 20 respondents were surveyed are included in the analyses. As a consequence, 79 areal units, representing 69 percent of the total 110 neighbourhood clusters, are included. This reduces the initial sample size by five percent (from 4911 to 4672 respondents). The administrative data on the social structure of Antwerp have for the most part been made available to us by the Antwerp Knowledge Center City Observation (in Dutch: *Studiedienst Stadsobservatie Antwerpen*). Administrative data on the social structure of Malmö have been provided by the City of Malmö and were originally gathered by Statistics Sweden. In both cities, neighbourhoods are rather homogeneous with regard to housing, thus the disadvantage measures in both cities correlate strongly with housing types.

4.2 A community survey to measure social trust and disorder in both cities

The measurement of community-level mechanisms has been improved since the survey-based measurement of community social organization started to flourish in the 1990s, and the importance of studying social structure (social cohesion and disorder) at the community level has been acknowledged (Sampson, Raudenbush and Earls, 1997). Neighbourhood research has clearly benefited from the ecometric approach to community mechanisms (Raudenbush and Sampson, 1999; Oberwittler and Wikström, 2009; Pauwels, 2006, 2007). The data on the social mechanisms of cohesion and disorder employed in the current study have been collected by means of an independent survey approach in both cities. The two surveys differ however in their selection of respondents. The expert survey of key informants in Antwerp used a very select group of key informants, i.e. persons whose structural position means that they are able to provide the researcher with the key information on a given neighbourhood that is required to measure community cohesion and disorder. These key informants were not residents but rather professionals active in Antwerp neighbourhoods (Pauwels, 2006). The principle of optimal selection, as described first and foremost by Campbell (1955), was applied. In total, 321 key informants were interviewed and asked to fill out a questionnaire on neighbourhood disorder, social cohesion and trust among neighbours (see Appendix 2C for information on the background characteristics of key informants). The information provided by these respondents was used to construct a social cohesion scale and a disorder scale. Reliability measures of both constructs were highly satisfactory and can be found in Appendices 2D and 2E. The ecological reliability of measures of social cohesion and disorder attained by means of the key informant technique has been previously established as satisfactory by Pauwels (2006).[3] In Malmö, survey data were provided by The Malmö Fear of Crime Survey, a cross-sectional study conducted in the police district of Malmö in 1998 (Torstensson, 1999). A total of 7,000 people between the ages of 16 and 85 living within the police district of Malmö, which comprises the two municipalities of Malmö and Burlöv, were randomly selected from the general population of Malmö and Burlöv respectively (254,904 on January 1st 1998). The members of the sample were asked to complete a questionnaire in the autumn of 1998. Two postal reminders were sent to the respondents. The overall response rate was 76 percent. The current study is based exclusively on data from the respondents living in Malmö, 4,911 persons (71, 6%). It is thus important to note that the key social mechanisms, social trust and disorder, have been measured in a different way across the two cities. We can do no more than conclude that both community surveys meet the criteria of reliability and validity. The key informant technique is probably less well-known than the resident survey technique, but tests conducted in Belgium suggest stable findings using professional key informants as social observers of neighbourhood processes (Pauwels and Hardyns, 2008).

5   Measurement of constructs

Factor analysis is a well-established technique in the study of the social ecology of crime. A latent variable approach of this kind is very useful for the purposes of the current study. It is common knowledge that indicators of disadvantage and disorder are highly correlated. While the classical social disorganization perspective pointed to the theoretical importance of separate structural determinants of disorganization, such as ethnic heterogeneity, instability and poverty, contemporary research often reveals intense segregation and thus very strong correlations at the census tract level, a problem that has been acknowledged before (Bursik, 1988; Bursik and Grasmick, 1993). This leads to multicollinearity and has consequences for the multivariate analysis of neighbourhood level data (Dolmén, 2002; Pauwels, 2001, 2002). Indicators of neighbourhood structural characteristics are measured slightly differently across the two cities, although they refer to the same content. We argue that latent variables represent the *"true scores of a neighbourhood"* on theoretical constructs in a better way than raw indicators. We recognise that the differences in the measurements of key concepts are a shortcoming in the present study but this shortcoming should be viewed in relation

---

3   Studies in which neighbourhood clusters are the unit of analysis sometimes have difficulties in conducting a survey at that same level. In the Antwerp study, questions that refer to small geographical units of analysis can be difficult for survey respondents to answer because they are not necessarily familiar with such administrative borders. This problem was technically solved in the original study (Pauwels, 2007) by conducting the expert survey at a slightly higher level of aggregation (larger administrative areas of clusters of adjacent neighbourhood clusters). The 210 Antwerp census tracts can also be divided into 42 larger areas, a division that is used by policy makers for the purposes of social policy. It was known which neighbourhoods comprised each and every larger administrative area. Thus it was possible to ascribe a score for social cohesion and disorder to each neighbourhood cluster. We could have easily aggregated the neighbourhood cluster administrative data to the larger administrative area, but this results in a substantial decrease in the sample size available for the purposes of multivariate analysis. This might lead to an increase in type 1 errors due to a serious reduction in sample size (N= 210 at the neighbourhood cluster level and 42 at the larger administrative area level). Nonetheless, we reanalysed the relationships at the larger administrative area level and did not find any aberrant results (Pauwels, 2006).

to the aim of the study, which is not that of conducting a strictly comparative study but rather that of testing a neighbourhood level theory in two independent samples simultaneously.

Neighbourhood disadvantage is a measure comprising the proportion of single person households, the proportion on social welfare, the proportion unemployed, residential stability, the proportion of immigrants, the proportion who own their own homes, and median income. In Antwerp, only official data are used, while both aggregated survey responses and official data are used in Malmö.[4]

Neighbourhood crime is measured in Antwerp using geo-coded police records of violence against persons, violent theft, car theft, theft from cars, theft of handbags, vandalism, and burglary and in Malmö using geo-coded police records of car theft, vandalism / violence against persons and the neighbourhood level victimization rate.[5] The Antwerp data were originally obtained from a research project on the applicability of hot-spot analysis in Belgium (Goeminne, Enhus and Ponsaers, 2002). Police recorded crime data in Malmö were provided by the Malmö police.

Social cohesion and trust is measured by three survey questions in both Antwerp and Malmö. The wording of the questions is somewhat different across the two surveys but both measures score similarly highly on internal consistency and ecological reliability (we refer the reader to Appendix 2D for an overview of the indicators and their factor loadings).

Neighbourhood disorder is measured using three and five indicators of social and physical signs of disorder in the neighbourhood in Antwerp and Malmö respectively. The internal consistency of the combined factors and the ecological reliability as measured by lambda are both acceptable. The indicators and their factor loadings can be found in Appendix 2E.

6   Methods of analysis

Initially we perform a series of block-wise OLS-regressions. This involves specifying a series of regression models in which variables are entered into the regression equation in blocks, with each block representing a theoretical variable.[6] The order in which these variables are subsequently entered is defined by theory. We perform block-wise models for crime and disorder as separate dependent variables, before proceeding to assess direct and indirect effects. Next, Structural Equation Modelling (SEM) is used to assess the direct and indirect effects of population density, disadvantage, social cohesion and crime and disorder. In order to evaluate the fit of the models, the Root Mean Square of Approximation (RMSEA) is preferred over the Chi-square value as a

---

4   We refer to appendix 2A for more details on the indicators used.
5   Neighbourhood victimization is an important additional indicator of neighbourhood crime that may not be measured using official data; therefore it was included as an indicator. Neighbourhood victimization loaded sufficiently highly on one latent variable, "crime", in Malmö. The indicators and their factor loadings can be found in Appendix 2B.
6   Essentially, some areas are geographically adjacent. This implies that observations are not independent and that spatial autocorrelation (i.e. a larger similarity between adjacent neighbourhoods than between distant neighbourhoods) may be present and bias the standard errors of the regression estimates. Spatial regression models (Anselin et al., 2000) can solve this problem, but we were not able to test for spatial autocorrelation.

measure of overall fit. Chi-square tends to be very sensitive to the size of the sample, resulting in accumulated high values of the statistic, i.e. the 'uncritical' rejection of models, in the case of large samples. RMSEA on the other hand is a measure of 'close fit', indicating that it takes into account the error of approximation in the population as well as the precision of the measure itself (Jöreskog and Sörbom, 1998, 1999). Models with a RMSEA <.05 are considered acceptable (Billiet and McClendon, 2000).[7] Although SEM is a powerful statistical tool in testing direct and indirect effects, the reader should bear in mind that even the most powerful techniques of causal modelling cannot truly establish causality. All analyses are conducted at the aggregate (neighbourhood) level, and thus all conclusions are drawn at the neighbourhood level.

7   Results from block-wise OLS-regression models

This section presents results from block-wise multivariate analyses that were conducted in order (a) to establish the strength of the relationships between population density, disadvantage, social trust and disorder and (b) to establish the nature of the relationship between population density, disadvantage, disorder and crime. Block-wise regression analysis provides an opportunity to develop a first insight into the independent effects of cohesion on disorder and of disorder on crime. These independent effects are assumed to exist independently of neighbourhood social structure, and the effects of social structure are viewed as being moderated by the mechanisms of community (dis)organization.

Table 1. Population density, disadvantage, social trust and disorder

Dependent: Disorder	Model 1 (A) Beta	Model 1 (M) Beta	Model 2 (A) Beta	Model 2 (M) Beta	Model 3 (A) Beta	Model 3 (M) Beta
Pop. density	.477***	.395***	.259***	.077 ns	.098 ns	-.006 ns
Disadvantage	--	--	.410***	.683***	.386***	.402***
Social trust	--	--	--	--	-.466***	-.464***
Model evaluation						
Rsq adj.	.224	.145	.342	.510	.528	.615
F-change	60.905***	14.235***	38.15***	58.22**	82.35***	21.73***

N: 210 Antwerp, N: 79 Malmö
***: p < 0.001 **: p < 0.01 * p<0.05

Table 1 presents the results of a series of block-wise regression analyses that explain neighbourhood-level differences in disorder. Only standardised regression coefficients are shown together with the significance level. To evaluate the models, the adjusted R-square value and the F-change values are presented.[8] From Model 1 it can be seen that population density has a strong positive effect on disorder and that this variable

---

7   Univariate tests of skewness showed that most of the variables are significantly skewed, and we therefore calculated normal scores prior to conducting the analyses of structural relationships between latent constructs, as suggested by Jöreskog and Sörbom (1998).

8   The adjusted R-square is similar to the R-square but is adjusted to take account of the number of parameters in the model. The F-change value shows whether the inclusion of a variable leads to a significant improvement in the model fit.

has a similar, positive effect on disorder in both cities; the effect of population density appears to be fully mediated by disadvantage in Malmö, whereas in Antwerp, the effect of population density merely decreases when disadvantage is included in the model (Model 2). Model 1 explains 22.4% of the variance in Antwerp and 14.5% in Malmö. Disadvantage has a strong positive effect on disorder in both cities. The effect of disadvantage is larger than the effect of population density. Model 2 explains 34.2% of the variance in Antwerp and 51.0% in Malmö. The final model includes social trust and reveals that social trust has the strongest direct effect on disorder in both cities, but that a direct effect of disadvantage on disorder still remains. This model explains 52.8 % in Antwerp and 61.5 % in Malmö. In congruence with the theoretical framework, social trust has the largest effect on disorder. The effect of population density is no longer significantly different from zero once social trust is entered into the equation.

Table 2. Population density, disadvantage, disorder and crime

Dependent: crime	Model 1 (A) Beta	Model 1 (M) Beta	Model 2 (A) Beta	Model 2 (M) Beta	Model 3 (A) Beta	Model 3 (M) Beta
Pop. density	.376***	.312***	.078 ns	.144ns	-.004 ns	.091 ns
Disadvantage	--	--	.560***	.363***	.429***	-.107 ns
Disorder	--	--	--	--	.318***	.687***
Model evaluation						
Rsq adj.	.137	.086	.360	.180	.424	.403
F-change	34.04***	8.32**	73.03***	9.806**	23.79***	29.49***

N: 210 Antwerp, N: 79 Malmö
***: $p < 0.001$ **: $p < 0.01$ * $p<0.05$

The next step in the analysis involves investigating the independent effect of population density on crime, independent of disadvantage and disorder. At the same time we get a first insight into the strength of the disorder – crime link, independent of neighbourhood structure. Model 1 shows that population density is positively related to crime in both cities, while Model 2 reveals that population density only has a significant effect on crime as long as disadvantage is not entered into the equations. Model 1 explains 13.7% in Antwerp and 8.6% in Malmö. Model 2 explains 36.0% in Antwerp and 18.0% in Malmö. The latter model suggests that population density has an indirect effect on crime which goes via its impact on disadvantage. Disadvantage seems to fully mediate this effect in both cities. Interestingly, Model 3 reveals that disorder has a strong direct effect on crime, independent of population density and disadvantage. Model 3 explains 42.4 % in Antwerp and 40.3 % in Malmö. When disorder is entered into the equation, the effect of disadvantage becomes insignificant in Malmö and decreases in Antwerp. If a common underlying neighbourhood structure were to be the reason as to why disorder and crime are so strongly correlated, then no direct effect of disorder should be found.

Block-wise regression analysis does not tell us how well a model fits the observed data in terms of direct and indirect effects. Block-wise regression tells us whether a parameter is significantly different from zero when other variables are entered into an equation, but it does not inform us about pathways and direct and indirect effects. As a means of better understanding the nature of the relationships between the theoretical variables, a number of confirmatory path analyses were fitted to the data using a structural equation modelling approach. The indirect effect hypothesis and the common cause hypothesis can thus be analyzed simultaneously and can be presented together with the best fitting model in both cities. This approach allows for a global evaluation of the models in both cities, while all variables are involved in a network of relationships.

8   Results of structural equation modelling

To briefly recapitulate the core assumptions of each hypothesis, the effect of social trust is at the heart of both hypotheses. *The "collective efficacy" derived model of crime and disorder* states that neighbourhood structure is related to rates of crime and disorder only to the extent that neighbourhood structure has an influence on neighbourhood social trust (Bursik and Grasmik, 1993). Population density is hypothesised to have direct effects on neighbourhood disadvantage, while neighbourhood disadvantage is hypothesized to have a direct effect on social trust. Social trust is hypothesized to have direct implications for both crime and disorder and thus disorder cannot be expected to have an independent direct effect on crime, since trust is the common cause.

Table 3. Testing the "collective efficacy" derived model of crime and disorder

Fit indexes	The "collective efficacy" derived model of crime and disorder (strong version) Antwerp	The "collective efficacy" derived model of crime and disorder (strong version) Malmö
RMSEA	0.31	0.20
CHI²	130.26	20.88
df	5	5
p-value	0.001	0.001
AGFI	0.50	0.72

N: 210 Antwerp, N: 79 Malmö

The results of our tests of the "collective efficacy" derived model of crime and disorder are summarized in Table 3. We performed a test of both the strong and weak versions of the model. The strong version is the one that was presented in Figure 1. Essential to both the strong and the weak versions of the collective efficacy derived model of crime and disorder is that we modelled the relationship between disorder and crime so that both are seen as outcomes of social trust, while allowing no direct effect of disorder on crime.[9] By doing so, we have empirically translated the theoretical assumption that all covariance between disorder and crime is due to a common cause (namely social trust) and that only correlated measurement error exists between the two constructs. After running these models, the model fits were interpreted. As can be seen from Table 3 the strong version of the collective efficacy model does not hold at all either in Antwerp (RMSEA: 0.31; AGFI:.0.50) or in Malmö (RMSEA: 0.20, AGFI: 0.72). A weak version of the collective efficacy model, in which room was allowed for independent direct effects of disadvantage on both crime and disorder, was also run. These models

---

9   This can be achieved by specifying that the relationship between disorder and crime is completely due to the common effect of social trust and by recognizing that the only covariance that still would exist between both constructs is due to error that might be correlated because both concepts share a common cause. This was achieved by freeing the error covariance between disorder and crime.

made the same assumption of the spuriousness of the disorder–crime link, of course, but allowed for additional direct effects of disadvantage on crime and disorder. These models did not lead to acceptable model fit indices. Let us now turn to interpreting the best fitting models, which to some extent were similar, but which also differed in some respects across the two settings.

Table 4. Best fitting models

Standardized Effects	Model 1: Best fitting model Antwerp				Model 2: Best fitting model Malmö			
	CRIME	DISOR-DER	TRUST	DISADV	CRIME	DISOR-DER	TRUST	DISADV
POP. DENSITY	--	--	-0.38	53	--	--	--	0.46
DISADVANTAGE	0.43	0.43	--		--	0.40	-.69	
TRUST	--	-0.49			--	-.46		
DISORDER	0.32				0.65			
R-SQUARE	0.43	0.52	0.14	0.28	0.42	0.63	0.48	0.22
FIT INDICES								
RMSEA	0.022				0.017			
CHI²	4.33				5.32			
df	4				5			
p-value	0.36				0.38			
AGFI	0.97				0.92			

This table summarizes all direct effects of the exogenous and endogenous variables of the best fitting models in both cities, as well as the fit indices and squared multiple correlations of all subsequent models. All path coefficients are statistically significant ($p < 0.01$). N: 210 Antwerp, N: 79 Malmö

From Table 4 we can see that population density does not have a direct effect on crime and disorder. Population density has a very similar positive effect on disadvantage in both settings. In accordance with the observations of Bursick and Grasmick (1993), the effects of disadvantage on disorder in Malmö and on crime and disorder in Antwerp, remain after adjusting for social trust. The analyses suggest that it takes more than social trust alone to explain neighbourhood differences in disorder. However, the best fitting models suggest that social trust is at least a key mechanism in explaining neighbourhood differences in disorder in both cities. It is also surprising how similar the standardized coefficients are. One difference between the two cities is the following: in Antwerp, social trust is only directly affected by population density, and not disadvantage, whereas the reverse is true in Malmö.

The statement derived from the Broken Windows derived model of disorder and crime, where disorder is allowed to have independent effects on crime, is confirmed, at least when considering the direct effect of disorder on crime. In Antwerp, however, direct effects of disadvantage are still observed. This finding suggests that the disorder-crime link may be different by context, and it is therefore important to study further other mechanisms that have not been taken into account in these analyses.

9   Conclusion and discussion

One thing is very clear from the analyses presented in this article: population density is only indirectly related to crime and disorder and its effect disappears when disadvantage and social trust are entered into the equations. This is true for both settings and suggests that this finding is rather stable. One key finding in this study is that social trust appears to be consistently and strongly related to neighbourhood levels of disorder in both cities. It cannot however be said that social trust fully mediates the effects of neighbourhood disadvantage equally in both settings. Furthermore, neighbourhood disorder is a strong predictor of crime in both cities. Structural equation modelling was used to establish the direct and indirect effects between these concepts and on the basis of the path models that were run and compared with one another, we cannot reject the hypothesis that the relationship between crime and disorder is spurious. We acknowledge that a cross-sectional study, such as the one presented here, cannot properly disentangle the temporal ordering issues involved, and thus leaves one critical element of the crime-disorder question unexamined. What is needed to this end are panel designs that follow up neighbourhoods over several years and that would allow for the examination of 'spirals of decline' (Schuerman and Kobrin, 1986). From the relationship examined in the current article, we can only conclude that the effect correlation between disorder and crime is strong in both Antwerp and Malmö, independent of important structural controls (population density and disadvantage) and social trust, which constitutes an essential concept in the collective efficacy-approach.

We would recommend that the relationships confirmed in this study be tested further. We have focused on social trust but this is not the only perspective that should be considered when testing the effects of neighbourhood-level social mechanisms. The key driving idea of the collective efficacy perspective is that area disadvantage determines different conditions for informal control (Oberwittler, 2001). This study was limited to a single dimension of collective efficacy, namely social trust. It is worth mentioning that environmental criminology posits a (complementary) view on the explanation of neighbourhood differences in crime and disorder that could not be examined in this study. Environmental criminology argues that concentrations of crime and disorder also depend on the differential 'opportunities' created on the one hand by the spatial differences in routine activity patterns that arise from different forms of land use and on the other by the physical features of neighbourhoods. This topic has not been addressed at all in this paper, but is one which also warrants attention when attempting to establish the strength of the independent effects of social trust. It might be that opportunity-based characteristics are equally or maybe even more strongly related to neighbourhood social trust, informal control and rates of crime and disorder than the characteristics described in the classic disorganization perspective. Opportunity structures might even account for some of the differences between the two settings that have been observed in the analyses conducted in this article.

We recognize that our data are subject to important limitations but we would nonetheless argue that the results are important, since this ecological study suggests that population density is only related to crime and disorder as long as disadvantage and particularly social trust are not taken into account. Social trust operates very similarly in shaping the uneven distribution of disorder in two settings, which are quite different with regard to population composition and the number of inhabitants. Unfortunately, comparative analyses are still rather scarce in ecological studies of crime and disorder (Eisner and Wikström, 1999; Sampson and Wikström, 2008). If one wants to assess how similarly social processes operate in explaining neighbourhood differences in crime and disorder, it is necessary to increase efforts to use identical operational measures of the same theoretical concepts and also to measure ecological units of analysis in a similar way. Another question raised here is that of the extent to which there is any justification to continue viewing the Collective Efficacy-approach and the Broken Windows-approach merely as opposing theories. A number of key hypotheses derived from each model were not falsified. It is therefore also important for future studies to assess how neighbourhood changes in social trust are related to neighbourhood changes in crime and disorder. The question previously raised by St.Jean (2007) also deserves more attention. Are we testing neighbourhood theories of crime and disorder at an inappropriate level? Although the performance of aggregate-level analyses is necessary to examine neighbourhood-level sociological theories of crime patterns, we must bear in mind that neighbourhoods are made up of individuals and future studies should therefore consider to study the micro-level consequences of macro-level variables. It is clear that the current study has not been able to meet all of these criteria, but we are nonetheless convinced that our work constitutes a valuable attempt at demonstrating that social mechanisms operate in a rather similar way across different structural and cultural settings. Despite the possible measurement errors that are present in a neighbourhood study that employs slightly to very different measures and methods of data collection to examine the effects of neighbourhood characteristics across different settings, the finding of very similar effect parameters remains. This should serve as a major stimulus to the development of optimal research designs for studying the relationships between neighbourhood structural characteristics, social mechanisms, and outcomes such as crime and disorder. From a theoretical point of view, the challenge is to search for additional mechanisms that may link disorder to crime, independently of mechanisms of informal control, such as social trust.

10   Bibliography

Anselin, L, Jacqueline Cohen, David Cook, Wilpen Gorr, and George Tita. 2000. Spatial Analysis of Crime. In Measurement and Analysis of Crime and Justice, ed. David Duffee. Rockville, MD: National Institute of Justice/NCJRS.

Billiet, JB, McClendon J. (2000). Modelling Acquiescence in Measurement Models for Two Balanced Sets of Items. *Structural Equation Modelling: A Multidisciplinary Journal* 7 (4):608-628.

Bursik, R. (1988). Social Disorganization and Theories of Crime and Delinquency: Problems and Prospects. *Criminology*. 26: 519-551.

Bursik,R. J Jr., Grasmick, H.G. (1993). *Neighborhoods and Crime*. New York: Lexington.

Campbell, D.T. (1955). The informant in quantitative research. *American Journal of Sociology*. 60 (4), 339-342.

Coleman, J.S. (1988). Social capital in the creation of human capital. *American Journal of Sociology.* 94: 95-120.

Dolmén, L., (2002). *Brottslighetens geografi. En analys av brottsligheten i Stockholms län.* Stockholm: Kriminologiska institutionen, Avhandlingsserie nr.6

Eisner, M., Wikström, P-O H (1999). Violent Crime in the Urban Community: A Comparison of Stockholm and Basel. *European Journal on Criminal Policy and Research.* 7 4: 427–442.

Goeminne, B., Enhus, E., Ponsaers, P. (2002). *Criminaliteit in de publieke ruimte. Een introductie tot de hot spot analyse.* Brussel: Politeia.

Granovetter, M.S. (1973). The Strength of Weak Ties. *The American Journal of Sociology.* 78 (6), 1360-1380.

Harell, A., Stolle, D. (2008). Reconciling Diversity And Community? Defining Social Cohesion. In: *Developed Democracies, Conference Proceedings of The International Conference on Social Cohesion.* Brussels, May 15th 2008, Brussels: The Royal Academy Press, (in press).

Jöreskog, K., Sörbom, D. (1998, fourth edition). *Structural equation modelling with the Simplis command language.* Scientific Software International, Chicago.

Jöreskog, K., Sörbom, D. (1999, third edition). *Prelis 2 User's reference guide.* Scientific Software International: Chicago.

Kasarda, J., Janovitz, M., (1974). Community attachment in mass society, *American sociological review,* Vol. 47, p427-433.

Kelling, G.L., Coles, C.M (1996). *Fixing Broken Windows: Restoring Order and Reducing Crime in Our Communities.* Free Press: New York.

Leighton, B. (1988). The Community Concept in Criminology: Toward a Social Network Approach. *Journal of Research in Crime and Delinquency.* 25 (4): 351-374.

Oberwittler, D. (2001). *Neighborhood Cohesion and Mistrust – Ecological Reliability and Structural Conditions* (working paper / No. 3).

Oberwittler, D., Wikström, P-O. (2009). Why small is better. Advancing the study of the role of behavioral contexts in crime causation. In Weisburd, W. Bernasco, & G. Bruinsma (eds.):Putting crime in its place. Units of analysis in geographic criminology. New York: Springer, 33-58.

Ouimet, M. (2000). Aggregation Bias in Ecological Research: How Social Disorganization and Criminal Opportunities Shape the Spatial Distribution of Juvenile Delinquency in Montreal, *Canadian Journal of Criminology.* 42 (2): 135-156.

Park, R. (1929). *Introduction to the science of sociology.* Chicago: University of Chicago Press.

Pauwels, L. (2001). Structurele contextualisering van de politionele registraties inzake intrafamiliaal geweld in Oost-en WestVlaanderen, *Panopticon,* 22: 220-238.

Pauwels, L. (2002). *De ene buurt is de andere niet, exploratie van mogelijkheden tot contextualisering van geregistreerde criminaliteit op buurtniveau.* Criminologische studies nr. 10, Brussel: VUBpress.

Pauwels, L. (2006). Ecologische betrouwbaarheid en constructvaliditeit van "sociale desorganisatieconstructen" op basis van de methode van "key informant analysis", *Panopticon,* 27: 34-55.

Pauwels, L. (2007). *Buurtinvloeden en jeugddelinquentie: een toets van de sociale desorganisatietheorie.* The Hague: Boom Juridische Uitgevers.

Pauwels, L., Hardyns, W. (2009). Measuring community (dis)organizational processes through key informant analysis, European journal of criminology, 6 (5): 401-418.

Raudenbush, S.W., Sampson, R.J. (1999). Ecometrics: Toward a Science of Assessing Ecological Settings, with Application to the Systematic Social Observation of Neighborhoods. *American journal of Sociology.* 105 (3): 603-651.

St. Jean, Peter K.B. 2007. *Pockets of Crime. Broken Windows, Collective Efficacy, and the Criminal Point of View.* Chicago: University of Chicago Press.

Sampson, R.J. (1985). Neighborhood and Crime: The Structural Determinants of Personal Victimization. *Journal of Research in Crime and Delinquency* 22(1):7–40.

Sampson, R. J. (2002). Transcending Tradition: New Directions in Community Research, Chicago Style. *Criminology* 40: 213-230.

Sampson, R.J., Groves, W.B. (1989). Community Structure and Crime: Testing Social Disorganization Theory. *American Journal of Sociology.* 94 (4):774-802.

Sampson, R.J., Raudenbush, S., Earls, F. (1997). Neighborhoods and Violent Crime: A Multilevel Study of Collective Efficacy. *Science* 277:918-24.

Sampson. R.J., Morenoff, J.D., Earls, F. (1999). Beyond Social Capital: Spatial Dynamics of Collective Efficacy for Children. *American Sociological Review.* 64 (5):633-660.

Sampson, R.J., Raudenbush, S.W. (1999). Systematic social observation of public spaces: A new look at disorder in urban neighborhoods. *American Journal of Sociology.* 105 (3):603-651.

Sampson, R.J., Wikström, P-O.H. (2008). The Social Order of Violence in Stockholm and Chicago Neighbourhoods: a comparative Inquiry. In: I. Shapiro, S. Kalyvas, T. Masoud (Eds). *Order, Conflict and Violence.* Cambridge, UK. Cambridge University Press.

Schuerman, L., Kobrin, S. (1986). Community Careers in Crime, 67-100. In: Reiss, A., Tonry, M. (Eds), *Communities and crime, Crime and Justice, Vol.8, Chicago, University of Chicago press.*

Shaw, C., McKay, H. (1969). *Juvenile Delinquency and Urban Areas.* Chicago: University of Chicago Press.

Skogan, W. (1990). *Disorder and Decline: Crime and the Spiral of Decay in American Neighborhoods.* Free Press: New York.

Smith, D.A., Jarjoura, G.R. (1988). Social structure and criminal victimization. *Journal of Research in Crime and Delinquency* 25 (1): 27-52.

Svensson, R., Pauwels, L. 2008. Is a risky lifestyle always "risky"? The interaction between individual propensity and lifestyle risk in adolescent offending: A test in two urban samples. *Crime & Delinquency.* OnlineFirst, published on September 24, 2008, DOI:10.1177/0011128708324290.

Taylor, R.B., Covington, J. (1993). Community Structural Change and Fear of Crime. *Social Problems.* 40 (3): 374-395.

Taylor, R.B. (1997). Social Order and Disorder of Street blocks and Neighborhoods: Ecology, Micro ecology, and the Systemic Model of Social Disorganization. *Journal of Research in Crime and Delinquency* 33:113-155.

Thrasher, F.M. (1927). *The Gang: A Study of 1.313 Gangs in Chicago.* Chicago: Chicago University Press.

Torstensson, M. (1999). *Lokala Problem, Brott och Trygghet i Polisområde Malmö – 1998 års Trygghetsmätning.* Polishögskolan och Avdelningen för Psykiatri. Umas, Malmö, Lunds Universitet.

Wikström, P-O. H. (forthcoming). *The Social Ecology of Crime. The Role of the Environment in Crime Causation: International Handbook of Criminology,* Volume 1: Fundamental Principles of Criminology.

Wilson, J. Q., Kelling, G. (1982). Broken Windows. *Atlantic Monthly.* 249 (3):29-38.

Appendices

Appendix 1. Descriptives

Antwerp	Min-Max	Malmö	Min-Max
Inhabitants	205-6486 +	Inhabitants	622-9370 +
Single person households	7.18-79.23 +	Single person households	8.00-67.74 '
Neighbourhood stability	32.23-83.20 +	Neighbourhood stability	5.00-89.47 '
Proportion immigrants	.75-38.33 +	Proportion immigrants	6.25-80.95 '
Proportion on social welfare	0-20.6 +	Proportion on social welfare	0.00-94.71 +
Proportion unemployed (18-64)	.45-24.52 +	Proportion unemployed (18-64)	1.70-16.6 +

Source: Knowledge Centre City Observation Antwerp (2002)
Source: City of Malmö (1999) + Official data, ' Survey data

Appendix 2. Measurement issues

Appendix 2A. Disadvantage

Antwerp indicators	Disadvantage factor loading	Malmö indicators	Disadvantage factor loading
Single households+	.751	Single households*	.451
Percent welfare+	.858	Percent welfare+	.906
Percent unemployed+	.867	Percent unemployed+	.914
Residential stability+	-.815	Residential stability*	-.614
Percent immigrants+	.788	Percent immigrants*	.701
Percent home ownership+	-.831	Percent not working*	.485
Median income+	-.898		

*: survey aggregates +: official data census

Appendix 2B. Neighbourhood crime

Antwerp indicators (2001)	Crime factor loading	Malmö indicators (1999)	Crime factor loading
Violence against persons / sq km	.983	Car theft / 1000 inh	.882
Violent theft /sq km	.938	Vandalism	.943
Car theft / sq km	.925	Violence against persons/ 1000 inh	.944
Theft from car / sq km	.905	Percent victimized*	.465
Theft of handbags / sq km	.845		
Vandalism buildings / 1000 buildings	.802		
Burglaries / 1000 households	.608		

* = Survey based (last year prevalence)

Appendix 2C. *Descriptives of key informants (N=321)*

Background Characteristics	No.	Percent
Professional Background		
Local shops and catering industry	131	40.8
Social work and medical doctors offices	67	20.9
Local governance	73	22.7
Local police and private security	50	15.6
Sex		
Man	174	54.2
Woman	147	45.8
Age		
18-25	20	6.2
26-35	77	24.0
36-45	108	33.6
46-60	95	29.6
60+	21	6.5

Appendix 2D. *Social trust*

Antwerp indicators (2001) (Alpha:.728; Lambda:.72)	Factor loading	Malmö indicators (1998) (Alpha:.693; Lambda:.867)	Factor loading
It is easy to get in touch with the inhabitants of this neighbourhood.	.741	People in my neighbourhood don't get along.	-.451
Most inhabitants of this neighbourhood are friendly.	.728	People in my neighbourhood help each other.	.796
Most inhabitants of this neighbourhood are prepared to help you when you ask them	.606	One can trust people in my neighbourhood	.776

Appendix 2E. *Neighbourhood Disorder*

Antwerp indicators (Alpha:.778; Lambda:.84)	Factor loading	Malmö indicators (Alpha:.821; Lambda:.902)	Factor loading
How many times do you see homeless people in the streets of this neighbourhood?	.569	Litter on the streets	.566
		Vandalism	.599
How many times do you see drunks in the streets of this neighbourhood?	.534	Annoying neighbours	.576
		Drunks	.704
How many times do you see visible signs of vandalism in this neighbourhood (e.g. broken windows, damaged public phone cells, graffiti on walls)	.713	People that quarrel	.735
		Apartments used by alcoholics and drug users	.591
How many times do you hear complaints about noise pollution?	.618	Youth hanging around in the streets	.628
How many times do you see garbage lying on the streets?	.590	Women that are harassed	.516
		Traffic problems	.325
How many times do you see people being harassed on the streets of this neighbourhood?	.630		

The continuum of conflicts of interest: from corruption to clubbing and the underlying risks at victimisation

Gudrun Vande Walle

1   Introduction

In recent years, many legal instruments have been developed and policy decisions have been taken in the field of public corruption. There are various reasons for this such as the fear of organised crime, the fear of terrorism, the economic losses, the bad reputation of the market and the loss of trust in the government (Beare, 1997, 66 ff; Williams and Beare, 1999, 113). Initiatives of the European Council, the UN and the OECD have caused a chain of corruption and anti-corruption initiatives at the national level.

The question about the importance of corruption is under discussion when studying recent political-economic changes and more specifically the blurring of boundaries between public and private power and the growing impact of private companies on the political decision making process. A market that gets more chances to pull up a chair at the decision making table maybe has less need to be corrupt in the sense of bribing a politician or a civil servant.

The main idea of this article is that corruption is a criminal phenomenon with the possibility of a high victimisation rate. For that reason, the political attention for corruption is reasonable. In the last decades, the relation between the public and the private level has changed from a top-down relation into a fragmentation and a privatisation of public power. The hypothesis is that the illegal act of corruption in the context of the western market is becoming less important than other legal mechanisms of conflict of interest, being for example the old boys network, lobbying, revolving doors,etc.. In other words corruption is part of a continuum of conflicts of interest, varying from legal to illegal activities but all with a high risk factor of victimisation.

If this hypothesis corresponds to reality, the question raises if the focus of criminologists is not too narrow and if it is not an obligation to study also legal injurious activities besides the criminal law environment. This consideration from a critical criminological point of view reminds us of an old criminological discussion that was held in the field of conflict criminology. Authors such as Quinney argued that *"the law represents an institutionalised tool of those in power which functions to provide them with superior moral as well as coercive power in conflict"* (Quinney, cited in Sumner, 1994). This question on the limitation of the study domain has logically been central to the debate on white collar crime since this is the domain of *those in power*. This issue has become less topical over the last decades until recently a new concept was introduced, namely state-corporate crime. This plain concept of Kramer and Michalowski brought back the older discussions on definition and limitation of the study domain.

However, this discussion is not restricted to academic circles; a government or an administration that disregards its duty of serving the public creates an environment *"where rules don't prohibit its own self-enrichment or that of the ruling class. The final concern is the well being of citizens and the equal "(Lambsdorff, 2007)*.

The remainder of this article is organised as follows: the first chapter motivates why the attention for corruption is probably out of balance when compared to other mechanisms of conflict of interest. This imbalance was one of the reasons why some criminologists have taken injurious effects as basic assumption for their criminological research (see e.g. Braithwaite, 1984; Croall, 2001). This feeling of uneasiness about the limitation of criminological research brought the researchers Kramer and Michalowki to the concept 'state-corporate crime'. A short explanation of the origin and meaning of state-corporate crime demonstrates the importance of this change in thinking about corporate crime. This can be illustrated by the domain of conflicts of interests in the sense that corruption is criminalised while other harmful mechanisms are widely accepted. We explain this with some examples, such as lobbying, networking and revolving doors. Some would call this point of view "conspiracy" while it is simply a completion of an until then incomplete explanation of some kind of injurious corporate activities (Pipes, 1997). Even if the comparison between corruption and legal mechanisms of political influence creates a feeling of uneasiness it is not totally unfounded since some exploratory studies have focused on a kind of interaction between corruption and lobbying. We end this contribution with a plea for the criminological perspective on the continuum of conflicts of interest in the interest of the victims.

This contribution is the first step in a research project that aims to study the anti-corruption strategies in the public and private sector to prevent public corruption of companies. More specifically, it is a reflection that we have made during the exploratory study of literature on corruption and a study of empirical data collected during an exploratory phase of observations and interviews with public officials and business representatives. Some documentary sources have been written by journalists or belong to the grey literature even if the work is well-founded.

2   How political-economic changes may have an impact on corruption

The domain of public corruption is an interesting laboratory to study the implications of a changed social landscape: the basic principle of public corruption is the illegal exchange relation between a decision maker (a politician or a civil servant) and an individual or organisation offering or promising an advantage in exchange for a desired decision outcome (Van Duyne, 2001, 2-3). It seems a reasonable hypothesis that changes in the political-economic ratio will directly influence the significance of corruption. What is meant by the alluded changes at the political-economic level? In the last decades, we have moved slowly from a central government to diversified governance with more responsibility for a field of private and semi-private organisations. Because of this diversification public and private actors are more than ever before working in partnerships and networks. In other words, society has evolved from a system of authoritarian regulation to public-private coregulation (see for example: Heere, 2004; Meehan, 2003). Because these changes have a direct impact on the heart of public corruption, namely at the power relation between the state and the company, the interest of corruption may change (see for example Kamminga, 2003, 11; Richter, 2001, 11). Why would a company bribe if it has been asked for advice in the decision making process?

The idea that gaining political power reduces bribery in the private sector is occasionally the subject of discussion on the prevention of corruption, most of the time from an economic point of view (Williams and Beare, 1999,90): *"privatization can reduce corruption by removing certain assets from state control and converting discretionary official actions into private, market-driven choices"* (cited in Rose-Ackerman, 1999, 35). Becker suggests that *"if we abolish the state we abolish corruption"* (Gary Becker cited in: Hodgson and Jiang, 2007, 1047). Other authors even relate the anti-corruption movement to deregulation. For example, Hopkins points out that the anti-corruption strategies advocated by mainstream economists largely consist of two main planks: reforming the state administration to minimize corrupt incentives and reducing the role of the state in economic life in order to leave as much economic activity as possible in the hands of the market (Lambsdorff, 2007, 2; Hopkins cited in: Hodgson and Jiang, 2007, 1048). In contrast to this economic point of view, Tillman and Indergaard consider the changed relation between the private sector and the state rather as a crime-inducing fact for corporate corruption. They studied the new economy and more specifically the corporate accounting scandals. The authors argue that *"analysts should focus on the creation of rules – both formal and informal – that govern markets, how those rules are the products of political projects and how those rules create opportunities for white-collar crime and corporate corruption"* They clearly believe that the type of market regulation may or may not create opportunities (Tillman and Indergaard, cited.in: Tillman, 2009, 74).

Maybe both points of view make sense. Deregulation probably means a reduction of corruption in the sense of criminal law but at the same time other harmful activities such as lobbying can create the same risks. Most conflicts of interest in the relation between the private sector and the state are not perceived as illegal even if the mechanisms are basically the same and even if the injurious effects could be much "higher". It seems that the underlying mechanisms of these conflicts differ depending on *when and where* these conflicts occur. The legal definition of corruption is a social construction of some activities of conflict of interest that are considered as criminal. This definition excludes other activities that are not considered as problematic but that are equally harmful. The first careful attempts of the EU and the UN to open the debate to other mechanisms of conflict of interest point at a growing consciousness that the risk factors can be completely hidden because of the institutionalised and legal character. We refer to article 12 of the EU Criminal Law Convention on Corruption that asks member states to criminalise:

> *"the promising, giving or offering (...) of any undue advantage to anyone who asserts or confirms that he or she is able to exert an improper influence over the decision-making of any person referred to in articles 2, 4 to 6 and 9(...) whether or not the influence is exerted or whether or not the supposed influence leads to the intended result."*

This article refers to the activity of paying bribes or giving advantages to lobbyists. Also the UNCAC offers some advice to prevent corruption, which is mentioned in article 12 as well. The advice focuses on the prevention of conflicts of interest by imposing restrictions on the professional activities of former public officials or on the employment of public officials by the private sector after their resignation or retirement. These functions directly relate to those held or supervised by the public officials in question during their tenure. In both regulations, not only corruption but also related mechanisms such as lobbying and revolving doors are considered.

Both points of view show the function of state intervention as a regulator: a state that exchanges its authority for a collaborative relation opens some problematic relations. This responsibility of the state authority has never been seriously questioned in criminology. Except for the conflict criminologists and marxist criminology, the state has never been considered as a real player. That is why the introduction of the concept 'state-corporate crime' by Kramer and Michalowski is a big step in the aetiology of organisational crime.

3   The added value of State-corporate crime

For decades, researchers of corporate crime have neglected the role of the state in committing crime (Kramer, Michalowski and Kauzlarich, 2002, 270). In theories explaining corporate crime, state responsibility was reduced to *a lack of state regulation* or *a lack of enforcement* (Box, 1983, 64) or, in the words of Sutherland, was conceived as *belonging to the same social class* (1961, 248). For most scholars of corporate crime, the state has never been perceived as a full partner in the commission of crime. The only exception is public corruption because it requires a governmental department, an individual civil servant or a politician as an indispensable participant. However, the statement: "no corporate crime without the state" holds water. If we consider cases such as Bhopal, Enron, Lernout and Hauspie, the Fortisgate or smaller ones, the state always has a certain responsibility: the state created a kind of opportunity because of the lack of enforcement, an unclear legal framework or a relation between the regulator and the regulated that is too close. To fill this gap, Kramer, Kauzlarich and Michalowski reintroduced the state as a participant in the commission of corporate crime, either as a facilitator or as the initiator. By introducing the concept of state-corporate crime, Kramer, Michalowski and their other colleagues reintroduced the state, not in the baseline as an element of explanation, but as an actor who is really responsible. The original concept has been further modified but the core of the definition is: *"state-corporate crimes are illegal or social injurious actions that occur when one or more institutions of political governance pursue a goal in direct cooperation with one or more institutions of economic production and distribution" (Kramer, Michalowski and Kauzlarich, 2002, 270).* The most important element of this definition is maybe the demarcation of state-corporate crime. The definition could be applied to illegal as well as *socially injurious* actions. The injurious character is the standard to decide if we are talking about a crime or not rather than the penal code. Other researchers such as Green and Ward specified the aspect of social injurious by referring to human rights violations (Green and Ward, 2004, 28). Especially in cases of socially injurious activities carried out with the help of the state, which is in the end the law maker and law enforcer, it is important to go beyond the legal definition and to look at the implications of these activities. Barak even went so far to say that injurious activities of the state are more threatening than harmful activities of the private sector because the first is making the rules in the name of common interest or national welfare.

The concept of state-corporate crime goes beyond the scope of the penal code to legal conflicts of interest that are equally harmful. The following questions are addressed: to what extent is the law-making process captured by industrial interests? Do the criminal definitions chosen encompass the socially injurious activities of companies? It is a debate that reminds us of the radical criminology of the Schwendingers who pleaded for a human rights definition of crime because "the legalistic definitions cannot be justified as long as they make the activity of criminologists subservient to the state" (Schwendinger and Schwendinger, 1972). In recent years, it has mainly been Van den Heuvel who has gone further than the legalistic delineation and has added collusion or *the Holland way of corruption*.[1]

We do not intend to debate the enlargement of the criminal stigma from corruption to other legal mechanisms. However, since we are mainly interested in the victims of conflicts of interest, this paper will narrow down its focus to all kinds of conflicts of interest between the "company" and "the state", legal or illegal, that carry the risk on victimisation. The concept of state-corporate crime enables us to have a complete understanding of the continuum of mechanisms of influence. The continuum consists of corruption, of a grey zone of collusion and of legal and legitimate but not unproblematic relations of interest between the private sector and the state.

In the following sections, we would like to draw attention to this continuum by offering three examples, namely networks, lobbying and revolving doors. We consider deregulation, coregulation and privatisation of regulation as important stimuli for the growth of this grey area around corruption.

4   "It takes two to tango": a framework for public corruption

A first mechanism of conflict of interest is public corruption, which is a rather vague concept because of its sociological origin. To avoid misunderstanding about our point of view, it is important to further explain this concept. The term 'public corruption' was defined by Petrus van Duyne as: *an improbity or decay in the decision-making process in which a decision-maker (in a private corporation or in a public service) consents or demands to deviate from the criterion, which should rule his decision making, in exchange for a reward, the promise or expectation of it* (van Duyne, 2001).

This decision making process can be situated at different professional levels, in different functions and with a difference in efficiency. The standard categories of public corruption are bureaucratic corruption and political corruption. Bureaucratic corruption attempts to influence the executive level, whereas political corruption is located at the level of law making. Some authors such as Moody-Stuart and Tillman added the category of grand corruption or "misuse of public power by heads of state, ministers and top officials for private profit", which in fact corresponds to a superlative of

---

1   Collusion is maintaining a secret relation with the aim of hindering the ability of illegal activities to be traced (Van den Heuvel, 1998, 12-13).

bureaucratic and political corruption (Moody-Stuart, 1996, 4; Tillman, 2009). The customs officer who allows to load an illegal freight in return for a bribe and the politician who inserts some exceptions to the environmental law in return for financial support in the elections are both committing corruption.

A fundamental feature of corruption is the need for at least two parties (van Duyne, 2001). "It takes two to tango". One party wants to have an impact on the decision making process and wants to give something in return (in the broad sense of the word); the other party has a certain power to answer the request of the first party because of its professional position. In contrast with other authors such as Larmour and Grabosky and also Tillman, we exclude fraud from the debate on corruption because it can be committed by one individual person or organisation (Larmour and Grabosky, 2000, 1; Tillman, 2009, 74). For example, for subsidy fraud, tax fraud or stock exchange fraud, no second party is required.

A third feature is the intention to exchange something: each party has something of interest for the other party because of his power position. The corruptee makes abuse of his professional power position to favour the corrupter who has the intention to give something in exchange, such as money, a material gift, a promise (for promotion, a job, friendship, love, etc.) and know-how. Mostly, the parties know each other well enough to avoid a written statement because they can trust the other party in respecting the deal.

In short, the main characteristics of corruption are: at least two parties, an exchange and a trust based relationship. It happens in the political, bureaucratic or private sphere, with a criminal intention and a misuse of power.

Although corruption is still considered as a major problem worldwide, it does not stem from a lack of regulation. At the international and national level, a wide range of regulations has been created to fight against corruption and bribery. Nowadays some researchers speak of a real anti-corruption movement supported by governments as well as regional and international organisations (Hindess and Sousa, 2006).

At the international level, the legal instruments with the most influence are the Foreign Corruption Practices Act, the OECD Convention on Combating Bribery of foreign public officials in international business transactions; the United Nations Convention against Transnational Organized Crime (UNTOC) (resolution 55/25, annex I) and the United Nations Convention against Corruption (UNCAC) (resolution 58/4 of 31 October 2003) and the Criminal Law Convention on Corruption and the Civil Law Convention on Corruption of the Council of Europe.

These examples of international initiatives reflect a diversity of definitions of corruption, a difference in objectives and a different appreciation of the injurious effects of corruption. Based on these diverse initiatives, we can draw some preliminary conclusions about the legal point of view on corruption. We could say that the first step in the fight against corruption is a fact: stimulated by international institutions, almost every country has an anti-corruption regulation. However, the second step, the implementation of this regulation in policy, is more problematic. The Global Integrity Index 2008 reflects the difference between the omnipresent anti-corruption regulation and the lack of implementation.[2] Only a limited number of countries which ratified the OECD convention are also enforcing it (Vogl, 2007, 179). A regulation that is not im-

---

2   (http://report.globalintegrity.org/globalindex/results.cfm)

plemented is nothing more than paying lip service to the anti-corruption institutions (Fubara-Anga, 2009).

A second conclusion we can draw when comparing the different initiatives relates to the differences in objective of each legal initiative. The OESO-convention was an answer to the question of the US to create a similar instrument as the FCPA in the aim to restore the equal competition of the American market. The idea was: "if we cannot benefit from corruption as we define it, then no one should" (Beare, 1997, 71). Limitations imposed by the OECD convention brought a new equilibrium on a world scale. On the other hand, the conventions of the Council of Europe have different objectives. A first objective is to protect the European member states and the European institutions against corruption. This means that the control of corruption is concentrated in the accession process.[3] A second objective is to protect Europe against organised crime (Huisman and Vande Walle, in review).

A third conclusion is the major difference in the scope of the application domain. The OESO and the FCPA are in the end rather tolerant of smaller bribes. Both conventions underline the importance of allowing facilitation payments or payments made to procure or speed up the provision of services. The limits of facilitation payments are not defined, which makes the authority of the OESO and FCPA conventions less coercive. In Belgium and in the Netherlands, facilitation payments are forbidden as minor bribery offence but some managers aknowledged that it is tolerated in practice (Conference, Ethical Corporation, May 27-28, 2009). Theoretically speaking, the European conventions and especially the UNCAC have a very broad application domain. In this context, we refer to the questioning of revolving doors and also of lobbying, which are both considered as risk factors for corruption. These references to legal activities in an anti-corruption legislation are not totally unfounded. Indeed, they clearly illustrate the changed relation between the private sector and the state as well as the growing consciousness about the risks involved in this new relation.

5   The grey zones of conflict of interest

Bearing in mind the rules of the European Conventions and the UNCAC on revolving doors and lobbying, we would like to question the continuum of conflicts of interest. One part of this continuum is well-known: corruption as a criminal phenomenon. In this part we focus on the legal mechanisms which are considered as a risk for harm.

5.1   Networking

A first phenomenon that is completely legal but that can hide conflicts of interest is networking. Networking can be defined as *initiating mutually advantageous new relationships with people or meeting people who can be of help to you while you are a help to them.* The aim of networking is to interact with individuals who have similar interests

---

3   The new member states of the European Union complain about the policy of double standards in the EU. The old member states are doing almost nothing to implement the anti-corruption regulation, while the new member states are subjected to a strict anti-corruption programme. This double standard policy gives the impression that the new member states are the most corrupt while the older ones are clean (Kos, 2009).

in an effort to build a relationship that will yield current and future benefits. At the private as well as at the public level, *networking* is today the key word for success in the job. Through networking a close relation between the state and private sector can be established, just as corruption may facilitate the work of criminal organisations. It could therefore be considered as a risk factor for social injurious conflicts of interest.

Networking can take different shapes. We present two different examples: clubbing in the society club and the professional network of interest. In an article on the murdered Belgian politician André Cools and the Agusta-affair, Marc Cools shows that society clubbing and the professional or institutionalised kind of networking are interrelated (Cools, 2009).

The first kind of networking is clubbing or being a member of a society club. The Belgian journalist Jan Puype has done research in what he called "clubland". During two years of field study, he tried to find out where journalists, politicians, captains of industry and top executives meet each other. He spent time in several Belgian clubs, including the A12 Business Club, Nieuw Economisch Appel, the European Round Table, De Warande, Cercle Gaulois and de Club van Lotharingen-Cercle de Lorraine. Even if these clubs are not the place for signing contracts, they are the first step in gathering information about the political and economic conditions, about future regulation. They are also an (ideal) opportunity to find business partners. There is nothing illegal in being a member of these and other similar clubs. Members pay a substantial enrolment fee (up to 170 EURO / year) but there is not really a problem of corruption in a criminal sense. In the early days of organisational criminology, Sutherland already mentioned that an organisation is able to commit white-collar crime without being perceived as a criminal or without being detected or prosecuted because of the social network of white-collar people. The social network was, according to Sutherland, related to the *cultural homogeneity* of people working for the government and business people, both being in the upper strata of the American Society, family and friendship relations and the mechanism of "the revolving door". M*any persons in government were previously connected with business firms as executives, attorneys, directors, or in other capacities, ...*(Sutherland, 1961, 248) Thus the initial cultural homogeneity, the close personal relationships, and the power relationships protect the businessman against critical definitions by the government.

Based on these observations, Puype was able to point out the obvious relationship between these clubs and the media. The membership list contains many journalists and one club is even using the infrastructure of a Belgian journal (Puype, 2005) This also reminds us of Sutherland's analysis, in which he argued that: *"the important newspapers, the motion picture corporations, and the radio corporations are all large corporations, and the persons who own and manage them have the same standards as the persons who manage other corporations. These agencies derive their principal income from advertisements by other business corporations and would be likely to lose a considerable part of this income if they were critical of business practices in general or those of particular corporations"*(Sutherland, 1961, 246). Society clubs as a kind of networking could be one of the examples of activities that make corruption redundant. The members of "clubland" belong to an elite of entrepreneurs who have alternative opportunities to influence law-making and regulation. It is a refined mechanism of conflicts of interest. For example, the engineer belonging to a club may not be privileged when submitting a public contract tender but the civil servant can give tips for being successful.

A higher level of networking is the institutionalised kind of networking: the professional network of interest. This is what happens, for example, in the pharmaceutical industry. It is a mistake to think/to believe that only the pharmaceutical companies are responsible for overmedication of the Belgian population, for undermedication of Third-World countries and for life-threatening side-effects of medical products. In the research project concerning conflict resolution in the case of injurious activities of the pharmaceutical industry (Vande Walle G., 2005), we found out that the governmental level had more responsibility than expected and that the government's role in the activities of the pharmaceutical industry is underestimated. The pharmaceutical industry is fundamentally based on a network with governmental institutions, doctors, medical researchers, and other parties involved. We have called this kind of institutionalised network 'a pharmaceutical complex', referring to the complexity of the network of all partners in the health sector. In this complex, different state institutions played an indispensable role for the profit motive of the pharmaceutical industry. These activities supporting the state range from inviting companies to coregulation to too readily approving the commercialisation of new medicines, tolerating hidden publicity for prescription drugs, etc. This kind of networks is more than initiated new relations between people for profit. These networks are deeply institutionalised in the system (Vande Walle, 2005). This institutionalised kind of networking is something Tillman and Indergaard referred to in their analysis of the Enron case: "the firm's leaders were able to build an elaborate network of interlocking connections to politicians, regulators, bankers, accountants, public relations experts and media heads". This is an example of a criminogenic institutionalised framework in which numerous institutions and organisational actors are considered as relevant in the construction of crime-facilitative environments (Tillman, 2009, 76).

In the white-collar crime study of Sutherland there is no question of corruption and we are not saying that every network in the pharmaceutical industry is a step to corruption. However, it cannot be denied that legal networks of business representatives, politicians and civil servants facilitate relations that are in the interest of one party and not in the interest of the public at large, even if these networks are legal. The power of the companies is the result of a more or less structured network of collaboration and conviviality that makes corruption redundant. Up till now there is little information about the relationship between networks and nepotism at the high political level. It would thus be an interesting research project to trace the impact of networks on the decision making process.

5.2   Conflicts of interest – Lobbying

A second risk of conflict of interest that is completely legal is lobbying. Blokland defines lobbying as *"promoting the interests by exerting pressure on decision makers"* (Blokland, 2006). A more neutral definition is: *giving information with the aim to have some impact on the decision making process* (Pauw, 1998, 130). Lobbying may take the form of campaign contributions or influence-buying through other means, as an activity that aims at creating rules or policies or changing those rules or policies. That lobbying is indirectly a lucrative activity is illustrated by the concentration of lobbyists in Brussels trying to influence the European politics. The exact number of lobbyists is not known but is estimated at approximately 15.000. While their employers vary from public authorities from cities, regions or countries, to NGOs, professional federations and companies, about two thirds of these lobbyists are working in the private sector (Schoofs, 2008).

In conversations, the comparison between corruption and lobbying often creates feelings of uneasiness because corruption is an illegal activity while lobbying is a widely accepted institutionalised mechanism of influencing power. The theoretical difference between lobbying and corruption is the question of "who is the winner of the game". While corruption is committed for personal gain and not in the public interest, lobbying is in the interest of a larger group of people, a sector or the environment. In practice, however, this distinction is not always that clear-cut. A first problematic issue of lobbying in practice is the importance of financial power. A lobby group with economic power has more chances to reach their aims than a low-budget and disorganised lobby group.

A second problem is the lack of transparency. Interest groups are considered as an advantage: it is a way of transferring expertise to the civil servants who have to regulate. But the lack of transparency gives the impression that the results of lobbying could be against the interests of the public at large. More transparency and a strict regulation of the practice of lobbying would guarantee more that lobbying happens in the interest of the general public and not in the interest of some people or one organisation (Dobovsek, 2009). So far, all attempts to create a more transparent system at the European level have failed. First, an obligatory registration system was proposed. However, this system has never actually been put into practice. Instead, a system of voluntary registration was implemented, which invites lobbyists to be open about the goals of their lobbying, the structure of the organisation, their network and an estimation of the budget assigned each year to representing interests.[4] On the 23th of June, the day that the voluntary register was launched, the *Financial Times* published a critique of different interest groups (Bounds and Tait, *Financial Times*, June 23th 2008). NGOs complain about the voluntary character of the regulation which is comparable to the regulation of dead letter. They predict that the new initiative will not offer information about the number of lobbyists nor about the customers they work for. On the other hand, corporate lobby associations protested about the official registration of lobbyists and fees, calling the initiative "an unnecessary burden", "discriminatory and unworkable", "voyeurism". They further argued that "money does not equate with influence and that financial disclosure therefore is irrelevant" (Corporate Europe Observatory, 2008). This debate about European transparency, which apparently also activated many lobbyists, has proved to be important for the interests of private com-

---

4 In constrast to the transparent system of the US, Europe has a closed policy of lobbying. In March 2005 Siim Kallas, European Commissioner, launched the European Transparency initiative to strengthen public trust in the EU through increased openness and accessibility. One of the highlights of this project was the transparency of interest representatives seeking to influence EU decision making and upholding minimum standards of consultation. The 3rd of May 2006, a Green paper was published which suggested registering the names of lobbyists, the name of their customer and the amount of money the customers paid, but on a voluntary basis. The 8th of May 2008, the European Parliament voted in favour of a mandatory, prescriptive system for the registration of lobbyists. The 23rd of June, the European Commission finally accepted a voluntary system of registration. Registrants also agree to abide by the Code of Conduct that flanks the registration procedure or declare that they have already signed a professional code with comparable rules.

For further information on the registration system, see: https://webgate.ec.europa.eu/transparency/regrin/consultation/listlobbyists/welcome.do).

panies. During a special session on lobbying, the Expert Group on Conflicts of Interest considered "vocal vested interests" over the "wishes of the whole community" in public decision making a major threat to public trust.

> *"While lobbying is often explicitly recognised as legitimate and essential, given the complexity of modern government decision making, assertions are often made that it too frequently borders on influence peddling. The conclusion is that though it may not be illegal, it is damaging to the integrity of democratic institutions".*[5]

5.3   The revolving doors

A third way of influencing political decision making is the conflict of interest caused by the mechanism of "revolving doors" or post-employment conflicts of interest. This same mechanism is considered in the UNCAC as a threat to corruption (art.7, 1b and art. 12, 2e)[6] In short, the term refers to the rotation of public officials to other positions related to the previous job. The mechanism of revolving doors is a phenomenon that is well known at the national level. As an example of non-administrative political corruption, Van Duyne refers to conflicts of interest between an entrepreneur and a politician. This conflict occurs when the entrepreneur begins a political career while the company continues and the business relations remain (Van Duyne, 1995, 124-125). The same happens when a politician picks up certain positions in companies after his political career. During my visits to pharmaceutical companies it happened a few times that a civil servant who worked for a public health service turned up later on as representative of that company (Vande Walle, 2005).

In the lobbying world at the European level, it happens on a regular basis that European commissioners and civil servants who worked in the domain of law making change their job for a lucrative job as European lobbyist, mostly to protect the interests of the audience they worked for at the government level. For the clients they represent, they are of special interest because of their know-how concerning European bureaucracy and their privileged access to decision makers *(Friends of the Earth press release – January 25th 2008).* The *Financial Times* reported in 2006 that about 100 civil servants went on sabbatical leave as a result of picking up a lucrative job as lobbyist. Friends of the Earth and Alter-EU warn for the revolving doors and ask for decisive regulation concerning the fast-moving revolving doors between the EU-Commission

---

5   OECD Ministerial meeting on Strengthening Trust in Government: What Role for Government in the 21st Century?" Electronic.copy: http://www.oecd.org/dataoecd/18/16/38944191.pdf)

6   Art. 12, 2 UNCAC: preventing conflicts of interest by imposing restrictions as appropriate and for a reasonable period of time on the professional activities of former public officials or on the employment of public officials by the private sector after their resignation or retirement where such activities or employment relate directly to the function held or supervised by these public officials during their tenure.

and lobbyists.[7],[8]. Equally problematic is the situation in which lobbyists with an expertise in a specific domain apply for a position in the European Commission. They may favour the company or sector the commissioner or civil servant used to worked for.

Similar to the other concepts of networking and lobbying, the mechanism of revolving doors in itself is not a criminal offence but the risk of being injurious without being noticed grows.

5.4   Are corruption and lobbying complementary?

A question that results from the previous reflection is the relation between the legal and illegal mechanisms on the continuum, such as lobbying on the one side and corruption on the other. Based on their study of corruption in the Balcan regions, Dobovsek and Kecanovic think that the relation between these two phenomena is a matter of losing trust. They argue that governance must be predictable, rational and transparent. When citizens become aware of the abuse of power and the conflicts of interests between the economic and political elite, they lose their faith in their government and public institutions. This loss of trust makes the step to corruption and fraud for citizens less difficult. Predictability, rationality and transparency are therefore the central reasons for regulation of lobbying. Transparency or openness about lobbying and conflicts of interest aims to prevent corruption (Kecanovic: in Dobovsek, 2009, 47)

Other researchers have formulated hypotheses about the complementary relation of lobbying and corruption (Campos and Giovannoni, 2007, 2). Campos and Giovanni elaborated "a conceptual framework that highlights how political institutions are instrumental in defining the choice between bribing and lobbying". They tested their predictions using survey data for about 6000 firms in 26 countries. The results suggest that (a) lobbying and corruption are fundamentally different, (b) political institutions play a major role in explaining whether firms choose bribing or lobbying, (c) lobbying is more effective than corruption as an instrument of political influence, and (d) lobbying is more powerful than corruption as an explanatory factor for enterprise growth, even in poorer, often perceived as highly corrupt, less developed countries Campos and Giovannoni, 2007) Harstad and Svensson think that corruption and lobbying complement and substitute each other, while Damania believes that both activities are only complementary (Campos and Giovannoni, 2007, 2).

Lobbying and corruption are certainly different from each other; firstly, lobbying does not require a kind of bribery because the expertise and information of the lobbyist could be more powerful as pressure tool. Secondly, lobbying is more directed towards the politician while corruption mostly involves bribery of public officers. Thirdly, lobbying is legal while corruption is illegal. This last difference may not be accepted as the main argument because powerful lobbyists can influence corruption policy, which

---

7   Friends of the Earth, Transparency in EU-decision making: reality or myth?, 13 (http://www.foeeurope.
     org/publications/2006/Transparency_in_EU_decision_making_May2006.pdf (17/01/2010)).
8   Alter-EU or the Alliance for Lobbying Transparency and Ethics Regulation is an alliance of about 160
     civil society groups, trade unions, academics and public affairs firms "concerned with the increasing
     influence exerted by corporate lobbyists on the political agenda in Europe, the resulting loss in democ-
     racy in decision-making in Europe: X, Critique of flawed study on regulating EU conflicts of interest.
     Alter-EU steering committee, February 2008.

has been argued by many other criminologists among others Sutherand, Quinney, Kramer and Michalowski.

The conventional wisdom is that lobbying is preferred in rich countries to exert political influence, while corruption is the preferred means in poor countries (Campos and Giovannoni, 2006). Campos and Giovannoni tried to discover why exactly companies choose for corruption or for lobbying. Based on an econometric exercise on data of 4000 firms spread over 25 countries, the authors concluded that: "lobbying and corruption are substitutes; firm size, age and ownership as well as political stability are important determinants of lobby membership and lobbying is a much more effective instrument for political influence than corruption, even in less developed countries" (Campos and Giovannoni, 2007). Similarly, Svensson and Harstad have suggested that the level of economic development could be a factor of explanation. More specifically, firms bribe if the level of development is low, but they switch to lobbying if the level of development is sufficiently high. These authors also consider lobbying and corruption as substitutes: the development of an economy can shift the means to change rules from corruption into lobbying. However, they warn for the corruption trap. Too large bribes discourage companies from investing in lobbying to "bend the rules" (Svensson and Harstad, 2005). While the authors tried to find out how corruption and lobbying interact and what the conditions are for an anti-corruption mentality, they have failed in the problematisation of state capture. In searching beyond a legalistic definition of state-corporate crime to the socially injurious activities of state-corporate collaboration, we meet a continuum of activities that are problematic. The authors mentioned above set out the first important steps to an overview and contextualisation. For the protection of democracy, it will be important to further explore the relationship between legal and illegal conflicts of interest.

6   The hidden risk of "conflicts of interest": regulatory capture and indirect victimisation

One of the features of corruption is that it happens unnoticed. Even if corruption is detected, it is a delicate and difficult job to find evidence. There are several reasons for this, including the relation of trust, the bribe that disappears in a construction of money laundering and the protected position of the corrupter. Even more difficult to detect are the legal mechanisms of conflict of interest. While the relationship between different means of conflict of interest must be further explored, the common risk of all means is obvious. Any kind of state-corporate collaboration is a step to regulatory capture or the capture of the regulator by the regulated. Richard Posner, economist and lawyer of the University of Chicago, was the first to speak about regulatory capture. "Regulation", he said, "is not about the public interest at all, but is a process, by which interest groups seek to promote their private interest. Over time, regulatory agencies come to be dominated by the industries regulated." Most other economists have a less extreme view, arguing that regulation is often beneficial but is always at risk of being captured by the regulated firms.

Judith Richter warns for this dangerous development. While regulatory capture was regarded as a major problem in the past, she says, the climate changed with the evolution to greater deregulation: "Today as policy *making has largely shifted from public mandatory regulation towards coregulation with or self regulation by industry; discussions about "due process" and "conflict of interest" and about whether, when and how public authorities should involve corporations in public regulatory processes are less common."(Judith Richter, 2001, 132)*

Networking and lobbying could be considered as signs of democracy because the public can participate in the decision making process through these mechanisms. However, the very same mechanisms risk becoming means of interest protection for the most powerful in society, usually at the expense of the powerless. Regulatory capture by means of lobbying or networking could be more dangerous than corruption because of three reasons: it happens without transparency; it is considered as legitimate; and it may indirectly hurt a large population.

7 Conclusion

It is known that corruption is a destructive criminal phenomenon, not in the least because it goes against fair competition, destroys democracy, sustains poverty in developing countries and deprives people of clean water, education and health care (Green and Ward, 2004, 18-19; Vogl, 2007, 173-174).Victimisation in the case of corruption is especially dramatic because of the unconscious character. Illegal waste dumping or the export of counterfeit medicines could cause thousands of victims who will never know that corruption is the source of their victimisation.

With the harmful character of corruption in mind and being aware of a changing political-economic relation, we have broadened our perspective to include conflicts of interest between private companies and public organisations. We have hypothesised that a too narrow focus on corruption has as a consequence that we go beyond the risks of some other harmful but legal mechanisms. In the last decades, the private sector has strengthened its position in the decision making processes; Several mechanisms, such as lobbying, revolving doors and co-regulation, have been introduced to exert pressure on the legal and bureaucratic decision making process. A legitimisation for this interference is the know-how and the stronger social engagement of the private sector to its stakeholders. This participatory idea has a serious downside as well: these mechanisms allow companies to support their own interests, sometimes against the well-being of the citizens and the environment. The central question for criminologists is: do we have to broaden our focus from corruption to the grey zone of *conflicts of interest for private gain?* An important step in this direction has been taken by Kramer, Michalowski and Kauzlarich, who have elaborated the concept of 'state-corporate crime'. The criterion for state-corporate crime is not the penal code but the socially injurious actions or for Green and Ward the violation of human rights.

The debate about the borders of organisational crime is as old as the study of this domain of criminology. In recent years this debate has faded into the background because it slows down the policy-oriented debate about enforcement and prevention. With this contribution on conflicts of interest of private companies we have tried to demonstrate that the debate on borders of crime is still important ... at least when someone cares about the harmful effects for people and planet.

8 Bibliography

Blokland H., Lobby in de VS ontspoort, *De Helling*, 2006, 3

Boekhout van Solinge T., Tropical deforestation and Greenhouse emissions as form of eco-crime, presentatie – NvK-congres Leiden, 19 and 20 June 2008

Boehm F., Regulatory capture revisited. Lessons from economics of corruption, *Working paper*, July 2007

Bounds A. and Tait N., Lobbying revamp under fire, *Financial Times*, 23 June 2008

Box S., Power, crime and mystification, London, Tavistock, 1983

Campos N. and Giovannoni F., Lobbying, corruption and political influence, *Public Choice*, 2007, 131, 1-21

Campos N. and Giovannoni F., Lobbying, corruption and political influence, September 2006, *IZA Discussion Paper* 2313

Cools M., André Cools en Agusta. Een Belgische affaire, *Justitiële Verkenningen*, 2009, no.3, 59 – 69

Croall H., Understanding white-collar crime, Buckingham, Open University Press, 2001

Dobovsek B., *Prevention of corruption* – Phd, Podgorica, Uprava za kadrove, 2009

Fubara Anga L., Case study: Anti-corruption legislation in Nigeria, presentation during the Ethical corporation conferences: the future of anti-corruption law and enforcement in Europe, Brussels, 27 and 28 May 2009

Friends of the Earth, Transparency in EU-decision making: reality or myth?, 13 (www.foeeurope.org/publications/2006/Transparency_in_EU_decision_making_May2006.pdf (17/01/2010))

Green P. and Ward T., State crime. *Governments, violence and corruption*, London, Pluto Press, 2004

Heere W., *From government to governance. The growing impact of non-state actors on the International and European legal System*, Cambridge University press, 2004

Hindess B. and de Sousa L., Workshop – The International anti-corruption movement, European Consortium for Political Research, General Conference, Nicosia, 2006

Hodgson G. and Jiang S., The economics of corruption and the corruption of economics: an institutionalist perspective, *Journal of Economic Issues*, 2007, 4, 1043-1061

Kamminga M., *Mensenrechtenschendingen door multinationale ondernemingen: wat valt er juridisch te doen?*, in: Brems E. en vanden Heede P. (eds), Bedrijven en mensenrechten. Verantwoordelijkheid en aansprakelijkheid, Antwerpen, MAKLU, 2003

Kecanovic, in Dobovsek (ed), the prevention of corruption – PhD, Podgorica, Uprava za kadrove, 2009

Kos D., presentation during the conference anti-Corruption in Europe, Ethical Corporation, Brussels, 19 and 20 March 2009

Kramer R., Michalowski R. and Kauzlarich D., The Origins and Development of the Concept and Theory of State-Corporate Crime, *Crime and Delinquency, April 2002,* 263-282

Lambsdorff J., *the institutional economics of corruption and reform: theory, evidence and policy,* Cambridge, Cambridge University Press, 2007

Larmour P. and Grabosky P., *Public sector corruption and its control,* Trends and issues in Crime and Criminal Justice, 2000, 143, Australian Institute of Criminology

Meehan E., From Government to Governance: New Opportunities for the Better Representation of Women, *ESRC Seminar run by Centre for the Advancement of Women in Politics (QUB),* 1 January 2003

Moody-Stuart, the costs of grand corruption (Column), *Economic Reform today,* 1996, 4

Pauw B., et al., *Lobbyen* – Dossier, Samson, 1998

Pipes D., *Conspiracy,* New York, Free Press, 1997

Puype J., *'De Elite van België, Welkom in de Club',* Leuven, Van Halewyck, 2005

Richter J., *Holding corporations accountable. Corporate conduct, international codes and citizen action,* London, ZED Books, 2001

Rose-Ackerman S., Corruption and Government. *Causes, consequences and reform,* Cambridge, Cambridge University Press, 1999

Schoofs N., De mist rond de Europese lobbyist, *Vacature,* Saturday 17 May 2008

Schwendinger H. and Schwendinger J., The continuing debate on the legalistic approach to the definition of crime, *issues in Criminology,* 1972, Vol.7, 71-81

Svensson J. and Harstad B., *Bribe or lobby? (It's a matter of development),* First draft – work in progress, June 2005.

Sumner C., *The sociology of deviance. An obituary,* Buckingham, Open University Press, 1994

Sutherland E., *White collar crime,* New York, Holt, Rinehart and Winston, 1961

Tillman R., Making the rules and breaking the rules: the political origins of corporate corruption in the new economy, *Crime, Law and Social Change,* 2009, 51, 73-86

van den Heuvel G., *Collusie tussen overheid en bedrijf. Een vergeten hoofdstuk uit de organisatiecriminologie,* Inaugurale rede, Maastricht, 23 January 1998

Vande Walle G., *Conflictafhandeling of risicomanagement. Een studie van conflicten tussen slachtoffers en ondernemingen in de farmaceutische sector,* Brussels, VUB-Press, 2005

van Duyne P., Will Caligula go transparent? Corruption in acts and attitudes, *Forum on Crime and Society,* 2001, vol.1, nr.2, 73-98 van Duyne P., *Machtsmisbruik in politiek en openbaar bestuur,* in: Van den Heuvel J. (ed), Ethiek in politiek en openbaar bestuur, Utrecht, Lemma, 1995

Vogl F., Global corruption: applying experience and research to meet a mounting crisis, *Business and Society Review,* 2007, 112:2, 171-190

Williams, J., Beare, M. (1999), "The business of bribery: globalization, economic liberalization, and the 'problem' of corruption", *Crime, Law, and Social Change,* Vol. 32,115-46.

Friends of the Earth press release "Bring lobbying out of the harbours" – 25 January 2008

OECD Ministerial meeting on *Strengthening Trust in Government: What Role for Government in the 21st Century?"* Electronic copy: *http://www.oecd.org/dataoecd/18/16/38944191.pdf*

X, Critique of flawed study on regulating EU conflicts of interest. Alter-EU steering committee, February 2008

http://ec.europa.eu/commission_barroso/kallas/transparency_en.htm

https://webgate.ec.europa.eu/transparency/regrin/consultation/listlobbyists/welcome.do).

Corruption as a judgement label

Arne Dormaels

1   Introduction

The relationship between people and the word 'corruption' is diverse and can be divided into the following three levels. On the first level the word 'corruption' is considered as a term which needs to be defined. This defining activity is the result of a conscious reflecting person. On the second level, the word 'corruption' is used as a label to demarcate a variety of situations within the ontological reality. It is our assumption that we are confronted on this level with a reflecting activity and that people use either an explicit or an implicit definition of the word corruption. On the third level the word 'corruption' is used to judge a concrete particular situation or a description of a concrete particular situation as corrupt or not. We assume that a judgement of such a situation might rather be a 'reflexive' (emotional) or a 'reflective' (rational) activity.

Within each of the three levels described above we observe an inter-personal or an inter-group variability. On the first – reflective – level we observe a manifest inter-personal variability. A good example to illustrate this inter-personal variability might be the academic definition debate on corruption. In the past, many researchers have tried to define the word corruption, which resulted in many (overlapping) definitions (Heidenheimer 1970; Gibbons 1990; Johnston 1991; Kurer 2005). In this context it is useful to cite van Duyne (2001), who argued that the definition of corruption is used in many overlapping contexts from a legal, economic, political or cultural comparative perspective. Based on the fundamental element of corruption – the behaviour of the individual deviant decision maker – van Duyne comes to the following definition of corruption: *"corruption is an improbity or decay in the decision-making process in which a decision-maker (in a private corporation or in a public service) consents or demands to deviate from the criterion, which should rule his decision making, in exchange for a reward, the promise or expectation of it"* (van Duyne 2001). Other researchers have argued, however, that a single universal definition which will fuse together the various theoretical understandings is hard to expect and that there exists little consensus about its definition (Werner 1983). Gardiner (1970) illustrates, for example, the range of definitions by grouping seven "broad" (e.g. *the exercise of governmental power to achieve nongovernmental objectives* (Scott, J. C. 1972, as cited in Gardiner and Lyman 1978) and "narrow" (e.g. *behaviour which deviates from the formal duties of a public officer for private wealth* (Nye, J. S. 1967, as cited in Gardiner and Lyman 1978)) categories (Gardiner and Lyman 1978). We observe that even the classification of different definitions of corruption into types is very numerous. Johnston makes a distinction between '*Behaviour-Classifying Definitions*' and '*Principal-Agent-Client Definitions*' of corruption (Johnston 1996). Peters and Welch (1978) introduced the following types of definitions: legal definitions, public interest definitions and public opinion definitions.

Inter-personal or inter-group variability on the second – reflective – level would be, for example, a situation in which two people use the word corruption in the same sense (with the same definition) but end up with a differentiated demarcation within the ontological reality. Take, for instance, a discussion on the incrimination of corruption in Belgian criminal law. Corruption is defined in article 246.§1 of the Belgian Criminal Code as follows: "*The action of a person performing public duties who requests or accepts, directly or through the intermediation of other persons, for himself or others, offers, promises or advantages of any kind, in exchange for committing the conduct referred to in Article 247 constitutes passive corruption*". One person or group might argue that it is necessary to read the whole of the expression used in §1, in which the notion of acceptance refers not only to an offer or promise, but also to an advantage, which implies the receipt of such an advantage. Another person or group might say that since acceptance of an offer is covered, then by extension the subsequent receiving of the advantage agreed must also be covered (GRECO 2009). This discussion of how to define the word corruption is by extension unmistakably related to the following question. Is it necessary that prior to a particular act or attitude on the part of the person performing public duties, he receives a request for an offer, promise or advantage of any nature (Flore 1999)? Some people will argue that in the case of no such offer being made to him there would consequently be no question of the offence of corruption. Others might defend the opposite interpretation and end up with a different demarcation.

Inter-personal or inter-group variability on the third – reflective and/or reflexive – level of judgment stands for any case in which a particular situation is judged as being corrupt by one person or one group and not by another person or group. An excellent case to illustrate the inter-personal variability on this level is a case in which a cabinet counsellor in the city of Antwerp, one of the major cities in Flanders, is responsible for the follow up of construction files for the city. Apparently, this counsellor rents for free a luxurious loft from a building contractor doing major city programmes. One day, a journalist raises the question of possible corruption in a newspaper article followed by many more news items covering this case (Van Der Aa 2009). It is revealing that at the outset local politicians and fellow cabinet counsellors did not label this case as corrupt. The media and the common public shared on the other hand a united opinion: this is obviously a corruption scandal. We found another illustration of inter-group variability in the work of Kjellberg, who discusses the contrast between publicly-perceived scandal and actual corruption within the municipality administrations of Oslo (Kjellberg 1995). A striking point in his analysis is the gap between the verdict in the media and public opinion about the alleged corruption scandals on the one hand, and the outcomes of three City Commissions exculpating the administration from the allegations on the other.

My interest goes in the present article to the third level, the level on which people judge a situation as being corrupt or not. We assume – on the basis of previous studies – that the judgement of such situation is influenced by two groups of characteristics (see for a discussion of these studies section 2 of present article). A first group contains situational characteristics (such as public officials, favours, payoffs, undue advantage, etc.). A second group consists of the personal characteristics of the observer (such as gender, social status, education, etc.). We assume that these characteristics influence, either separately or in combination, the judgement of a situation as corrupt or not.

In this article, we will describe a method – a so-called scenario-based questionnaire – to assess to what extent the aforementioned two groups of characteristics, either themselves or in combination, influence the judgement of a situation as corrupt or not. The elaboration of our scenario-based questionnaire, which is the core subject of this article, will be discussed in more detail in section three.

In the past, a handful of scholars conducted studies developing a methodology to study the influence of characteristics on the judgement process of a situation as corrupt or not. These methods – assumptions, procedures, results and understandings – form the basis of my methodology. These studies will be presented together with their methodologies in the first introductory section of this article. Subsequent sections will address the elaborations of our scenario-based questionnaire.

2   Understanding judgement of a situation as corrupt

2.1   Historical overview of relevant research

As mentioned above, researchers have constructed methods to observe in a more meticulous way the influence of particular characteristics on the judgement of a situation as corrupt or not.[1] The subdivision between these two groups – situational and personal characteristics – forms the basis for structuring this chapter in which we will present previous research on judgments of situations as corrupt. Within each group we will address the researches in a chronological way. Some of the studies we will introduce in our article focused on the situational characteristic, others addressed the personal related characteristics.

2.1.1   Situational characteristics

Our starting point is the work of Gardiner (Gardiner 1970a; Gardiner 1970b) who introduced a behavioural and normative perspective in the context of analysing the judgement of a situation as corrupt or not. An important methodological improvement is that Gardiner introduced the application of scenarios describing situations. The respondents had to judge in this way whether or not, and to what extent, a scenario was corrupt. As we will discuss further on, his constructivist method has since been evolved (Bezes and Lascoumes 2005).

At the same time, Heidenheimer opened new horizons to study the judgement of situations by appreciating the existence of different normative dimensions in this

---

1   We should mention that 'judgement of a situation as corrupt or not' is addressed in most of the above discussed researches as the 'perception of corruption'. For that reason we must make clear the difference between two ways in which the phrase 'perception of corruption' is generally used. The first one relates to contexts in which the perception of corruption – or in other words the judgment of situations as corrupt – is analysed. These studies try to isolate values, norms and characteristics that might influence a judgement. The latter is what should be understood within the framework of our present study.

'Perception of corruption' might refer however to another meaning: measuring the scale of corruption. This is measuring the scale of corruption by perception based indicators. In other words, not measuring corruption itself but opinions about its prevalence (Treisman 2007). The underlying idea is that these perception-based-indices can 'predict' the 'real' levels of corruption.

process (Heidenheimer 1970). His approach resulted in the familiar black-white-grey classification of judgements of situation as corrupt. A political action that is not only condemned by public opinion but also demands punishment is evaluated as 'black corruption'. Acts judged corrupt by both politicians and citizens, but not seen as severe enough to be punished is evaluated as 'white corruption'. 'Grey corruption' indicates that either public officials or citizens expect to see an act punished, while the other group does not. Rationalizing the judging process of situations as corrupt proceeding from this continuum broadens the scope of the study subject. Several studies analysing different indexes of 'situations judged as corrupt' validate the presumption that the judgment of a situation as corrupt is a multi-dimensional process with different evaluative criteria intersecting to produce multiple perspectives on these situations (Gibbons 1990; Treisman 2000; Anderson and Tverdova 2003).

In 1978 Peters and Welch conducted a scenario-based questionnaire study amongst 978 senators in 24 US states to explore Heidenheimer's theory (Peters and Welch 1978). Their questionnaire was built out of a set of 10 scenarios. Each scenario had to be rated on a five-point scale according to whether the senators believed the scenario was corrupt, whether they believed most members of public would condemn this act and whether they believed most public officials would condemn it. Innovatively, they varied the scenarios' significant characteristics, the so-called 'salient characteristics'. They successfully unravelled the judging process of corruption into four significant dimensions. The 'public official' involved, the actual 'favour' provided by the public official, the 'payoff' gained by the public official and the 'donor' of the payoff. Peters and Welch concluded in their study that judgments of corrupt behaviour do not only vary between different social groups but also vary in relation to the nature of the acts. The methodological framework elaborated by Peters and Welch presented new perspectives in the empirical study of corruption. This methodology allows us to observe in a very meticulous way the influence of significant characteristics of different acts and to compare the judgment of scenarios as corrupt between different social groups.

Gorta and Forell explored the public judgement of situations by determining what types of activities are labelled corrupt by random samples of 1,978 NSW public sector employees (Gorta and Forell 1994). A central question in their study was how people decide whether an activity is corrupt and how this relates to their willingness to take action about that conduct. The core of their methodology consists of brief descriptions of 12 scenarios, depicting different types of conduct which could potentially occur in any public sector organisation. For each scenario respondents were asked to rate, on a six-point scale, how *desirable* they believed the behaviour to be, how *harmful,* and how *justified* they considered it to be. They were also asked to judge whether the conduct was corrupt or not.

Twenty years after the publishing of Peters and Welch's article, Jackson and Smith revealed empirical evidence that supports the hypothesis that judgements of scenarios as corrupt are affected by four particular dimensions (Jackson and Smith 1996). They used empirical evidence from 100 interviews with Australian politicians and 500 voters using the scenarios elaborated by Peters and Welch. The politicians' and publics' judgments of corruption are affected by the broad dimensions of corruption – the natures of the official, payoff, favour and donor – alike. Activities embodying a greater number and strength of salient characteristics were condemned by both the elites and the citizens. Activities with fewer and weaker number of salient characteristics turned out to be judged more ambiguously, in which the politicians were the more lenient group.

Mancuso illustrated the enduring utility of Peters and Welch's dimensional scheme analysing data from a large-scale Canadian national survey on political ethics (Mancuso 2005). The core of his questionnaire was a series of hypothetical scenarios describing political behaviour. In his study Mancuso examined the effects of changing contexts – public, private or something in between – of an activity. Placing the judgment of a scenario along the spectrum from public to private proved to be useful in understanding how corruption is perceived. Scenarios are more likely to be judged as corrupt when they involve illegal activities, larger payoffs and more direct and immediate benefit. The dimensional analysis defined by Peters and Welch holds true in the context of Mancuso's study in 2005. What's more, the identification of the context in which an activity takes place seemed to be significant in the prediction of whether or not it would be judged as being corrupt.

Redlawsk and McCann (Redlawsk and McCann 2005) presented a questionnaire to 6,829 voters. This questionnaire included a group of questions where voters had to rate the extent to which various hypothetical actions by government officials or citizens were judged corrupt. Examining popular judgments of what constitutes political corruption in the United States, they found two distinct judgment dimensions: corruption understood as lawbreaking, and corruption as favouritism. Secondly, it became clear that these judgments are heavily conditioned by the voter's socioeconomic background and are politically consequential. The latter should be seen as independent variables – personal characteristics – interfering in the judgement process as addressed in the following part.

2.1.2   Personal characteristics

Certain studies pointed out the importance of personal characteristics interacting and influencing the judgment of situations as corrupt. More concrete judgements of situations are probably influenced by independent variables such as demographic, socioeconomic characteristics, trust in political or governmental organizations and gender (Jackson and Smith 1996; Swamy, Knack et al. 2001; Park 2003; Redlawsk and McCann 2005; de Sousa 2008).

Based on a multiple correspondence analysis of data deriving from a nationally representative sample survey carried out in France in February 2006 on citizens' perceptions of professional politicians, public office holders and political corruption, Lascoumes and Tomescu-Hatto empirically illustrated citizens' ambiguities regarding the evaluation of corruption (Lascoumes and Tomescu-Hatto 2008). Personal information – sex, age and level of education – on each respondent has been introduced in the analysis. One of the outcomes was that a highly qualified professional turned out to be more tolerant of favoritism whereas less well-educated people and those in lower-status occupations are less tolerant of favoritism.

Analysts have long sought to understand whether women and men have different ethical orientations. Some researchers have argued that women and men consistently make fundamentally different ethical judgments, especially of corruption. When it comes to gender, for example, it seems to be the case that women, when individually surveyed, appear more public-spirited or less tolerant of corruption (Swamy, Knack et al. 2001). Others, however, have found no such disparities. Other studies have emphasized the interaction between age and gender (Pharr 1998; Aldrich and Kage 2003). Aldrich and Kage statistically investigated the interactive effects between gender and age in a nationally representative data set from Japan. The concluded that the interaction between gender and age functions better as a predictor of moralism than the independent variables of education or gender alone. Older individuals of both sexes were found to have similarly strict moral perceptions.

Furthermore there is sturdy empirical evidence that somebody's socioeconomic status conditions their judgment of a situation and whether they condemn behaviour as corrupt. The higher a person's social status, the more he will approach corruption from a legalistic point of view and vice versa (Jackson and Smith 1996; Redlawsk and McCann 2005).

2.2   Concluding remarks

Presenting the aforementioned studies on the judgement of situations as being corrupt or not, it becomes clear that this phenomenon has been mainly studied in the field of political science. From this point of view research on judging corrupt situations is limited to (disjunctions between) democracy and 'political' corruption and consequently restricted to the relationship between elites and voters. This relationship should be understood as the aforementioned inter-personal or inter-group variability in the judgement of situations as corrupt. Although corruption has gained more and more attention in the last decades, it is surprising to see that this subject has not yet entered into the broader field of socio-criminological research addressing the judgement of day-to-day scenarios possibly labelled as corrupt.

Secondly, Heidenheimer's hypothesis about categorizing the judgment of situations into black-grey-white corruption has been discussed. This discussion is, in our opinion, mainly focused on defining corruption and should be located at our first identified level in which the word corruption is used. Nonetheless we found several empirical studies which supported Heidenheimer's 'polychromatic' approach to the judgment process of situations as corrupt (Heywood 1997; Heidenheimer 2004). For that reason it is surprising that only a handful of studies analysed the judgement of situations from a multidimensional perspective involving both personal and situational characteristics. The same goes for the thesis that judgments of corruption vary between groups – the aforementioned intergroup variability – and over time, which is ignored by many scholars.

The empirical approach introduced by Peters and Welch has been evolved and scrutinized by other scholars and proved to be a reliable and sturdy method to investigate opinions about corruption in relation to the context of the social group to which the respondent belongs to. Nevertheless we should draw attention to our observation that researchers have struggled in the past to cope with a multidimensional approach to the judgement of situations as corrupt and are in some way implicitly bound to either the criminal law or to the widely used academic definitions. What really is judged by the public as being corrupt in the broadest sense is in this respect not captured.

3 A scenario-based questionnaire to understand judgements of corruption

The question is how to capture the above discussed methodological points of attention in our research design. Hereafter we will present an alternative approach to studying the judgement process of situations by evolving the Peters and Welch method. To do so we will elaborate a questionnaire constructed out of a set of scenarios. First, we will draw attention to the similarities and differences between our methodological concept and the methods used within the above discussed researches. Second, we will make clear how we selected meaningful situations to be judged as corrupt or not. Next, the introduction of different dimensions covering various characteristics which may possibly influence the judgement of a situation will be introduced. To conclude this chapter we will present the blueprint of our questionnaire.

3.1 Analogy with previous research methods

Our proposed method relies on a rich research tradition focused on understanding the judgement process of situations as corrupt or not. Similarly to those studies, we introduce the assumption that salient characteristics are involved in judging a situation. As explained above we make a clear distinction between situational and personal characteristics. Our project has the aim of contributing to the question of which dimensions and salient characteristics are involved in judging a situation as corrupt or not. To do so, we will work with a comparable research design. On the basis of a questionnaire we will organise a quantitative survey and statistical analysis of the results. In other words, we will organise an experimental design to examine individual judgement of a series of hypothetical scenarios.

We will, however, add some new elements into our research method and design. In line with recent findings we will involve both situational and personal characteristics in our present study. In this way we will try to combine different dimensions and their salient characteristics into the judgement process.

Secondly, we elaborated a method to vary more than one salient characteristic regarding a particular situation. Our questionnaire is constructed out of different scenarios describing in a neutral way a meaningful situation. We will successively bring in salient characteristics during the judgment of the scenarios as corrupt or not. In this way we will try to observe the shift of a respondent's opinion. Our thesis is that in this way we will be able to differentiate transgressive variables through which we expect to detect the bascules in the judgement process.

In our research we will give very broad interpretation of situation to be judged as corrupt or not. So far, studies dealing with the question of understanding the judgement process used a set of scenarios illustrating situations within a political context to be assessed as corruption. Additionally, some of these studies addressed the question of how these judgements affect the ballot box (Fackler and Lin 1995; Karahan, Coats et al. 2006; McCann and Redlawsk 2006; Birch 2008; Ferraz and Finan 2008; Lascoumes and Tomescu-Hatto 2008). Outlining the relationship between political corruption and democracy was the main focus of these studies (Gardiner 1967; Redlawsk and McCann 2005). As already mentioned above it is our intention to broaden the scope and to study the phenomenon corruption *as such*.

Last but not least we recall our observation that inter-personal or inter-group variability exists within the judgement of situations as corrupt. In line with this observation, we will include samples from different populations (police, municipalities, Flemish citizens) to help us to better understand the judgement situations.

3.2   Selection of the scenarios

We conducted an in-depth search for articles reporting corruption within the Flemish Digital Press Database '*Mediargus*' over the period January 1995 to March 2009. The Dutch language is rich in terms and phrases related to corruption. Although the use of different terms may vary depending on the context, the basic word referring to 'corruption' is understood in Dutch as '*corruptie*'. Nevertheless a vast range of alternative synonyms exist, such as '*omkoping*' understood as bribery, '*gesjoemel*' translated as fiddling, '*steekpenning*' to be understood as bribe or '*smeergeld*' understood as slush money. Based on a query with the aforementioned search terms we received over 15,000 articles reporting corruption in the broadest sense and covering different continents and regions. On the basis of this empirical information, we observed that a vast range of different situations is judged to be corrupt.[2] Moreover we observed that some cases are judged corrupt by only one individual person, while other situations are condemned by more than one person or a group of people (for example citizens, journalists, or profession groups such as the police, judicial service, or politicians). On the other hand, we observed that some situations have never been judged as corrupt (e.g. rapes, theft, assault and battery, etc.). This observation provided the opportunity to select meaningful situations which were rich on diversity in such a way that they included all kinds of stories which came to the attention of the general public and have been judged in some way as corruption by at least one person or a group of people.[3]

3.3   Dimensions covering different characteristics that influence judgement

Different dimensions involved in the judgement of a situation as corruption resulted from the analysis of the newspaper reports. These dimensions were on the whole congruous with the components of a potentially corrupt act and the salient characteristics introduced by Peters and Welch (Peters and Welch 1978). In what follows we will describe the four dimensions – public/private actor, undue favour, relationship corruption and bribery, and payoffs – in which we will bring in salient characteristics to observe shifts in the judgement of a situation.

---

2   We obtained the impression that 'smeergeld' is in essence associated with the context of political and governmental corruption. Private corruption or the concept of active corruption is often referred to by the word 'steekpenning'. Financial fraud and corruption in a financial and economic context is often expressed as 'gesjoemel' (see Bové 2009 for a more detailed discussion of the influence of newspapers on the perception of corruption).

3   Our first idea was to come to a selection of hypothetical scenarios based on a qualitative case study using court cases. However, court cases are often the result of a complaint or an investigation following alleged violations of the law. Consequently, we had to admit that the resulting scenarios would illustrate the criminal law definition and be consequently incongruous with our research objective to study situations as corrupt or not. In fact, we would be using our survey methodology as a method to validate in some way the criminal law definition. Or in other words the scope of our project would be situated at the second level in which the word 'corruption' is used – in a reflective way – to label a variety of corrupt situations.

3.3.1   Public versus private actor

The component 'public official' and 'the role that a public official performed' introduced in the study of Peters and Welch is transformed in the present study into the continuum 'public' versus 'private' actors.

A public actor has to be understood as any role relating to a public or a political position. A private actor means any private branch or personal position. Doing so, we bring in more differentiation into potentially corrupt relationships. This provides us with the opportunity to test whether a person in a private role, for example, is less severely judged than a public role when it comes to corrupt behaviour. In other words, is the general public more forgiving of private-to-private individual corruption as opposed to corruption involving public officials or politicians? We believe that the position of the party concerned will determine the dimension of the judgement to a large extent. Someone engaged in a public function will be judged more harshly than a private employer abusing his position for personal gain. The latter is often called 'just doing business'. Secondly, it seems interesting to investigate to what extent the position of the 'initiator' determines one's judgement. In other words, does it matters who requests or receives an undue advantage? Is a public servant who demands business-class tickets in exchange for influencing the allocation procedure more blameworthy than a corporate manager engaged in hanky-panky via gift giving during the bidding procedure?

3.3.2   Favour

The second dimension to be considered is the favour or the undue advantage intrinsically linked to the process of corruption. The favour and the perceived corruptness of the favour will vary strongly from one situation to another. Nevertheless we will try to cover the following two axes in our scenarios: the nature of the conduct that brings about the favour and secondly the beneficiary of it. To illustrate the vast nature of the conduct we refer to the following scenario that has been included in the questionnaire: '*Bart is an engineer and he is responsible for the production process in an automobile plant. A supplier of airbags wants to increase his deliverance of airbags*'. When it comes to judge a scenario as corrupt there are different possible situations to test. Firstly, does it matter if the favour rendered by Bart means a breach of his duties? This has been translated into the following question: '*Do you consider it corrupt if the supplier of airbags rewarded Bart with a holiday trip before Bart optimized the production?*' What about the situation in which the undue advantage is linked to a routine part of Bart's job? The question would then be '*Do you consider it corrupt if the supplier of airbags rewarded Bart with a holiday trip after Bart optimized the production?*' To what extent does it affect the general public opinion if Bart has to break rules to render the favour? This has been translated into the following question: '*Do you consider it corrupt if the supplier of airbags rewarded Bart with a holiday trip before Bart organized errors in the production*'.

3.3.3   Relationship corruptor and corrupted

Who or what is the beneficiary of the undue advantage and does it affect the judgement of potentially situation as corrupt? To test our assumptions we introduced different situations in which we alternate the characteristics of this dimension trying to influence the bascule in the verdict. The most obvious way to catch the dynamics of this dimension is to introduce a gradual transition in the characteristics of the beneficiary from the public interest to personal gain into particular scenarios. For that reason we give the scenario a purposeful twist as follows: '*A local politician visits the office of a fellow party member to discuss and bring about the settlement of a construction licence*'. Particular to this case is that the Belgian general public is very reluctant when it comes to doing politics. We do not want to run ahead of the research findings but we expect that playing with the salient characteristics of this dimension we might get a switch in the evaluation of the described situation. Does the judgement for example vary if the local politician is doing politics and tries to promote the building of a home of rest in his town? What if he goes about promoting the setup of factory grounds? These first two situations are strongly related to the common good and will probably be accepted by the most part of the respondents. Let us enter the grey zone via the following situation. How will the general public perceive the situation in which the politician goes to a fellow party member to bring about the construction of a golf course? This is less linked up to the common good and we expect therefore more diversification in the respondents' opinions. Moreover, this grey zone would be for example very informative if we could interpret the respondent's response in relation to his or her demographic, socio-economic characteristics, age and gender characteristics or in relation to the social group the respondent belongs to. Finally we refer to a situation in which the politician tries to influence the renovation of his ranch (farmhouse) out of self interest.

3.3.4 Larger/smaller and direct/indirect payoff

In our study we will also take into account research findings to date that suggest that situations are more likely to be judged as corrupt when they involve larger payoffs, more direct payoffs and more immediate payoffs (Peters and Welch 1978; Johnston 1991; Welch and Hibbing 1997). Does it mitigate judgements when less direct and less immediate gifts are involved? In that way we will make, for example, a distinction between:

- *Settling the invoice by paying x Euros...*
- *Depositing an amount x Euros into one's own account...*
- *Depositing an amount x Euros into the account of political party...*
- *Demanding a free laptop...*
- *Receiving a free laptop...*

Furthermore, we believe that the nature of the relationship between the 'giver' and the 'receiver' affects the judgment. When the connections between giving and getting are more clearly within the sphere of interest peddling we expect the judgments to be less tolerant. Does it involve, for example, a local builder or an international construction firm? Will opinions change when it involves just a business meeting in Brussels or a business trip to Mallorca?

3.4 Blueprint of the survey instrument

All the aforementioned dimensions have been processed to a guiding scheme to cata-logue situations as potentially corrupt acts based on the combination and interaction of different dimensions and their salient characteristics (see table 1). Although this scheme is a draft and is the result of an evolving scientific process we think it is useful to discuss the dimensions of hypothetical situations as corrupt acts.

On the basis of the following scenario we will illustrate the application of the scheme: *'Jean works as a local government councillor and is responsible for the follow up of construction files for the city. Jean rents a luxurious loft from a building contractor doing major city programmes'*. We expect that this scenario will be categorized in a certain sense as neutral, meaning that most people would not condemn the described act. Through the combination of some significant dimensions we will try to capture the bascule in the judgments of this situation from a tolerated to a condemned situation. For that purpose, we present successively four situations following the neutral scenario in or-der to shift the respondent's opinion. In this way we will try to differentiate transgres-sive variables through which we expect to detect the bascules.

- *How corrupt do you consider this situation?*
- *How corrupt would you consider it if Jean's sister rented the loft at a reduced tariff?*
- *How corrupt would you consider it if Jean rented the loft at a reduced tariff?*
- *How corrupt would you consider it if Jean rented the loft for free?*

We expect that the majority of the respondents will judge the first situation accept-able. It is plausible that the second situation would enter the grey zone and that re-spondents will state that they reject the third and fourth situation. On the basis of the scheme we disentangle the situations into the following characteristics:

Table 1. Characteristics potentially influencing judgement

	Position	Payoff	Favour/gift	Relationship
Situation 1	Private benefit from public act (?)	Ambiguous (+/-)	Unknown (-) Not direct (-)	Yes +
Situation 2	Private benefit from public act (++)	Third party (+/-)	Unknown (-) Not direct (-)	Yes +
Situation 3	Private benefit from public act (++)	Large (+)	Unknown (-) Not direct (-)	Yes +
Situation 4	Private benefit from public act (++)	Large (++)	Unknown (-) Not direct (-)	Yes +

+ More corrupt
– Less corrupt

The favour remains in all three situations unknown and probably has no significant influence affecting the bascule in the judgment. When it comes to the payoff we dis-tinguish a clear shift in the characteristics of this dimension. Starting from an am-biguous situation (just renting) it goes over to a more objectionable payoff (reduced rent for Jean's sister) up to the culpable payoff understanding as free rent. Important in this scenario is that even in the first situation it might very well be the case that respondents disapprove of the fact that Jean is renting the loft from a building con-tractor. This is because the ambiguous payoff interferes with a strong closeness, i.e.

'building contractor' doing city jobs. Testing the influence of the dimension 'closeness' through other scenarios with other characteristics will probably provide interesting information. Last but not least we have the position of the people concerned. In this case it concerns a public official who collects a big favour by misusing his function. This is a classic dimension in the judgement of situations as corrupt and we believe that this will be judged more harshly compared to for example a purely private business.

3.5   Format of the scenarios

The formulation of our scenarios is crucial for our research. Each scenario and the forced choices have to be drafted in such a way that they can be easily understood by the respondents. Secondly, the wording of our scenarios has to be simplified as much as possible in order to be able to use the survey to be read by the respondents themselves but also to use it in interviewing by telephone as we aim to reach different social groups.[4]

For that reason we tested our questionnaire on a focus group.[5] The organizational set-up of this focus group imitated a telephonic interview. Nobody received a paper version of the questionnaire and all scenarios have been read to the participants. Let us illustrate the major outcomes with the aforementioned question about Jean working for the local government. We learned that it seemed to be more proper to insert all the information that remained unaffected throughout this situation into the neutral scenario itself:

- Jean is a local government councillor
- Works on construction dossiers
- Rents a loft from a prominent building contractor.

Whereas the situations following the neutral scenario have to incorporate the changing characteristics and background information:

- His sister gets a reduced tariff
- Jean gets a reduced tariff
- Jean gets it for free.

Otherwise the respondents get confused in distinguishing the variable versus the invariable information and not knowing anymore what has changed compared to the previous situations. Secondly, it came to our attention that this way of presenting our scenarios demanded less effort and concentration, which is a benefit when it comes to a telephone interview.

We observed that it is of utmost importance that we refrained from inserting different meaningful variables at the same time. This would make it impossible to draw valid conclusions. We can make this clear on the basis of an illustration. Let us recon-

---

4   The elaborated questionnaire will be used within the framework of a six-year PhD study on corruption. The project consists of two distinct phases: (1) a quantitative empirical study involving different social groups; and (2) a qualitative study. A telephonic survey based on the CATI method with will be organized, together with postal surveys within different samples of the population (different profession groups) during this quantitative phase. Our questionnaire has to serve in other words for a telephonic as well as a postal survey.

5   This focus group took place on May 20, 2009. The group consisted of college students at Ghent University.

sider again the scenario in which a politician tries to influence the renovation of his ranch and therefore visits the cabinet of a fellow party member. Via the question '*Do you consider this corrupt?*' we try to find out if this way of 'doing politics' is denounced. If we for example would speak about a 'ranch located within a nature reserve' we run the risk that respondents would in the first place condemn the lack of respect for nature reserves instead of rejecting a possible act of political corruption.

3.6   Measuring the judgement

Last but not least we have to discuss the selection of our question format. Will we use forced-choice questions or apply rating-scale items? Previous studies have used a point scale asking their respondents to answer 'how corrupt' they perceive the behaviour to be. This leads, however, to problems when we want to dichotomize responses into 'corrupt' versus 'not corrupt' (Gorta and Forell 1995).

Seeing that we want to understand the bascules in the judgement of corruption this implies that we will have to force respondents to take a certain position: whether they condemn a hypothetical corrupt scenario or they have no strong feelings about it. In doing so it is less useful to let our respondents take a position on, for example, a five-point scale of whether they condemn the scenario or not. In this case directionality is all that is necessary and in that case the best approach is to use a simple two-point or four-point scale: do you consider it appropriate or not (and to what extent)? But there is much more to choosing a question format.

When it comes to increasing the response degree, experimental research has shown that respondents to forced choice questions produce higher endorsement than, for example, to the web check-all format (Smyth, Christian et al. 2008). More important is that experiments found that respondents spent significantly longer on the forced-choice formatted questions. This suggests that all respondents more deeply process the response options in this question format. In relation to the organization of a telephonic organized survey we must take into account that it will be almost impossible that respondents will readdress a question or demand additional information, unlike in in-depth interviews. For that reason we must increase their attention on processing the questions.

3.7   Surveys and samples used

Surveys are an important means for assessing corruption and surveys always go together with samples. According to Langseth, a common error in the study of political corruption is to over-sample urban areas, where people are more accessible, and to under-sample more rural areas (Langseth 2007). One of our methodological assumptions is that our sample not only has to be representative, but also has to cover various social groups.

A general public survey, for example, can show that a great number of citizens do not tolerate it that a local politician visits the cabinet of a fellow party member to discuss and bring about the settlement of a construction licence. On the other hand it may very well be the case that a sample of elites does not experience this particular scenario to be corrupt at all. Studies on political corruption revealed that opinions of corruption can vary over different social groups. Jackson and Smith subjected Heidenheimer's theory about political corruption to a significant interview-based study amongst Australian politicians and voters (Jackson and Smith 1996). Reckoning by the results of their survey – that the perceptions of government insiders differ from the perceptions of outsiders – it is likely that the outcome of our study on the perception of corruption will be strongly determined by our sample choice.

In 2005 Bezes and Lascoumes made a pointed analysis of 12 identified empirical studies on the perception and judgement of political corruption (Bezes and Lascoumes 2005). They came to the conclusion that a significant number of studies applied Peters and Welch's methodological design (Peters and Welch 1978) to scrutinize Heidenheimer's hypothesis of black-grey-white corruption. However, most of these studies investigated the perception and judgement of corruption from one single setting: either they investigated the citizens or the elite's perspective of corruption. Heidenheimer's thesis was however that whether a corrupt activity was tolerated (white corruption) or condemned (black corruption) depended upon the type of social group the observer belonged to (Heidenheimer 2004). Only the study of Jackson and Smith (Jackson and Smith 1996) analysed the perception and judgement of corruption by comparing the citizens' and the politicians' perceptions of corruption (Bezes and Lascoumes 2005). This methodological framework presented by Heidenheimer and followed up by a select number of scholars proved to be useful in gaining new insights into the judgements and opinions of unethical behaviour.

As opposed to the empirical findings which suggest that it may be that members of some subgroups of the population share a common understanding of corruption which differs from the understanding shared by other subgroups, Gorta and Forell ascertained that the existing literature has not revealed clear cut distinctions (Gorta and Forell 1995). The differences in findings are the result of studies being conducted in different decades and involving different social groups. In line with this, it is our contention that if we want to deduce a more comprehensive understanding of the perception and judgement of corruption we must analyse scenarios about corrupt behaviour involving different social groups within a restricted time span.

As far as we know no one has examined the judgement of situations using empirical evidence in a direct comparison of different social groups and elucidated the inter-personal or inter-group variability in the judgment of situations as corrupt or not. As discussed above, the equation of the judgments of different social groups brought significant findings into the study field of political corruption. So, why should this not be the case in researching corruption in a broad sense?

4   Conclusion

The perception of a particular situation or a description of a situation as corrupt (or not) is not a linear judgement. Our assumption is that one's judgement of a situation shifts along multiple dimensions in relation to the particular activity and the context of the observer in relation to the one being observed.

We assume – on the basis of previous studies – that the judgement of a situation is influenced by two groups of characteristics. A first group contains situational characteristics (such as public officials, favours, payoffs, undue advantage, etc.). A second group consists of the personal characteristics of the observer (such as gender, social status, education, etc.). Our assumption is that these characteristics influence, either separately or in combination, the judgement of a situation.

In this article, we described a method – a so-called scenario-based questionnaire – to assess to what extent the aforementioned two groups of characteristics, either in themselves or in combination, influence the judgement of a situation as corrupt or not. For that reason we evolved Peters and Welch's methodology and elaborated a questionnaire composed of different scenarios. Each scenario is followed by descriptions of different successive situations in order to shift the respondent's opinion. In this way we will try to differentiate transgressive characteristics through which we expect to detect the bascule in the judgments of a situation from a tolerated to a condemned situation.

The purpose is to more precisely elucidate these shifts in order to enhance our understanding of corruption. In other words we will try to find situational and personal characteristics or combinations of characteristics which determine the judgement of a scenario as corrupt or not.

5   Bibliography

Aldrich, D. & Kage, R. (2003). Mars and Venus at twilight: A critical investigation of moralism, age effects, and sex differences. *Political Psychology, 24*(1), 23-40.

Anderson, C. J. & Tverdova, Y. V. (2003). Corruption, political allegiances, and attitudes toward government in contemporary democracies. *American Journal of Political Science, 47*(1), 91-109.

Bezes, P. & Lascoumes, P. (2005). Percevoir et juger la 'corruption politique'. Enjeux et usages des enquêtes sur les représentations des atteintes à la probité publique. *Revue française de science politique, 55,* 757-786.

Birch, S. (2008). Electoral institutions and popular confidence in electoral processes: A cross-national analysis. *Electoral Studies, 27*(2), 305-320.

Bové, L. (2009). Tegen de perceptie in. De Vlaamse dagbladpers en de beeldvorming over corruptie. In A. Dormaels, G. Vande Walle, P. Ponsaers, & M. Easton (Eds.) *Verkocht of omgekocht? Een criminologische benadering van corruptie.* Themanummer Orde van de dag, 47, 24-27. Mechelen: Kluwer.

de Sousa, L. (2008). I don't bribe, I just pull strings: Assessing the fluidity of social representations of corruption in Portuguese society. *Perspectives on European Politics and Society, 9*(1), 8-23.

Fackler, T. & Lin, T. M. (1995). Political corruption and presidential elections, 1929-1992. *Journal of Politics, 57*(4), 971-993.

Ferraz, C. & Finan, F. (2008). Exposing corrupt politicians: The effects of Brazil's publicly released audits on electoral outcomes. *Quarterly Journal of Economics, 123*(2), 703-745.

Flore, D. (1999). *L'incrimination de la corruption: les nouveaux instruments internationaux. La nouvelle loi belge du 10 février 1999.* Bruxelles: La Charte.

Gardiner, J. A. (1967). Public attitudes toward gambling and corruption. *Annals of the American Academy of Political and Social Science, 374,* 123-134.

Gardiner, J. A. (1970). The politics of corruption in an American city. In A. J. Heidenheimer (Ed.) *Political corruption. Readings in comparative analysis* (pp.167-175). New Brunswick, NJ: Transaction Books.

Gardiner, J. A. (1970). *The politics of corruption. Organised crime in an American city.* New York: Russel Sage Foundation.

Gardiner, J. A. & Lyman, T. R. (1978). *Decisions for sale. Corruption and reform in land-use and building regulation.* New York: Praeger Publishers.

Gibbons, K. M. (1990). Toward an attitudinal definition of corruption. In A. J. Heidenheimer, M. Johnston & V. T. Le Vine (Eds.) *Political corruption: a handbook* (pp.165-171). New Brunswick, NJ: Transaction Publishers.

Gorta, A. & Forell, S. (1994). Unravelling corruption. A public sector perspective. Survey of NSW public sector employees' understanding of corruption and their willingness to take action. *Research Report No. 1* (p. 183). Independent Commission Against Corruption.

Gorta, A. & Forell, S. (1995). Layers of decision: linking social definitions of corruption to a willingness to take action. *Crime, Law and Social Change, 23,* 315-343.

GRECO (2009). *Evaluation report on Belgium on incriminations* (p. 31). Strasbourg Cedex: Council of Europe.

Heidenheimer, A. J. (2004). Disjunctions between corruption and democracy? A qualitative exploration. *Crime Law and Social Change, 42*(1), 99-109.

Heidenheimer, A. J. e. (1970). *Political corruption: Readings in comparative analysis.* New Brunswick, NJ: Transaction Books.

Heywood, P. (1997). Political corruption: Problems and perspectives. *Political Studies, 45*(3), 417-435.

Jackson, M. & Smith, R. (1996). Inside moves and outside views: An Australian case study of elite and public perceptions of political corruption. *Governance – an International Journal of Policy and Administration, 9*(1), 23-42.

Johnston, M. (1991). Right and wrong in British politics – fits of morality in comparative perspective. *Polity, 24*(1), 1-25.

Johnston, M. (1996). The search for definitions: The vitality of politics and the issue of corruption. *International Social Science Journal, 48*(3), 321-335.

Karahan, G. R., Coats, R. M. et al. (2006). Corrupt political jurisdictions and voter participation. *Public Choice, 126*(1-2), 87-106.

Kjellberg, F. (1995). Conflict-of-interest, corruption or (simply) scandals – the Oslo case 1989-91. *Crime Law and Social Change, 22*(4), 339-360.

Kurer, O. (2005). Corruption: An alternative approach to its definition and measurement. *Political Studies, 53*(1), 222-239.

Langseth, P. (2007). Measuring corruption. In A. S. Sampford, C. Connors & F. Galtung (Eds.) *Measuring corruption*. Aldershot: Ashgate Publishing Limited.

Lascoumes, P. & Tomescu-Hatto, O. (2008). French ambiguities in understandings of corruption: Concurrent definitions. *Perspectives on European Politics and Society, 9*(1), 24-38.

Mancuso, M. (2005). Contexts in conflict: Public and private components of assessment in ethical judgements. *Journal of Canadian Studies – Revue D'Etudes Canadiennes, 39*(2), 179-203.

McCann, J. A. & Redlawsk, D. P. (2006). As voters head to the polls, will they perceive a 'culture of corruption?' *Ps-Political Science & Politics, 39*(4), 797-802.

Nye, J. S. (1967). Corruption and political development: A cost-benefit analysis *American Political Science Review, 61*, 417-427 (as cited in Gardiner and Lyman 1978)

Park, H. (2003). Determinants of corruption: A cross-national analysis. *Multinational Business Review, 11*(2), 29-48.

Peters, J. G. & Welch, S. (1978). Political corruption in America – search for definitions and a theory, or if political corruption is in mainstream of American-politics why is it not in mainstream of American-politics Research. *American Political Science Review, 72*(3), 974-984.

Pharr, S. J. (1998). 'Moralism' and the gender gap: Judgments of political ethics in Japan. *Political Psychology, 19*(1), 211-236.

Redlawsk, D. & McCann, J. (2005). Popular interpretations of 'corruption' and their partisan consequences. *Political Behavior, 27*(3), 261-283.

Scott, J. C. (1972). *Comparative political corruption*. Englewood Cliffs: Prentice-Hall (as cited in Gardiner and Lyman 1978).

Smyth, J. D., Christian, L. M. et al. (2008). Does 'Yes or No' on the telephone mean the same as 'Check-All-That-Apply' on the web? *Public Opinion Quarterly, 72*(1), 103-113.

Swamy, A., Knack, S. et al. (2001). Gender and corruption. *Journal of Development Economics, 64*(1), 25-55.

Treisman, D. (2000). The causes of corruption: a cross-national study. *Journal of Public Economics, 76*(3), 399-457.

Treisman, D. (2007). What have we learned about the causes of corruption from ten years of cross-national empirical research? *Annual Review of Political Science, 10*, 211-244.

Van Der Aa, J. (2009). Kabinetsadviseur woont gratis in loft. *Het Laatste Nieuws*, 18.

van Duyne, P. C. (2001). Will Caligula go transparent? Corruption in acts and attitudes. *Forum on Crime and Society, 1*(2), 73-98.

Welch, S. & Hibbing, J. R. (1997). The effects of charges of corruption on voting behavior in congressional elections, 1982-1990. *Journal of Politics, 59*(1), 226-239.

Werner, S. B. (1983). New directions in the study of administrative corruption. *Public Administration Review, 43*(2), 146-154.

Towards an integral and integrated drug policy: pearls and pitfalls

Liesbeth Vandam
Charlotte Colman
Freya Vander Laenen
Brice De Ruyver

1   Introduction

In most countries in the European Union, including Belgium, the approach of the drug problem is an integral and integrated one, in which several policy levels and policy domains are involved in the drug policy (EMCDDA, 2008). This approach requires a vertical policy concurring between the different competency levels (e.g. national and local level) and a horizontal policy concurring between the different policy domains (e.g. prevention, harm reduction, treatment, health and well-being and repression).

Since the 1990's, evaluation of policy has become increasingly important in western societies (Leeuw, 2005). In accordance with this tendency, evaluation has been executed in several domains involved in approaching the drug phenomenon. In scientific literature, studies are found evaluating interventions in the prevention (Stevens, Trace & Bewley-Taylor, 2005; Stevenson & Mitchell, 2003; Botvin & Griffin, 2003; Kumpfer & Alder, 2003), the treatment (Autrique, Vanderplasschen, Pham, Broekaert, & Sabbe, 2007; Rigter, van Gageldonk, Ketelaars, & van Laar, 2004; van Gageldonk, Ketelaars, & van Laar, 2006) and the law enforcement domain (Stevens et al., 2005; Mazerolle, Soole, & Rombouts, 2007). Studies evaluating the cooperation between several domains involved in approaching the drug phenomenon are rare though. When developing an evidence based drug policy, more insight in best practices in an integral and integrated drug policy are absolutely necessary. With this contribution, we try to answer the following research questions: first, in developing an integral and integrated drug policy, which international best practices can be identified (i.e. *what* works)? Second, how must we implement these interventions (i.e. *how* does it work)? Third, to what extent have these effective interventions and best practices been implemented in the Flemish Community?

The aim of this chapter is to identify international best practices in developing an integral and integrated drug policy. No (new) evaluations of cross-cutting interventions were conducted, but by means of a study of literature, we present best practices in the context of developing an integral and integrated drug policy and the essential preconditions involved. These cross-cutting interventions and essential preconditions are better illustrated with reference to specific examples. To this end, we analyse to what extent these best practices have indeed been implemented in the Flemish

Community[1]. We will give two examples of best practices and two examples of less favourable practices in this Community of Belgium.

In the first part of this contribution, we define the scope of this paper: we describe what is meant with an integral and integrated drug policy and we identify the actors involved. After a presentation of the methods used, we present the international best practices in an integral and integrated drug policy. Finally, we analyze how these international best practices are implemented in the Flemish Community.

This chapter is based upon the research project "Do's and don'ts in an integral and integrated drug policy" (De Ruyver, Lemaître, Ponsaers, Born, Pauwels, Vander Laenen et al., 2009). This research project started in October 2007 and ended in April 2009. The research project is part of the Research programme in support of the federal drug policy document and is funded by the Federal Science Policy[2].

2   Defining the scope of this paper

2.1   An integral and integrated drug policy

In (scientific) literature the concept 'integral and integrated drug policy' is used as a container concept. Policy makers and scientific researchers apply the concept to various projects, interventions and policies. It has become a fashionable term which is often randomly used without the exact interpretation being entirely clear. The concepts 'integral and integrated' thus need a clear definition. In this chapter, *integral* signifies 'global.' The drug phenomenon is multidimensional and, therefore, all facets must be tackled. The drug phenomenon is related to health, social and economic aspects. It also has an international dimension where both the supply and demand side can determine policy (De Pauw, 2007; De Ruyver, 2007; Decorte, De Ruyver, Ponsaers, Bodein, Lacroix, Lauwers et al., 2004). Where demand reduction derives from interventions with the goal of raising personal resistance to drug consumption, supply side reduction is aimed at reducing the access to and consequently the use of drugs (Pentz, 2003). For this reason, an integral drug policy is related to several policy domains as a result of which, welfare, health, prevention, treatment and repressive elements are brought together.

The aspect *integrated* has a direct link with an integral approach although integral and integrated are two separate concepts that cannot be used interchangeable. An approach whereby all aspects of the drug phenomenon are addressed (integral) requires the involvement of all relevant actors and services which represent the different domains (integrated). Co-operation and concurring between actors are therefore required (Heed, 2006). Both a horizontal concurring between domains and a vertical concurring between all competency levels is required to tackle the drug phenomenon from several angles. The various competences (welfare, public health, policing and

---

1   Belgium is divided in three communities: the Flemish Community, the French Community and German-speaking Community. The Communities have powers for amongst others health policy and assistance to individuals.

2   More information can be obtained on: http://www.belspo.be/belspo/fedra/proj.asp?l=en&COD=DR/33

judicial authorities, environmental services,...) involved in the approach of the drug phenomenon are located on federal, regional, provincial and local level.

In the European Union, the drug phenomenon is in the first instance viewed as a public health problem, whereby the distinction between legal versus illegal drugs is only relevant in a legal-criminological context (EMCDDA, 2006). Therefore, we consider drug policy to include the approach of the drug phenomenon for both illegal and legal drugs.

2.2   Identification of domains involved

Since an integral and integrated drug policy covers several policy areas, we distinguish four domains: welfare and health, prevention, treatment and law enforcement. This classification reflects the different domains involved in drug policy. Alongside these four domains, we further include a number of sub-categories in order to introduce an enhanced degree of specificity (cf. table 1).

Table 1. Domains involved in an integral and integrated drug policy (De Ruyver et al., 2009)

Domains and subcategories				
Domain	Welfare and health	Prevention	Treatment	Law Enforcement
Subcategory	Not drug-specific	Drug-specific harm reduction Early intervention	Drug-specific	Drug-specific (interventions situated at the various levels of the criminal justice system)
Pretext	Demand side			Supply side

We list under welfare and health those services which aim to promote general well-being or which have health promotion aims. More concretely, this domain includes non-drug specific initiatives which relate to different life domains of the drug user such as housing, return to work training, education, financial assistance and leisure activities.

Prevention includes interventions which are specifically focused on the drug phenomenon and on specific risk groups or risk settings (selective prevention) or on specific high risk individuals (targeted prevention). Prevention activities that are not specifically targeted at the drug phenomenon but rather at the general population (universal or general prevention) (Botvin & Griffin, 2003) are not included as part of prevention but in welfare and health. Such interventions are aimed at the promotion of general welfare and commonly have a health promotion objective. Harm reduction is included under the prevention heading and is not classified as a distinct domain within the research since needle exchange programmes or safe injecting rooms e.g. have as their primary aim the prevention or limitation of drug-related harm for both the individual (drug user) and society. Early intervention is included under the prevention umbrella as well since such interventions are also aimed at reducing the risks for harm of drug use for users at risk for problematic drug use (Brook, Brook, Richter, & Whiteman, 2003).

Treatment includes drug specific services or specialised interventions such as substitution therapy, day care treatment or therapeutic communities. Non-drug specific treatment interventions or interventions concerning general health and welfare are listed under welfare and health.

Law enforcement includes interventions that are situated at the various levels of the criminal justice system: the investigation level, the prosecution level, the sentencing level and the level of the execution of sentences.

2.3 An integral and integrated drug policy: more than one domain involved

Based upon the previous definition of an integral and integrated drug policy, the emphasis of the study lies on cross-cutting interventions, i.e. interventions between several domains involved in drug policy. Therefore, we did not select projects or studies focusing on one domain.

3 Methodology

The following methodology was used in order to reach the research goals. Firstly, a literature search was conducted to identify international and national best practices. Secondly, in order to analyse to what extent these best practices had indeed been implemented in the Flemish Community, all provincial and local drug coordinators in the Flemish Community were contacted. The possibilities and problems associated with the implementation of an integral and integrated drug policy in this region of Belgium were mapped by means of interviews and one focus group.

3.1 Literature search

An intensive literature search was carried out aimed at the identification of best practices in an integral and integrated drug policy. We searched for studies that evaluate cross-cutting interventions in an integral and integrated drug policy. First, these studies allow to examine 'what works'; enabling to see if there are interventions which appeared to be effective on the basis of international evaluations. Second, an answer is given to the 'how' question; how an intervention must be implemented and which pitfalls there are. The search for literature started in December 2007 and ended in June 2008.

The following online scientific databases were screened: Web of Science, Pubmed, Drugscope, Alcohol Concern, Archido, Toxibase, EDDRA (Exchange on Drug Demand Reduction Action), EIB (Evaluation Instruments Bank) and the database of the Cochrane Collaboration. The following websites were screened: Website EMCDDA (European Monitoring Centre for Drugs and Drugs Addiction), Website Home Office, Website WODC (Research and Documentation Centre), Website European Commission, Website of ECAD (European Cities Against Drugs), Website World Health Organization Regional Office for Europe. Health in prisons project, Website Campbell Collaboration, Website Beckley Foundation Drug Policy Programme, Website The Dutch Society, Security and Police Foundation, Website VAD (Organization Alcohol and other Drug problems). The following Reitox-national focal points[3] were contacted by telephone and e-mail in order to find additional literature: Belgium, United Kingdom, the Netherlands, Ireland, Luxembourg, Finland, France, Spain, Italy and Portugal. The following research centres, study centres and organisations were contacted: Crime Concern (UK), Centre for crime and justice studies (UK), Centre for criminal justice studies (UK), North Inner City Task Force (Ireland), European forum for urban safety, Pompidou Group, Drug Info (Australia), Australian Drug Information Network (Australia), Netherlands Institute for the Study of Crime and Law Enforcement (the Netherlands), the Trimbos Institute (the Netherlands) and Intraval (the Netherlands). Many other national and international experts, both scientific researchers and policy makers in this field were contacted[4].

When screening the scientific data bases the search terms 'multidisciplinary', 'multi-agency', 'transdisciplinary', 'evidence-based', 'coordination', collaboration', 'linkages', 'integrated', 'integral', 'integral' were used in combination with 'drug policy', 'approach', 'drug related nuisance' and 'nuisance'.

We focused our search upon cross-cutting interventions in Europe, Canada and Australia. We did not include interventions evaluated in the United States, because of the difference in policy aims between the United States and the countries included in the study. The policy in the United States is aimed at prohibition, while the policy in the countries included are aimed at normalization (De Ruyver, Vermeulen, Vander Beken & Vander Laenen, 2003).

The primary focus of the search was on *effect evaluations*. However, a preliminary literature review learned that effect evaluations measuring the effect of interventions involving multiple domains, were frequently unavailable. Most of the studies measure the effect of an intervention within one domain, for example the effect of a specific treatment program. Therefore, *process evaluations*, namely studies focusing on the factors stimulating or hampering the implementation of an intervention were also included (Swanborn, 1999).

In sum, effect and process evaluations of cross-cutting interventions approaching the drug phenomenon in Europe, Canada or Australia were included in the inventory. After the literature review, 27 evaluation studies were selected. Twelve studies evaluate the effect of a cross-cutting intervention, eight studies evaluate the process of an intervention and seven studies evaluate both the effect and the process of an intervention.

Based upon the Maryland Scientific Method Scale (MSMS), we are able to determine the effectiveness of a cross-cutting intervention. The MSMS is a simple 5-point scale of methodological quality. This scale makes it possible to compare the effectiveness of various cross-cutting interventions in various settings (Sherman, Farrington, Welsh & Mackenzie, 2002). The majority of the identified effect evaluations are located at level 1 or level 2 of the MSMS. Only four studies were located at level 3 of the MSMS. No studies were located at levels 4 or 5 of the MSMS. These findings are not surprising. First, 'first measurements' are often missing in social scientific stud-

---

3   The EMCDDA coordinates a network of National focal points set up in the 27 EU Member States, Norway, the European Commission and in the candidate countries. Together, these information collection and exchange points form Reitox, the European Information Network on Drugs and Drug Addiction. This human and computer network links the national information systems of the 27 Member States and Norway and the key partners to the EMCDDA.

4   For a complete overview of all organisations contacted, we refer to: http://www.belspo.be/belspo/fedra/proj.asp?l=en&COD=DR/33

ies. Second, social scientific researchers seldom use control groups to compare them with the experimental group. Third, in social scientific research, groups are seldom randomly composed (Loosveldt, 2001). Process evaluations cannot be evaluated by the MSMS. Because few of the process evaluation studies are included in Web of Science, it is difficult to estimate the international scientific quality of these studies. These findings confirmed our hypothesis that evaluations of interventions – that are often locally implemented – are seldom reported on in international databases or distributed internationally. These findings highlight the importance of contacting experts and fieldworkers to gain insight into the existence of specific interventions. Finally, despite our intensive efforts, we acknowledge that the list of studies collected, is probably not an exhaustive one.

3.2   Identifying best practice in the Flemish Community

As a specific example, we analyse to what extent the identified best practices were indeed implemented in the Flemish Community.

All provincial (N = 6) and local drug coordinators (N = 13) in the Flemish Community were interviewed in order to identify cross-cutting interventions in this region. Furthermore, we contacted the VAD (Organization Alcohol and other Drug problems), an organisation coordinating the majority of organizations working in drug prevention and treatment. The interviews were done using a semi-structured questionnaire. In the beginning of the questionnaire, important terms, such as 'cross-cutting interventions', were clearly defined. After an explanation of the subject of investigation, the coordinators were asked for information about cross-cutting interventions in their sphere of action. In the questionnaire we asked for information about the existence of cross-cutting interventions. The name, aim, organization level and geographical area of the initiative and actors involved, were collected.

We became aware of the limitations of our interviews when initiatives outside the scope of the research were mentioned by the interviewees more than once. Moreover, specific cross-cutting interventions might not be mentioned, e.g. initiatives with a broader scope than the drug phenomenon.

Interviews and one focus group were conducted with the actors involved in order to identify essential preconditions when implementing cross-cutting interventions in the Flemish Community in the context of developing an integral and integrated drug policy. A semi-structured questionnaire was used in order to identify the 'do's and don'ts' when implementing cross-cutting interventions in an integral and integrated drug policy. A focus group was organized with actors performing a coordinating role in cross-cutting interventions in order to test the results from the interviews. All interviews and focus group data were analysed with Maxqda allowing to efficiently organize and analyse the qualitative information.

The results presented in part 4.1 are based upon the international literature review. The results presented in part 4.2 (To what extent are these best practices implemented in the Flemish Community?) are based upon the data collected in the interviews and the focus groups.

4   Results

4.1   Best practices in an integral and integrated drug policy

When evaluating cross-cutting interventions, it is difficult to define which domain involved is responsible for the effect (Brook et al., 2003; Ritter, Bammer, Hamilton, Mazerolle & The DPMP Team, 2007; Stevens, 2008). Nevertheless, Kibel and Holder (2003) found that interventions where several domains cooperate and therefore several policy objectives are pursued are the most promising. Based upon the evaluation studies of cross-cutting interventions, we present in the following part best practices and the preconditions involved. In this presentation of the best practices, we will make a distinction between interventions that were proven to be effective and interventions that hold promise.

4.1.1   General findings

Based upon our list of evaluation studies, we conclude that certain countries are more represented than others in producing evaluation studies on our topic of interest. The United Kingdom, closely followed by the Netherlands, produces a large amount of the evaluation studies. These findings are in line with what was expected, since the United Kingdom and the Netherlands have a long tradition in policy evaluations (Leeuw, 2005). While other countries may produce effective or promising interventions, these interventions may remain unobserved due to the lack of policy evaluation.

4.1.2   Cross-cutting interventions in the context of diversion to treatment by law enforcement

In drug policy, the collaboration between law enforcement and treatment has become increasingly important. Drug users are orientated to treatment by the law enforcement actors in order to divert (drug) offenders away from crime and drugs (De Wree, De Ruyver, Verpoest & Colman, 2008). In our search for studies evaluating interventions of diversion to treatment by law enforcement, we found six process evaluations (Barton, 1999a; Barton, 1999b; Institute for Criminal Policy Research, 2007; Hunter et al., 2005; De Ruyver et al., 2008; Harman & Paylor, 2005) and eleven effect evaluations. Three of these studies are at level three of the MSMS (Holloway, Bennett, & Farrington, 2005 ; Koeter & Bakker, 2008; Heale & Lang, 2001) and eight are at level two of the MSMS (Seeling, King, Metcalfe, Tober, & Bates, 2001; De Ruyver, Ponsaers et al., 2007; van Ooyen-Houben, 2008; Barton, 1999a; Institute for Criminal Policy Research, 2007; Skobdo et al., 2007; Mazerolle et al., 2007; Scottish Executive Effective Interventions Unit, 2004).

In each stage of the criminal justice system, best practices aimed at the diversion to treatment by law enforcement are identified.

At the moment of *arrest*, "arrest referral schemes" are a form of orientation to treatment (Scottish Executive Effective Interventions Unit, 2004; Seeling et al., 2001; Hunter et al., 2005). When a drug user is arrested, police officers and treatment workers are brought together to motivate the offender to have a consultation with a social worker. This cross-cutting intervention is without obligations for the drug offender.

The effect studies show hard to reach drug users are reached by means of the arrest referral schemes and motivated to contact treatment. A decrease in drug related crime is noticed when the intervention takes place at an early stage in the criminal justice system (Scottish Executive Effective Interventions Unit, 2004; Seeling et al., 2001). The process evaluations show that the practical organization of the intervention is important: a central office is installed where social workers and police officers can meet on a daily basis in order to enhance the dialogue between the social workers and the police officers. Communication and consultation are important in order to work with a common aim. Training and supervision are essential preconditions that allow for a successful cross-cutting intervention. With regard to treatment, insufficient capacity and client focus are the biggest pitfalls (Hunter et al., 2005).

Another good practice at the *prosecution level* is the 'Pilot care-project' (Proefzorg) (De Ruyver, Colman, De Wree, Reynders, Van Liempt & De Pauw, 2008). The 'Pilot care project' orientates individuals that have committed offences due to their drug dependence to treatment. The process evaluation showed that the installation of an intermediary manager enables both law enforcement and treatment to respect their aims and responsibilities (De Ruyver et al., 2008).

At *the sentencing-level*, the drug courts are identified as a best practice. Drug courts aim to refer clients to treatment, they aim to guarantee a satisfactory completion of treatment, and to diminish the extent of re-offending The drug courts are effective as they refer a growing amount of drug users to treatment at the sentencing level (Heale & Lang, 2001). The process evaluation shows that the information exchange on drug tests between treatment and law enforcement is an important pitfall, because more consequences are at stake for the drug user at this level of the criminal justice system. Moreover, given the distinct aims of treatment and law enforcement, this cross-cutting intervention runs the risk that both domains interpret 'success' in various ways. A clear demarcation of aims and boundaries is therefore even more important at this level of the criminal justice system (Barton, 1999b).

At *the level of execution of sentences*, best practices were identified as well, e.g the CARAT (Counselling, Assessment, Referral, Advice and Throughcare) -schemes. The CARAT-schemes want to increase the support available to drug-using prisoners both during custody and on release (Harman & Paylor, 2005). No effect evaluations were found; two process evaluations refer to pitfalls when implementing this type of intervention. The most important pitfall refers to the lack of coordination and cooperation with facilities outside of prison (Harman & Paylor, 2005; Mair & Barton, 2001). When a clear demarcation of the division of roles and responsibilities is missing, this leads towards a fragmented treatment offer for the (former-)prisoners. Precise protocols and procedures are necessary to realise a productive cooperation. Training sessions for prison personnel as well as for extramural social workers are necessary to enhance cooperation between the treatment and law enforcement domain at the level of execution of sentences (Mair & Barton, 2001).

To conclude, evaluations show that the above mentioned types of orientation for drug users to treatment who come into contact with the police or justice at various echelons of the criminal justice system are promising in order to diminish the use of drugs and drug-related crime. Besides this, hard to reach drug users, who never were in contact with treatment before, are referred to treatment for the first time.

Based upon the process evaluations included, the following preconditions are identified for cross-cutting interventions in the context of diversion to treatment by law enforcement: the availability of sufficient financial means, the availability of sufficient capacity, communication and consultation between the several domains involved, training and supervision of the actors involved and a clear demarcation of aims, boundaries and responsibilities (including an agreement on information-exchange). Respecting each other's aims, boundaries and responsibilities can be facilitated by the installation of an intermediary manager between treatment and law enforcement (Barton, 1999a; Barton, 1999b; Institute for Criminal Policy Research, 2007; Hunter et al., 2005; De Ruyver et al., 2008).

4.1.3 Cross-cutting interventions in the context of returning to or starting work

We found one study that evaluated both the process and the effectiveness (MSMS 1) of an intervention providing assistance to drug users in returning or starting to work in the Netherlands (Michon, Rondez, & van Weeghel, 2000). Welfare, health and drug treatment are working together in this intervention. After controlling the drug problem, the client starts an education or work training in order to start to work. This project seems promising because the majority of the clients start a job and an improvement in other life domains is noticed. 43% of the participants had a job after the completion of the programme and maintained this job for a period of at least 6 months.

An essential precondition for this type of cross-cutting interventions is coordination so that continuity of care is guaranteed throughout the project. A formal cooperation structure between the domains involved, is considered a factor of success, because each domain recognizes its own responsibility in the project. The biggest pitfall of the initiative is the fact that the project is only moderately embedded in drug treatment so that few drug users are referred to this project (Michon et al, 2000).

Thus, cross-cutting interventions in the context of returning or starting to work require the following preconditions: coordination and a formal cooperation structure between the domains involved and the guarantee that the project is fully embedded in the drug treatment domain.

4.1.4 Cross-cutting interventions in the context of day and night care and referral to housing projects

The literature-review resulted in one study that evaluated both the process and the effectiveness of this type of intervention (Wits, Biesma, Garretsen, & Bieleman, 1999). *Day and night care interventions* aim to provide drug users with an alternative way of spending their time instead of using drugs on the street. These programmes also offer (basic) supplies to users, such as showers and washing machines. Coffee, tea, sandwiches, a television, table football etc. is offered. There is cooperation with the police authorities and with drug treatment that can both refer drug users to this project. The effect evaluation shows that this intervention ensures a reduction of nuisance on the street and a reduction in heroin use in favour of methadone use. This results in an improvement in the various life domains and a reduction in drug-related crime (Wits et al., 1999).

With *regards to housing projects* for drug users, help is provided for accommodation and for assisting users to address their problematic drug use and by offering assistance in their search for work. In order to reach its aim, the project works together with welfare and health in the context of the housing facilities and with drug treatment in the context of the individual drug problem. Again, the effect evaluation shows that this intervention ensures more structure for the drug user and less nuisance on the street (Wits et al., 1999).

For these types of interventions it is an essential precondition to have regular meetings and a clear communication with the various partners involved. A clear distribution of tasks and responsibilities is needed. These preconditions guarantee the intake and straight flow of clients and avoid that one of the partners involved takes over the entire project (Wits et al., 1999).

4.1.5 Cross-cutting interventions in the context of harm reduction

We were able to identify six studies for interventions with harm reduction goals, evaluating user rooms. Four studies were measuring the effectiveness of user rooms, three of which were at level one of MSMS (Spijkerman et al, 2002; van der Poel et al, 2003; Zurhold, 2003) and one at level two of MSMS (Bieleman, et al, 2007). Two of these studies were evaluating the process of the implementation of the intervention (Spijkerman et al, 2002; van der Poel et al, 2003). In these interventions, law enforcement is not directly involved as a partner although these initiatives started from both a health and a security perspective. These interventions start from an integral approach, but are not strictly 'integrated'. This practice came about to tackle drug-related nuisance and to improve the health situation of the drug user by reducing the presence of drug users on the street and providing them with prescribed drug supplies and safe spaces in which they can use. Drug users can call upon general medical supplies but also drug specific assistance such as opiate substitute prescription with accompanying advice and guidance. The effect evaluations show that user rooms guarantee tackling drug nuisance (decrease in overt use and decrease in number of needles left behind by drug users), contributing to an improvement in the health of drug users and harm reduction. Hard-to-reach drug users are reached with these projects (Spijkerman et al, 2002; van der Poel et al, 2003; Zurhold, 2003; Bieleman, et al, 2007).

The most important pitfall for this type of intervention relates to the resistance of people in a neighbourhood where the user rooms are located. This resistance can prevent the installation or continued existence of such user spaces. Nevertheless, acceptance by local residents is possible if they experience the advantages of user spaces (e.g. a reduction in drugs nuisance in public space).

4.1.6 Cross-cutting interventions in the context of prevention and early intervention

In the framework of early intervention, cross-cutting interventions are set up with various domains in the community, together with the local government.

One effect evaluation at level 3 of the MSMS (Crow, France, Hacking, & Hart, 2004) and three process evaluations were found (Crow et al., 2004; Kellock, 2007; Toumbourou, 1999). These projects are referred to as 'multi-agency programmes', whereby health and welfare, prevention, treatment and if necessary, law enforcement, cooperate. These cross-cutting interventions stimulate professionals to work in the framework of early intervention and prevention and are located in the "Communities that Care" project. "Communities that care" originated in the United States, but was also set up in Australia, the UK and the Netherlands. These interventions start from the growing belief in criminological sciences that the approach of risk and protective factors is the best way to prevent drug use and other social problems. By means of school surveys and other research methods, risk and protective factors are mapped and a community profile is made up around which practical initiatives can be set up (Crow et al., 2004; Horn & Kolbo, 2000; Arthur & Blitz, 2000).

The study of Crow et al. (2004) could not draw conclusions about the effectiveness of the Communities that Care projects, but gives an overview of essential preconditions when implementing this type of projects.

Several preconditions are identified (Crow et al., 2004; Kellock, 2007; Toumbourou, 1999). An early intervention project consists of several phases in order to develop specific actions. After the screening of risk and protective factors in a specific local entity, the actors involved belonging to various domains are identified. Actors may be juvenile and family services, schools and judicial authorities involved with youth. The local government should also be included in order to create a political basis, enabling financial resources. The next step includes developing a plan of action with the actors and domains identified. The local problem defines the development of specific actions. It is important to evaluate the interventions in order to adjust the interventions when necessary. Another precondition refers to the presence of contacts and collaboration initiatives between domains. Using cross-cutting interventions are in itself an essential precondition when implementing new interventions in the context of early intervention. Another precondition refers to the initial phase of analyzing the risk and protective factors, namely that all potential partners must be involved. When specific actors are missed in the beginning, these actors might not be motivated to develop a specific plan of action. The employment of a drug coordinator is another important precondition for this type of cross-cutting intervention. Communication and the organization of meetings is important because several partners are working together. Another precondition refers to the involvement of both executives and field workers in order to reach an agreement about the possible interventions. Furthermore, the plan of action must be based upon consensus. A lack of consensus leads to a lack of motivation and even resistance to implement the action plan. To increase the credibility and support for the interventions within the "Communities that Care" project, multiple cross-cutting interventions should be set up. When multiple interventions are set up, the insolvency of one intervention should not be insurmountable, because other interventions are still feeding the project.

4.1.7 Cross-cutting interventions with the involvement of other domains

Besides the 'traditional' domains involved in drug policy, other fields are also important partners if an integral and integrated approach to the drug phenomenon is to be realised. We found three effect evaluations at level two of the MSMS that include other domains (Snippe, Bieleman, Kruize & Naayer, 2005; Snippe, Naayer & Bieleman, 2006; Mazerolle, Soole & Rombouts, 2007). The property market is an important player in the treatment of drug-related nuisance (Snippe, Bieleman, Kruize, & Naayer,

2005; Snippe, Naayer, & Bieleman, 2006). Spatial management of places where (drugs) nuisance is identified can bring about a fall in such nuisance and reduced feelings of insecurity by itself. Besides repressive action with respect to drug dealers and users who cause nuisance, the local administration must also be involved. Home-owners and renters must be sensitized to avoid that their houses become dilapidated. They must invest in the renovation and renewal of their neighbourhood. Administrative sanctions are also a means to counter neighbourhood degeneration and can be used to close premises where anti-social behaviour takes place (Mazerolle et al., 2007).

4.1.8 Cross-cutting interventions within the framework of policy development and concurring

Some interventions are aimed at the development of a local integral and integrated drug policy irrespective of the type of partnership arrangements envisaged. Two studies evaluating the effectiveness of this type of interventions were identified, both at level 1 of the MSMS. (Connolly, 2002; Doherty, 2007).

Several steps must be followed so that an effective integral and integrated policy can be delivered. The first step is the identification of partners to be involved in drug policy. Key persons involved in the development of an integral and integrated drug policy should be included. In big cities one can start from existing networks. The inclusion of citizens to inform them about local problems is considered as a best practice. The second step relates to leadership and coordination. A steering group should identify means and aims of the local integral and integrated drug policy. The coordinator of the steering group should mobilize and bring together the identified actors involved. Thirdly, a local integral and integrated drug policy can only be developed when based upon a local problem analysis. Monitoring is essential because each local entity has its own history and its own drug problems. Based upon this analysis a local strategy and action plan are developed. As a last step, the local initiatives should be evaluated regarding the process of their implementations and regarding their effectiveness (Connolly, 2002; Doherty, 2007).

4.2 To what extent are these best practices implemented in the Flemish Community?

In the study of literature, we have presented effective cross-cutting interventions and essential preconditions in the context of developing an integral and integrated drug policy. These interventions and essential preconditions are better illustrated with reference to a specific example. To this end, we analyse to what extent these effective interventions and best practices have indeed been implemented in the Flemish Community. Our analysis is based upon the interviews and focus group conducted within the framework of the research project 'do's and don'ts in an integral and integrated drug policy' (De Ruyver et al., 2009).

4.2.1 The lack of effect- and process evaluations in Belgium

Monitoring and evaluating is necessary to realize and to adjust an evidence-based policy. With this, it is important to have a high quality level of evidence itself and a high quality of the processes through which this evidence is transformed into policy options (Defra, 2006). An important ingredient for an evidence-based policy are the evaluation data. However, from effect studies it can be indicated that the application of evidence-based policy (especially in drug prevention) is still an uphill battle (Botvin, 2004). So we can identify a gap between academic research and practice (Arthur & Blitz, 2000; Vander Laenen, 2008). Furthermore, in the Flemish Community, effect evaluations measuring the effect of specific interventions involving multiple domains, are absent[5]. Even process evaluations of the various cross-cutting interventions meeting the standards and principles of an academic evaluation are rare. In the Flemish Community, only one process evaluation within this context was found.

In this way, we cannot speak of a high-grade evidence based drug policy in the Flemish Community.

However, we do have information, available from experiences from practice, (based upon the interviews and focus group mentioned before), allowing us to presume that the project at least holds promise and therefore can be mentioned, with some caution, as a "best practice".

4.2.2 Presumptions of "best" and "less favorable practices"

Based upon the best practices identified in the literature review, we can state that a lot of work remains to be done in Flemish Community concerning setting up and improving cross-cutting interventions. In some areas there are more projects than in others. For example, we have various cross-cutting interventions in the context of diversion to treatment by law enforcement and only a few interventions in the context of day and night care and referral to housing projects, in the context of harm reduction, in the context of prevention and early intervention and within the framework of policy development and harmonization. However there is a lack of cross-cutting interventions where other domains are involved and in the context of returning or starting to work.

From the identified process evaluations, the interviews and focus group with practitioners, we already have a general view of the quality of the cross-cutting interventions with regard to the preconditions (identified in the literature review) that are present or lacking. In the context of diversion to treatment by law enforcement, frequently mentioned absences of preconditions are: availability of sufficient (financial) means, availability of sufficient and diverse treatment capacity. In the context of day and night care and referral to housing projects the following preconditions are frequently absent: the existence of a protocol, communication and consultation between the several actors involved and the presence of a uniform financial organization system. Concerning the interventions in the context of prevention and early intervention following shortcomings are mentioned: the absence of sufficient (financial) means, not reaching all the identified actors involved, a lack of selective and indicated prevention, the lack of a clear interpretation of the term 'early intervention', the absence of more and diverse actions within this framework (in most cases the police is one of the partners, although scientific research shows that the provision of information on drugs by the

---

5   Effect evaluations do exist of general interventions (e.g. De Ruyver et.al., Effects of alternative measures and sanctions) but there are no effect evaluations available of specific projects.

police is not appropriate (EMCDDA, 2006) nor effective when vulnerable groups are involved (Hammersley, Ditton & Main,1997).

In this part we will give two examples of "best practices" and two less favourable examples of cross-cutting projects in the Flemish Community. Furthermore we will verify whether the previously mentioned preconditions in the literature review are fulfilled in these interventions.

Best practices and presumptions of cross-cutting "best practices"

As examples of best practices or presumptions of best practices we discuss *Pilot care* (Proefzorg) and a *local Steering group drugs*. Based upon the study of literature, we indicate for these projects in which way they can be identified as a "best practice".

The first intervention that can be identified as a good practice is *Pilot care*, an intervention in the context of diversion to treatment by law enforcement. This intervention was started as a pilot project in the judicial district of Ghent in 2005. The judicial authorities of Ghent decided to respond to the hiatus at prosecution level in the drug policy field, by diverting drug offenders, who confessed to having committed offences stemming from dependency problems, to treatments services. The legal basis for this alternative is based upon the principle of discretionary powers of the prosecutor. The project has three central objectives: 1) the diversion of drug offenders of victimless crimes to treatment rapidly (early intervention), 2) efficiently (with minimal practical obstacles and maximal cooperation between judicial authorities and treatment services) and 3) effectively (with positive results).

In order to implement the project, two new actors were created. Firstly, a "proefzorgmanager" (Pilot care manager), who handles the judicial part of the diversion, was engaged. This Pilot care manager represents the bridge between the criminal justice system and the treatment services. Secondly, two coordination centers were created, where the intake procedures for the treatment centers take place. The coordination centers refer the client to one of the treatment centers in the region (from low threshold centers to residential centers). There are two versions of a Pilot care procedure. The short-version of Pilot care contains one interview in the coordination centre. Mainly non-problematic users who have no problems in other life spheres are received in the short version. The long-version Pilot care lasts six months and contains three interviews in the coordination centre. During this period, a thorough treatment proposal is developed. Especially problematic users, who have problems in other life spheres and previous drug related offences are included in the long version. When there is a positive outcome, the case will be dismissed. A negative outcome will lead to the prosecution of the offender or a diversion towards the drug treatment court.

The pilot-project was evaluated in 2005-2007 (De Ruyver, et al., 2008). This process evaluation comprised of a quantitative and a qualitative part. The qualitative study consisted of an impact- and a process evaluation. The process evaluation found strong evidence of a successful cooperation between the criminal justice system and the treatment services within this project. The majority of the respondents were pleased with their role and had a positive attitude towards the project. The critical elements (the standard form for feedback, the role of the Pilot care manager,...) were recommended as an example for developing or optimizing other alternatives to sanctioning. We can state that, based upon the preconditions derived from the literature review, this project can be indicated as a "best practice". After all, Pilot care anticipated to the known obstacles in an interaction between the criminal justice system and the treatment services and in particular to the problems concerning the professional secrecy and role definition. Furthermore, this project is characterized by a strong bottom-up impulse. During this project, this turned out to be a very important aspect of the success of the cooperation. Before the start of the project, agreements were made between the several actors involved in the project. The judicial authorities and the treatment centers agreed upon the formal aspects of the diversion procedure. Furthermore, the form and content of the feedback the treatment services were to give to the judicial authorities, was discussed. There was also a proper role-definition. The two bridges of the project (the Pilot care manager and the coordination centers) are working independently from one another and both actors have respect for each other's working principles and philosophies.

However, some bottlenecks could be identified namely the legal insecurity (the conditional dismissal is not legally embedded), the absence of sufficient financial means in particular for treatment, the lack of availability of sufficient treatment capacity (especially in crisis and residential centers) and a diversity of treatment centers to give an appropriate response towards every (different) problematic drug user.

The second example of a good practice is the local steering group on drugs in Ghent, which is a cross-cutting intervention within the framework of policy development and policy harmonization. The basic conditions for a best practice in an integral and integrated drug policy have been created in Ghent. Firstly, the drug policy of the city of Ghent operates from a bottom-up approach, by which goals and priorities are set by field workers. Secondly, through a "two-tracked-policy", the drug policy is developed both on a policy as well as on a practical level. The drug phenomenon is dealt with both on a short term in order to answer the ever changing social reality as well as on a long term in order to develop a sound and well substantiated drug policy.

The starting point and the goal of the local drug policy in Ghent is the place and the position of the Municipality and the partners in the field. The Municipality has a pioneer role in the coordination of the local drug policy. This means that the Municipality creates the conditions for the development of the local drug policy. Consequently, the local drug policy is more than the policy of the Municipality: it is a policy of all the different partners in Ghent, developed for the entire population. In this sense, the Municipality is merely one of the partners in the debate (Drug policy plan city of Ghent, 2000).

In December 1998 the steering committee for the drug policy was founded. In this way, the drug policy of the city of Ghent is developed in a systematic and coordinated way. This steering committee is composed of representatives from the domains that are involved in the development and/or execution of the drug policy in Ghent (drug prevention, drug treatment, local police, street corner work, general treatment, the Municipality and provincial drug coordinator). The drug coordinator is chair of this committee. The members of the committee are appointed by the domains they represent.

The steering committee has a vital role in the preparation of the local drug policy, in close cooperation with the drug coordinator. Participation and involvement are at the core of its work. The strategy for the local drug policy and the policy plan are prepared by this committee. To further guarantee the participation and commitment of all domains involved, an open meeting is organized, each time before a crucial step is taken in the local drug policy. During these meetings, all the domains involved in the drug phenomenon are represented. The open meeting has several goals; namely to inform the domains, to get feedback on the preparatory work done by the steering committee and to improve and stimulate the communication between the different domains. The open meeting establishes working groups aimed at the realization of the fixed priorities. These working groups are strictly goal-oriented, and operate within a present time-schedule. The coordinator of the drug policy is the interlink between the working groups, the steering committee and the open meeting. (S)he acts also as the liaison for the communication between the work field, the local government and the supra-local government. During the open meeting, the partners from the work field define the priority actions for the drug policy in Ghent for that working year. The priority actions are these actions aimed at realizing the essential basis conditions to be able to take the next step in the process of developing the drug policy. Communication and cooperation are an essential part of an integrated policy and the inventory of the wants and the needs is a necessary condition to develop a policy based on well-founded choices (Vander Laenen, 2001).

This initiative complies with the postulated preconditions. Moreover, similarity is found with the previous mentioned preconditions of the "communities that care-projects": there is communication and cooperation between all the domains involved in the drug phenomenon. Furthermore, an interlink is created namely the drug coordinator. In this way there is a proper guidance of the different partners.

Unfortunately, there is no participation of neighbours and drug users in the development of the drug policy in Ghent. After all, from the literature review it became clear that drug users are an important target group because they experience the drug problem directly and they receive signals concerning new developments.

Towards some presumptions of unfavorable examples...

We discuss two less favorable examples of cross-cutting initiatives, including the reasons why these are identified as such. The first example is the intervention aimed at treatment for (former) detainees in the prison of the city of Mechelen, the second intervention is the working group "Early intervention".

The first example is an intervention aimed at treatment for (former) detainees in the prison of Mechelen, an intervention in the context of diversion to treatment by law enforcement. At the level of execution of sentences, it is possible to divert detainees to treatment in preparation of their release. This project is called "Centraal aanmeldingspunt" (CAP) (Central registration point). This CAP includes ambulant and residential drug treatment centers in the region. On the basis of intake-interviews, organized in prison, the social worker checks which treatment centre is the best solution for the drug problem of the detainee. After release from prison, the former detainee can start with the treatment program. However CAP is a promising project, a CAP-unit is not installed yet in every prison in Flanders, although extension to all prisons is foreseen. In the prison of Mechelen, for example, there is no CAP. To partially fill this gap, a collaboration exists between an ambulant drug treatment centre and the prison. A social worker of this centre visits the prison regularly at certain points in time to conduct the intake-interviews.

The participants of the interviews and the focus group did not indicate strengths of this project, while several bottlenecks were stressed. Both criminal justice actors and treatment providers indicated that the collaboration between the partners involved is very limited. Only one treatment centre is organizing intake-interviews aimed at referral to its own centre. It is a solo-collaboration of this centre with prison staff. It is clear that extension of the network towards other treatment centers is appropriate so that the most appropriate treatment program can be offered to the (former) detainee. Furthermore, there is a lack of appropriate (financial) means to realize the diversion to treatment. The result is that the time between the intake in prison and the actual admittance to treatment is too long and that waiting lists for intakes are growing. The absence of sufficient financial means is also a reason why other centers do not organize intake-interviews in prison. Diversion to other, especially residential centers is insufficient as well, when ambulant treatment is not appropriate. Residential centers resist treating this target group because they are former detainees, resulting in even less referrals to residential treatment. The absence of a structured collaboration network like the CAP is one of the reasons why (former) detainees do not enter treatment after detention. After all, there is no collaboration structure that puts into operation and structures the diversion of detainees towards treatment. Moreover, treatment (drug related but also for other life domains) during detention is lacking.

To conclude, we can state that (former) detainees with a drug problem are a forgotten target group in which no means are invested. This is not only the case for the prison of Mechelen, but also for Belgium and other countries. Moreover, treatment services are often reluctant to treat those clients (Seal, Eldridge, Kacanek, Binson & MacGowan, 2007; Freudenburg, Daniels, Crum, Perkins, & Richie, 2005; Freudenburg, Wilets, Green & Richie, 1998).

The second example of a less favourable intervention is the Working group "Early intervention" in the city of Antwerp, an intervention in the context of prevention and early intervention. The working group "early intervention" in Antwerp is an initiative of *SODA* (City Platform on Drugs, Antwerp) the local drug coordination unit of Antwerp. This working group was set up with two goals: 1) to answer to the gaps in the provision of early-intervention initiatives 2) to centralize all organizations working around this theme in order to eliminate an existing overlap of similar initiatives (in particular concerning the training of intermediaries).

Positive elements of the intervention are that the working group brings together actors from welfare and health and from treatment. SODA is the initiator of the initiative and has the mandate to centralize all the organizations in that field. As such, SODA facilitates cross-cutting collaboration.

However for this project, several bottlenecks can be identified. There is some tension between the organisations considering the (financial) means. At present, no financial means are available for early-intervention initiatives so it becomes impossible to develop new projects. This leads to a decline in the motivation of the partners involved in the Working group to participate. Furthermore, no collaboration has yet been set up with the schools in the region, although they are an essential partner for this project.

To conclude, there is no clarity of the concept of "early intervention". Early intervention is still a relatively new domain with few realizations. Each partner needs to look within his/her own organisation which possibilities concerning collaboration are possible. Currently, early intervention is limited to trainings for intermediaries. It is still unclear for the involved actors if they have to create new ideas in this context of early intervention or if the existing projects within this context can overcome this.

5   Conclusion

At the European Union level the focus is upon an integral and integrated approach of the drug problem. Both in scientific literature and in policy documents, the concepts of 'integral' and 'integrated' are frequently used without the exact interpretation being entirely clear. Because we aimed to identify international best practices in developing an integral and integrated drug policy, our first goal was to develop a clear and precise interpretation of an 'integral and integrated drug policy'.

We defined *integral* as 'global.' The drug phenomenon is multidimensional and, therefore, all facets must be tackled. The drug phenomenon is related to health, social and economic aspects. It also has an international dimension where both the supply and demand side can determine policy. Where demand reduction derives from interventions with the goal of raising personal resistance to drug consumption, supply side reduction is aimed to reduce access to and consequently use of drugs. For this reason, an integral drug policy is related to several policy domains as a result of which, welfare, health, prevention, treatment and repressive elements are brought together. The aspect *integrated* has a direct link with an integral approach although integral and integrated are two separate concepts and may not be used interchangeable. An approach whereby all aspects of the drug phenomenon are addressed (integral) requires the involvement of all relevant actors and services which represent the different domains (integrated). Both a horizontal concurring between domains and a vertical concurring between all competences is required to tackle the drug phenomenon from several angles. The various competences (welfare, public health, policing and judicial authorities, environmental services,...) involved in the approach of the drug phenomenon are located on federal, regional, provincial and local level.

This definition, applied to search for our evaluation studies, can be used by policy makers and scientific researchers when developing and evaluating cross cutting interventions in the context of an integral and integrated approach.

When we searched for international best practices of cross-cutting interventions in the context of an integral and integrated approach of the drug phenomenon, we found few evaluation studies. After the literature review, 27 evaluation studies were selected. Twelve studies evaluate the effect of a cross-cutting intervention, eight studies evaluate the process of an intervention and seven studies evaluate both the effect and the process of an intervention.

These findings confirmed our hypothesis that evaluations of interventions – that are often locally implemented – are seldom reported on in international databases or distributed internationally. This highlights the importance of evaluating cross-cutting interventions as well as publishing and reporting about the evaluation studies.

Based upon the effect evaluations, we could identify various *effective* cross-cutting interventions, both in the context of diversion to treatment by law enforcement (e.g. the arrest referral-schemes), in the context of returning to or starting work, in the context of day and night care and referral to housing projects, in the context of harm reduction (e.g. user rooms), in the context of prevention and early intervention (e.g. Communities that care-project), with the involvement of other domains and cross-cutting interventions within the framework of policy development and concurring.

Based upon the process evaluations, we could identify the following similar pre-conditions across the different cross-cutting interventions: coordination between the partners involved, a formal agreement between the partners involved, including a clear demarcation of tasks and responsibilities and communication, including regular meetings between the partners involved.

The results of the study of literature can be used to modify or rationalize cross-cutting interventions in drug and other policy domains. Research into effective cross-cutting interventions and interventions meeting essential preconditions is important for each policy domain to meet the requirements of an evidence-based policy.

When developing an integral and integrated drug policy, is has become clear that more evaluation studies of cross-cutting interventions are definitely needed, both at national and international level.

When developing an evidence based policy, it is absolutely necessary that we obtain more effect and process evaluation studies, especially for a societal delicate subject as drugs (Agar & Reisinger, 2000).

Although research has shown that policy makers seldom use scientific data to guide their decisions (Gorman, 1998; Waddell et al., 2005), this should not restrain both policy makers and scientific researchers to take their proper responsibilities in order to overcome this bottleneck.

6 Bibliography

Arthur, M.W. & Blitz, C. (2000). Reaching the gap between science and practice in drug abuse prevention through needs assessment and strategic community planning, *Journal of community psychology*, 28(3), 241-255.

Autrique, M., Vanderplasschen, W., Pham, T., H.,, Broekaert, E., & Sabbe, B. (2007). *Evidence-based werken in de verslavingszorg: een stand van zaken*. Gent: Academia Press.

Barton, A. (1999a). Breaking the crime/drugs cycle: the birth of a new approach? *The Howard Journal, 38*(2), 144-157.

Barton, A. (1999b). Sentenced to treatment? Criminal justice orders and the health service. *Critical Social Policy, 19*(4), 463-483.

Bieleman, B., Biesma, S., Hoorn, M., & Kruize, A. (2007). *Overlastmetingen in de Gravinnesteeg en omgeving*. Groningen: Intraval

Botvin, G., J, & Griffin, K., W. (2003). Drug abuse prevention curricula in schools. In Slobada, Z. &

W. Bukoski, J (Eds.), *Handbook of drug abuse prevention. Theory, science, and practice.* New York: Kluwer Academic/Plenum Publishers.

Brook, J., S., Brook, D., W., Richter, L., & Whiteman, M. (2003). Risk and protective factors of adolescent drug use: implications for prevention programs. In Z. Slobada & W. Bukoski, J. (Eds.), *Handbook of drug abuse prevention. Theory, science, and practice.* New York: Kluwer Academia.

Connolly, J. (2002). *Community policing & drugs in Dublin.* Dublin: The North inner City Community Policing Forum.

Crow, I., France, A., Hacking, S., & Hart, M. (2004). *Does Communities that Care work? An evaluation of a community-based risk prevention programme in three neighbourhoods.* York: Joseph Rowntree Foundation.

Coppel, A. (2008). *Drug use, front line services and local policies. guidelines for elected officials at the local level*: European Forum for urban safety. Democracy, cities & drugs projects

Decorte, T., De Ruyver, B., Ponsaers, P., Bodein, M., Lacroix, A. C., Lauwers, S., et al. (2004). *Drugs en overlast. Drogues et nuisances.* Gent: Academia Press.

De Maeseneire, I., & Rosiers, J. (2003). *Evaluatie. Een thema in de kijker.*: Vereniging voor Alcohol- en andere Drugproblemen.

De Pauw, K. (2007). Een veiligheidsnetwerk rond druggebruik te Geraardsbergen en Lierde. In W. Bruggeman (Ed.), *Cahiers Integrale Veiligheid. Thema Regie.* Brussel: Politeia.

De Ruyver, B., Casselman, J., Meuwissen, K., Bullens, F., & Van Impe, K. (2000). *Het Belgisch drugbeleid anno 2000: een stand van zaken drie jaar na de aanbevelingen van de parlementaire werkgroep drugs.* Gent: Onderzoeksgroep Drugbeleid, Strafrechtelijk beleid en Internationale criminaliteit.

De Ruyver, B., Vermeulen, G., Vander Beken, T., Vander Laenen, F. (2003), *International Drug Policy, Status Quaestionis, Compendium of Articles,* Antwerpen-Apeldoorn, Maklu.

De Ruyver, B. (2007). Integraal drugbeleid. In J. Casselman & H. Kinable (Eds.), *Het gebruik van illegale drugs. Multidimensionaal bekeken.* (pp. 325-341).Heule: UGA.

De Ruyver, B., Ponsaers, P., Lemaître, A., Macquet, C., De Wree, E., Hodeige, R., et al. (2007). *Effecten van alternatieve afhandeling voor druggebruikers.* Gent: Academia Press.

De Ruyver, B., Colman, C., De Wree, E. Vander Laenen, F., Reynders, D., Van Liempt, A., et al. (2008). *Een brug tussen justitie en drughulpverlening. Een evaluatie van het proefzorgproject.* Antwerpen: Apeldoorn.

De Ruyver, B., Lemaître, A., Ponsaers, P., Born, M., Pauwels, L., Vander Laenen, F., et al. (2009). *Do's and don'ts in een integraal en geïntegreerd drugbeleid* Gent: Academia Press.

De Wree, E., De Ruyver, B., Verpoest, K., Colman, C. (2008). All in favour? Attitudes of drug users and stakeholders towards judicial alternatives, *European journal on criminal policy and research*, 14, 431-440.

Doherty, G. (2007). *Report of the community policing forum for 2006-2007*. Dublin.

European Monitoring Centre for Drugs and Drug Addiction. (2006). *Annual report 2006: The state of the drugs problem in Europe*. Lisbon: European Monitoring Centre for Drugs and Drug Addiction.

European Monitoring Centre for Drugs and Drug Addiction. (2008). *Annual report 2008: the state of the drugs problem in Europe*. Lisbon: European Monitoring Centre for Drugs and Drug Addiction.

European Monitoring Centre for Drugs and Drug Addiction. (2008). *Selected Issue, National drug-related research in Europe*. Lisbon European Monitoring Centre for Drugs and Drug Addiction.

Federale Regering. (2001). *Beleidsnota van de Federale Regering in verband met de drugproblematiek*. Brussel: Federale Regering.

Freudenburg, N., Daniels, J., Crum, M., Perkins, T., & Richie, B. E. (2005). Coming home from jail: the social and health consequences of community reentry for women, male adolescents, and their families and communities. *American Journal of Public Health*, 95(10), 1725-1736.

Freudenburg, N., Wilets, I., Green, M., & Richie, B. E. (1998). Linking women in jail to community services – factors associated with rearrest and retention of drug-using women following release from jail. *Journal of American Medical Women Association*, 1998, 89-93.

Hammersley, R., Ditton, J. & Main, D. (1997) Drug use and sources of drug information in a 12-16-year old school sample. *Drugs: Education, prevention and policy*, 4, 231-241.

Harman, K., & Paylor, I. (2005). An evaluation of the CARAT initiative. *The Howard Journal* 44(4), 357-373.

Heale, P., & Lang, E. (2001). A process evaluation of the CREDIT (court referral and evaluation for drug intervention and treatment) pilot programme. *Drug and Alcohol Review*, 20, 223-230.

Heed, K. (2006). If enforcement is not working, what are the alternatives? *International Journal of Drug Policy*, 17(2), 104-106.

Hunter, G., McSweeney, T., & Turnbull, P., J. (2005). The introduction of drug arrest referral schemes in London: A partnership between drug services and the police. *The International Journal of Drug Policy*, 16, 343-352.

Holloway, K., Bennett, T., & Farrington, D. (2005). *The effectiveness of criminal justice and treatment programmes in reducing drug-related crime: a systematic review*. London: Home Office, Research Development and Statistics Directorate.

Institute for Criminal Policy Research. (2007). *National evaluation of criminal justice integrated teams.* London: The Institute for Criminal Policy Research, University of Bristol.

Kellock, P. (2007). *Communities that Care. Review of implementation in three Australian communities.* Brunswick: The Asquith Group.

Kibel, B., M., & Holder, H., D. (2003). Community-focused drug abuse prevention. In Z. Sloboda & W. Bukoski, J. (Eds.), *Handbook of drug abuse prevention. Theory, science, and practice.* New York: Kluwer Academia.

Koeter, M. W. J., & Bakker, M. (2007). *Effectevaluatie van de Strafrechtelijke Opvang Verslaafden (SOV).* Den Haag: WODC – Ministerie van Justitie.

Kumpfer, K., L., & Alder, S. (2003). Dissemination of research-based family interventions. In Z.

Sloboda & W. Bukoski, J. (Eds.), *Handbook of drug abuse prevention. Theory, science, and practice.* New York: Kluwer Academic.

Leeuw, F. L. (2005). Trends and developments in program evaluation in general and criminal justice programs in particular. *European Journal on Criminal Policy and Research, 11,* 223-258.

Loosveldt, G. (2001). Experimentele designs. In J. Billiet & H. Waege (Eds.), *Een samenleving onderzocht. methoden van sociaal-wetenschappelijk onderzoek* (pp. 157-179). Antwerpen: Standaard Uitgeverij.

Mair, G., & Barton, A. (2001). Drugs Throughcare in a local prison: a process evaluation. *Drugs: education, prevention and policy, 8*(4), 335-345.

Mazerolle, L., Soole, D., & Rombouts, S. (2007). Drug law enforcement. A review of the evaluation literature. *Police Quarterly, 10*(2), 115-153.

Michon, H., Rondez, M., & van Weeghel, J. (2000). *Een werkend middel. Evaluatie van*

Baanberekend. Arbeidstoeleidingsproject voor mensen met een verslavingsachtergrond. Utrecht: Trimbos-instituut.Ministerie van Justitie (2006).

Pentz, M., A. (2003). Anti-drug-abuse policies as prevention strategies. In Z. Slobada & W. Bukoski, J. (Eds.), *Handbook of drug abuse prevention. Theory, science, and practice.* New York: Kluwer Academia.

Rigter, H., van Gageldonk, A., Ketelaars, T., & van Laar, M. (2004). *Hulp bij probleemgebruik van drugs. Stand van wetenschap voor behandelingen en andere interventies, 2004. Achtergrondstudie Nationale Drug Monitor.* Utrecht: Trimbos Instituut.

Ritter, A., Bammer, G., Hamilton, M., Mazerolle, L., & The DPMP Team. (2007). Effective drug policy: a new approach demonstrated in the Drug Policy Modelling Program. *Drug and Alcohol Review, 26,* 265-271.

Scottish Executive Effective Interventions Unit. (2004). *Reducing the impact of local drug markets.* A research review.: Scottish Executive Effective Interventions Unit.

Seal, D. W., Eldridge, G., Kacanek, D., Binson, D., MacGowan, R. J., & The Project START Study Group (2007). A longitudinal, qualitative analysis of the context of substance use and sexual behavior among 18- to 29-year-old men after their release from prison. *Social Science & Medicine 65*(2007), 2394-2406.

Seeling, C., King, C., Metcalfe, E., Tober, G., & Bates, S. (2001). Arrest Referral-a proactive multi- agency approach. *Drugs: education, prevention and policy, 8*(4), 327-333.

Sherman, L. W., Farrington, D. P., Welsh, B. C., & Mackenzie, D. L. (Eds.). (2002). *Evidence-based crime prevention.* London: Routledge.

Skodbo, S., Brown, G., Deacon, S., Cooper, A., Hall, A., & Millar, T. (2007). *The Drug Interventions programme (DIP): addressing drug use and offending through 'Tough Choices'.* London: Home Office, Research Development and Statistics Directorate.

Snippe J., Bieleman, B., Kruize, A., & Naayer, H. (2005). *Eindevaluatie: inspanningen, proces en resultaten 2001-2004. Hektor in Venlo.* Groningen: Intraval.

Snippe, J., Naayer, H., & Bieleman, B. (2006). *Hektor in 2005. Evaluatie aanpak drugsoverlast in Venlo.* Groningen: Intraval.

Spijkerman, R., Biesma, S., Meijer, G., van der Poel, A., van den Eijnden, R., & Bieleman, B. (2002). *Vier jaar verantwoord schoon. Evaluatie van Verantwoord Schoon: programma voor aanpak van drugsoverlast in Rotterdamse deelgemeenten.* Rotterdam: Instituut voor Onderzoek naar Leefwijzen & Verslaving (IVO) /Stichting INTRAVAL.

Stevens, A., Trace, M., & Bewley-Taylor, D. (2005). *Reducing drug related crime: an overview of the global evidence.* Report Five. Oxford: The Beckley Foundation.

Stevenson, J. F., & Mitchell, R. E. (2003). Community-level collaboration for substance abuse prevention. *The Journal of Primary Prevention, 23*(3), 371-404.

Swanborn, P. G. (1999). *Evalueren.* Amsterdam: Boom.

Toumbourou, J. W. (1999). *Implementing communities that care in Australia: a community mobilisation approach to crime prevention.* Retrieved 31/03/08. from http://www.aic.gov.au/publications/tandi/ti122.pdf.

Vander Laenen, F. (2008). *Drugpreventie bij kwetsbare groepen? Jongeren met een gedrags – en emotionele stoornis aan het woord.* Den Haag: Boom.

Vander Laenen, F., 'The drug and cannabis policy of the city of Ghent, Belgium', in European City Conference on Cannabis Policy, 6 – 8 December 2001, Utrecht, The Netherlands, *Drug Policy in European countries and cities, Reference book*, Ministry of Justice, ES&E, the Netherlands, p. 41 – 62.

Van der Poel, A., Barendregt, C., & van de Mheen, D. (2003). Drug consumption rooms in Rotterdam: an explorative description. *European Addiction Research, 9*, 94-100.

Van Ooyen-Houben, M. (Ed.). (2008). *Quasi-compulsory treatment in the Netherlands: promising theory, problems in practice.* Oxford/Brighton: Pavilion Publishers UK.

Van Ooyen-Houben, M. (2008). Usage de substances illicites et politique néerlandaise en matière de drogues: vue d ensemble et évaluation exploratoire. *Déviance et Société*, Vol. 32, 3, 325-348.

Wits, E. G., Biesma, S., Garretsen, H. F. L., & Bieleman, B. (1999). *Evaluatie SVO-projecten. Rotterdam. Eindrapport.* Rotterdam/Groningen: IVO/Intraval.

Zurhold, H., Degkwitz, P., Verthein, U., & Haasen, C. (2003). Drug consumption rooms in Hamburg, Germany: Evaluation of the effects on harm reduction and the reduction of public nuisance. *Journal of Drug Issues*, 663-688.

Explaining Violence and Aggression on Public Transport – Literature on Typology and Etiology Applied

Neil Paterson
Patrick Moreau
Gert Vermeulen
Marc Cools

1   Abstract

Questions concerning crime, safety and security have become and continue to be a hot topic in many western European countries with Belgium being no exception. A number of high profile incidents, although atypical in their severity, have focused attention on problems of violence and aggression on public transport in Belgium. As part of a wider research project aiming to improve knowledge of violent incidents in this area from the offender's perspective and thus contribute to their prevention, this article explores a number of related questions. What is the extent of the violent crime problem on the Belgian public transport system? Are there differing forms of violence on public transport and if so, must we search for differing etiological explanations? Does academic literature and criminological theory provide us with any helpful explanations as to the causal factors – both personal and situational – for violence committed in the public transport arena? In light of the above, will situationally based crime prevention initiatives prove sufficient to address the problems of violent crime on public transport or do we need to complement them with other types of intervention?

2   Introduction

Questions relating to safety and security have assumed increasing importance during recent decades. In many west European countries, the topic of public security in various guises now occupies a prominent place on the public and political agenda. In Belgium, a number of factors have been cited as drivers for this development including the rise of right wing political parties in the early 1990's, the 'Dutroux' crisis, the ensuing mobilisation of public opinion and consequent reforms of the country's policing system.[1] More recently, a number of high profile violent crime cases such as the

---

1   Maesschalck, J & Ringeling, A, (2008). What goes up must come down? The career of 'safety and security' as a policy issue in Governance of Security in the Netherlands and Belgium, Cachet, I., De Kimpe, S, Ponsaers, P & Ringeling, A (Eds), Den Haag, Boom Juridische Uitgevers, pages 316-322

murder of Luna Drowart and her child minder Oulemata Niangadou in an Antwerp street in 2006 have served to maintain this momentum.

Issues concerning violence and aggression have also been construed as a problem in and around public transport. The murder of Joe Vanholsbeeck and the death of Guido De Moor, although atypical of the general problematic in terms of their severity, served to highlight to the public at large that public transport was not immune from problems of violence and aggression.[2] Consequently, the Belgian Home Office provided funds to Gent University in 2008 for a research project concentrating on the perpetrators of violence on public transport. Perpetrators were identified from judicial case files in Belgium's five largest cities: Antwerp, Brussels, Charleroi, Ghent and Liège. Thereafter, semi-structured interviews with perpetrators were conducted which aimed to explore the processes and mechanisms surrounding violence on public transport. The specific focus of the project was not coincidental. It has been noted that there is 'very little research aimed at improving our understanding of the origins and escalation of violent interactions. There is also little well-founded knowledge concerning the prevention and control of violence'.[3] The research aimed to fill this gap at least in part.

A broad based international literature review was undertaken during the preparatory phase of the project. Drawing upon the findings of this literature review, this article explores a number of related questions:-

• What is the extent of the violent crime problem on public transport in Belgium?
• Are there differing forms of violence on public transport and if so, must we search for differing etiological explanations?
• Does academic literature and criminological theory provide us with any helpful explanations as to the causal factors – both personal and situational – for violence committed in the public transport arena?
• In light of the above, will situationally based crime prevention initiatives prove sufficient to address the problems of violent crime on public transport or do we need to complement them with other forms of intervention?

3   The extent of violent crime on public transport in Belgium

Precise up to date information on the extent of violent crime on the Belgian public transport system is not easy to come by. Federal Police statistics show that some 73,000 offences of intentional assault were recorded in 2008 (the most recent year for which full statistics are available) and that this represented an increase of 19% in the five year period since 2003. Unfortunately, no breakdown is provided of the various locations in which these crimes were committed thereby making it impossible to

---

2   In the wake of the murder of Joe Vanholsbeeck in Brussels Central station in 2006, increased feelings of insecurity among young people aged 15-24 (the victim's own age group) were recorded. Interestingly, the murders of Luna Drowart and Oulelata Niangadou appeared to have no perceptible impact on feelings of insecurity amongst the population at large. See Van Den Bogaerde, E, Van Den Steen, I, Klinckhamers, P & Vandendriessche, M (2007). Veiligheidsmonitor 2006, Federale Politie – Directie van de nationale gegevensbank

3   Adang, O. M. G. (2000). "Jonge Mannen In Groepen – Een Geweldige Combinatie?." Justitiële Verkenningen(1), page 79

establish whether these increases also applied to crimes of violence within the public transport arena. Federal Police statistics do however indicate that some 2,100 recorded crimes in the general category of 'violence against the person' were committed in locations directly related to public transport during 2008 (train, bus or metro stations, on trains, buses or the metro or at bus and tram stops). This represented some 2.5% of overall recorded violent crime during the year. 33% of incidents took place in the Brussels region, 39% in the region of Flanders and 28% in the region of Wallonia. The highest number of offences were committed in or around train, bus or metro stations with 45% of these taking place in Brussels. It is not possible to discern from these statistics, however, the differing types of violent crime from which the global total is comprised: nor is time series comparison information available.

Police statistics also provide us with a picture of the extent and distribution of theft within the public transport system. In 2008, approximately 19,000 such offences were recorded representing 5% of overall recorded thefts in Belgium. 52% of incidents occurred in Brussels, 37% in Flanders and 11% in Wallonia. Again, the highest number of incidents took place in or around train, bus or metro stations although a significant number of thefts were also committed on trams, trains and the metro itself. As with violent crime against the person, the greatest incidence of theft on public transport can be found in Brussels where 52% of the total crimes were recorded. It is not possible to extrapolate from these figures the proportion of thefts on public transport which actually involved violence or the threat thereof. Interestingly, the incidence of both violent crime and theft on public transport in Brussels is significantly higher than one might expect given the city's total population. As indicated, in 2008 33% of violent crimes and 52% of thefts on public transport in Belgium occurred in Brussels whilst the city's population comprises approximately 10% of the country's total.[4]

In common with many other forms of criminality there are, of course, a number of difficulties regarding the use of official crime statistics to obtain an accurate picture of violent crime on public transport. It has been highlighted, for example, (Kyvsgaard 2003) that judicial sources should not in themselves be automatically considered as a reliable source of data owing to, for example, inaccuracies or inconsistencies in police recording, offence categorisation and information provided by victims and witnesses. Equally, the definition of violent crime deployed will have an impact on the type of incidents included and, therefore, the kinds of interventions which are considered to address the problem. The definition of violent crime on public transport employed by the Police in Belgium has two components. Specifically, an incident must concern a reported crime of violence or a crime involving the direct threat of violence and have been committed in a place relating to public transport (in or around train, bus or metro station or, a bus or tram stop or, on the following forms of public transport: train, tram, metro or bus). Whilst such a definition may be satisfactory from legal perspective, a more accurate picture of crime and safety on public transport would be gleaned by adopting a 'whole journey approach' and thus extending the definition to include incidents which occurred victims were walking to, from and between bus and tram stops or train, bus or metro stations (Newton, Johnson and Bowers 2004).

---

4   Politiële Criminaliteitsstatistieken – België 2000/2008, Federale Politie – CGOP – Beleidsgegevens, 2009.

Maklu   265

There is, furthermore, a strong possibility that a significant amount of violent crime on public transport remains unreported and recorded by the police or other responsible authorities. This issue was highlighted in a 2004 study concerning violence against staff on railways in the United Kingdom which found that only 52% of physical assaults on staff were reported to an appropriate authority. Reasons cited included a belief that nothing would happen, that the incident was not serious enough to report and that assaults were an accepted 'part of the job. [5] On the part of passengers, reluctance to delay one's journey, lack of confidence that the offender would be apprehended, the physical absence of someone to actually report an incident to and the belief that a reported incident will not be taken seriously have been cited as reasons for non-reporting behaviour in the specific context of public transport.[6] In addition, the most recent Belgian Crime Survey estimated that only 34% of all crimes were reported to the police and that only 25% resulted in the production of a dossier requiring further action. Reporting levels were particularly low for crimes involving the threat of violence and for sexual offences each of which were reported by less than 10% of victims.[7] There appears no reason to suppose that these trends will not impact on the reporting of crime within the Belgian public transport system.

Finally, we should note that the three regional public transport authorities in Belgium (De Lijn in Flanders, MIVB in Brussels and TEC in Wallonia) and the national rail service provider NMBS do not operate a uniform recording system for violent incidents. Moreover, only one of these providers, De Lijn, actually publishes information on violent incidents which occur within its jurisdiction.

4   Typologies of violent crime on public transport in Belgium

In light of the difficulties in effectively ascertaining the extent of violent crime on public transport in Belgium, it should probably not surprise us that relatively little information is readily available concerning the typologies of incidents. Information from public transport providers, academic literature and analysis of legal dossiers does, however, indicate that the nature of violence on public transport in Belgium encompasses a relatively diverse range of scenarios, behaviours and actors. For example, the Safety and Security survey undertaken in 2008 by the Flemish public transport provider *De Lijn* highlighted that violent incidents can revolve around a variety of participants – passenger(s) and staff member(s), non-passengers and staff members and incidents between passengers themselves. Such incidents involved a range of behaviours including physical aggression, sexual intimidation and assault, spitting, threatening behaviour (both with and without the use of weapons) and thefts (including thefts with violence).[8]

By studying dossiers concerning anti-social behaviour committed by young people on and around public transport, De Wree, Vermeulen and Christiaens (2006) also

---

5   Summerell, J, Shorrocks, T & Mitchell, K (2004). Reducing Assaults On Railway Staff – A Report For Railway Safety, Rail & Safety Standards Board, page 17

6   Cozens, P, Neale, R, Hillier, D, & Whitaker, S (2004). Tackling Crime and Fear of Crime While Waiting at Britain's Railway Stations, Journal of Public Transportation, volume 7, no. 4, University of South Florida, page 24

7   Van Den Bogaerde et al

8   Veiligheidsmonitor 2008 http://www.delijn.be/over/veiligheid/eerste_resultaten_veiligheidsmonitor. htm?ComponentId=7566&SourcePageId=6028

identified that a wide range of incident types were recorded by both the Police and public transport providers. These ranged from thefts involving violence to assault, threats of assault, sexual intimidation and brawls. As above, these incidents involved varying configurations of actors – violence amongst young people themselves, violence directed at public transport personnel, violence directed toward adult passengers and violence perpetrated for material gain.[9]

Finally, information gleaned by the authors themselves from a review of 49 legal dossiers in Antwerp for the year 2008 found that a significant majority (73%) of registered cases involved violence where no financial or material motive was apparent. In the majority of cases, 83%, the offender acted alone. 86% of offenders were public transport passengers a status shared by 61% of victims: the remaining victims were employees of public transport providers. 77% of offenders were male. 83% of offenders were adults at the time the offence was committed. 20% of all the analysed cases centred around violence committed during checks of tickets or other travel documents by bus, tram or train inspectors. Interestingly, a parallel analysis of legal dossiers carried out for the central police district in Brussels unearthed some differing characteristics: approximately 70% of cases registered in 2008 and the first part of 2009 involved violence with a financial or material motivation whilst in the remainder no such motive could be discerned. Incidents were, furthermore, committed by a more even mixture of offenders acting alone and in groups than those in Antwerp.

5   Personal and situational causal factors

In the following paragraphs, an overview of literature which casts some light on both the personal and situational causal factors of violent behaviour with relevance to public transport is provided. The themes selected are based on the diverse typologies highlighted above and include consideration of:-

• the differing motivational drivers contributing to the commission of an offence
• the status and selection of the victim
• the role of alcohol and drugs
• whether the offence was committed individually or in a group
• situational factors relating to violent crime on public transport

It should be noted at the outset, however, that the exercise was of necessity limited in scope being constrained by the lack of dedicated research into the causes of violence from the offender's perspective and the fact that within a relatively broad canon of research into generic criminal behaviour, very little attention has been paid to the specific problem of violence and serious violence in particular.[10]

It has been noted that 'violence, like any complex behaviour, has multiple and heterogeneous etiologies, and there is no unitary type of person who is violent.'[11] By way

---

9   De Wree et al. pages 25, 309-310
10   Guerra, N.G. (1998), Serious & Violent Juvenile Offenders: Gaps in Knowledge & Research Priorities, in, Loeber, R., Farrington, D.P., Ed. (1998). Serious & Violent Juvenile Offenders – Risk Factors & Successful Interventions, Sage Publications. See also Levy & Maguire, page 832
11   Reiss, A.J, Roth, A (ed), Understanding & Preventing Violence, Panel on the Understanding and Control of Violent Behaviour, National Research Council, 1993, page 361

of illustration, Toch (1992), writing from a social psychology perspective, conceptualised several distinct types of violent offender: the 'self-image demonstrator', someone who uses violence as a means of demonstrating toughness to gain the admiration of his peers; the 'self-image defender', someone who is easily slighted or insulted and will react violently in order to defend their ego and the 'reputation defender', who acts to defend the values of the group to which they belong when they believe it to be under threat. The form of violent behaviour exhibited alongside the situation in which it occurs and potential trigger points are likely to differ for each of these types of offender.[12]

A distinction has also been drawn in research between what is described as instrumental and expressive forms of violence. The former is thought to be characterised by a degree of rationality and might include obtaining money or re-saleable goods through violent robbery. Such incidents may be more likely to occur in situations where supervision and oversight is low. Expressive violence, by comparison may be construed as satisfying or functional from an offender's perspective but without necessarily being underpinned by a clear rationale or economic motive. As such, illustrative examples of expressive violence would be street fights, football hooliganism or, in an institutional setting, violence committed against prison officers by inmates. Interestingly, such violence can still occur (or may even be more common) in certain settings and situations where supervision and oversight is high in that this can sometimes act as a trigger to potential offenders.[13] Such situations could plausibly include violence committed by passengers against public transport employees during, for example, the ticket control process.

If the distinction between instrumental and expressive violence was clear cut and valid, two differing and distinct forms of prevention strategy could evidently be devised. Such a relatively crude bifurcation of motive and motivation has, however, been called into question in a number of studies. In their research into violent street robbery, Jacobs and Wright (2007) explored offenders' motivations and decision making processes in the run up to the commission of an offence. Whilst many offenders cited an ostensibly rational need for money to meet financial and addiction problems, their selection of targets was often more opportunistic and instinctive than premeditated. Financial need was, furthermore, somewhat unconventionally defined and included the ability to maintain consumption of non-essential status enhancing items. Interestingly however, the choice of robbery over other offences appeared to be considerably more rational in that it was seen as an easier and less risky method of making money than other legal or illegal alternatives.[14] Similarly in their research with carjackers, Topali and Wright (2007) found that offences triggered by 'pure' opportunity or 'pure' need were relatively rare. More usually, the motivation or trigger lay somewhere between these two extremes.[15] The implied specialism in offence selection referred to

12 Toch, H (1992). Violent Men – An Enquiry Into The Psychology Of Violence, American Psychological Association, pages 135-162
13 Gadon, L., Johnstone, L., Cooke, D. (2006). "Situational Variables And Institutional Violence: A Systematic Review Of The Literature." Clinical Psychology Review 26, page 524. This finding relates specifically to 'expressive' violence within a prison setting.
14 Jacobs, B.A, & Wright, R, Stick-Up: Street Culture & Offender Motivation in, Pogrebin, M., Ed. (2007). A View Of The Offender's World About Criminals, Sage Publications, pages 63-67
15 Volkan, T. and Wright.R, Dubs and Dees, Beats and Rims: Carjackers and Urban Violence in, Pogrebin, M., Ed. (2007). A View Of The Offender's World About Criminals, Sage Publications, pages 72-76

above appears, however, to run counter to other studies which have found that violent offenders also tend to commit non-violent crimes: in other words, they are versatile rather than specialised. Linkages between differing forms of violent behaviour have also been found, in that people who commit one form of violence have a relatively high probability of committing another.[16] The frequency and amount of violent offences committed by individual offenders is not easy to determine. Farrington (1989) found that while nearly all of the offenders in his longitudinal study were convicted of non-violent as well as violent crimes, only a quarter of their crimes were violent. Moreover, whilst 70% of offenders were convicted of only one violent offence, they were frequent general offenders who appeared to 'graduate' to violence after property crime.[17]

Various studies have also been conducted which attempt, in one way or another, to cast some light on the role of alcohol and drugs in violent crime. By way of example, the 2005 European Crime and Safety Survey identified a positive correlation across 18 European Union countries between rates of violent crime and the consumption of alcohol.[18] Similarly, crime survey data from the United Kingdom has established that in over half (53%) of violent incidents where the perpetrator and victim were previously unknown to each other and in over a third (36%) of acquaintance violence incidents, the victim described the perpetrator as being under the influence of alcohol[19] whilst American research (Frieze and Browne 1989) established a clear relationship between alcohol use and violence in general and against partners in particular. 'Alcohol-related violence' is often associated with places where young people meet and where disputes may arise both in and around drinking establishments. Such disputes often extend into the public transport arena as people attempt to make their way home following a night out.[20] It is, however, overly simplistic to regard alcohol alone as a sufficient precursor to violent behaviour in that not all heavy drinkers are prone to violent behaviour and even violent offenders who drink heavily are not violent every time they drink. Nonetheless, the opportunities for certain forms of interaction combined with the kinds of fragile self-respect described by Toch above, would appear to implicate alcohol, at the very least, in the process of becoming violent.[21] In contrast, there is very little evidence to suggest that the pharmacological effects of cannabis, hallucinogens or opiates make people violent (at least when they are consumed alone). Clearer linkages have, however, been found between violent behaviour and the use of amphetamines or solvents albeit that these findings need to be treated with same degree of

16 Reiss, A.J, Roth, A (ed), Understanding & Preventing Violence, Panel on the Understanding and Control of Violent Behaviour, National Research Council, 1993, pages 359-361. See also Pease, K. (2005) No Through Road Closing Pathways To Crime, in Crime Reduction & The Law, Moss, K. S., M., Ed. Routledge, page 55 and Whyte, B. (2001). Effective Intervention for Serious and Violent Young Offenders, Criminal Justice Social Work Centre For Scotland, page 1

17 Farrington, D. (1989) Early Predictors Of Adolescent Aggression & Adult Violence, Violence & Victims 4, Springer Publications, pages 307- 331

18 Van Dijk, J, Van Kesteren, J, Nevala, S, Hideg, G. (2005) The Burden of Crime in the EU: Research Report – A Comparative Analysis of the European Crime and Safety Survey (EU ICS 2005), EUICS Consortium, page 2 and appended tables

19 Mattinson, J, (2001). Stranger And Aquaintance Violence: Practice Messages From The British Crime Survey, Briefing Note 7/01, London: Home Office, page 3

20 (2007). Youth Violence Alcohol & Nightlife – Late Night Transport. W. H. O. G. C. F. V. Prevention. Liverpool, John Moores University

21 Levy & Maguire, pages 827-828.

caution as those which emphasise the causal link between alcohol and violence. [22] In both instances, however, it is probably an exaggeration to consider alcohol and drug related violence as distinct forms of violence: rather, we should note they role that they can play alongside other factors in contributing to violent behaviour in the public transport arena as elsewhere.

Research has also attempted to explore the characteristics of violence committed in groups much of which starts from the presumption that 'exceptional' behaviour is somehow easier to carry out in a group setting with its associated features of relative anonymity, group solidarity and peer pressure. A piece of observatory research conducted with football supporters in the Netherlands by Adang (2000) concluded that the existence of a group can in itself sometimes act as an escalatory factor which leads to violence because communication between potential aggressors, victims and bystanders which might mitigate the situation is harder to hear and interpret. He saw, furthermore, that the greater the prospect of participants remaining anonymous, the greater the probability that violence would be used. Such violence was not, however, employed by all gang members as even in sharply escalating situations, only some 10% of the group became actively involved with the rest acting as onlookers. Other factors influencing the use of violence were primarily physical: the availability of weapons, knowledge of a particular area or territory and the associated possibility of a safe escape. The number of people in a group was not seen to be important as a contributory factor in violent behaviour. Rather, the characteristics of the group and the relationship between its members were of greater significance. Interestingly, Adang does not subscribe to the view which sees gangs and gang members as essentially apart from mainstream society pointing out that many of the factors which appear to contribute to violence in group situations are also features of everyday life: obedience, conformity, prestige and solidarity with peers. [23]

A small number of studies have concentrated specifically on the factors which can contribute to violence on or around public transport. In a research project focusing on the public transport system of Birmingham, a major British city, Burrell (2007) highlighted how public transport hubs can act as both generators and attractors for crime. A crime generator is characterised as a place where significant numbers of people are attracted for reasons entirely unrelated to criminal motivation. Consequently, large numbers of criminal opportunities become manifest as potential offenders and targets are brought together both spatially and temporally. Crime attractors, on the other hand, are places which offer many criminal opportunities that are well known to offenders. People with criminal motivations are attracted to such locales to commit crime. Within the context of public transport, Burrell encourages us to view large stations or bus terminals as generators because they create opportunities for crime due to the amount of people using them. The concept of an attractor would be more appropriate for particular stations or bus stops which are poorly lit, unstaffed or have gained a bad reputation for crime. She also makes an interesting assertion based on previous research (Newton 2004) that the incidence of crime at a particular bus stop, station, bus, tram or train route is likely to be higher in areas where the incidence of crime is higher in the local community. Burrell's research found evidence that violent crime

---

22   Ibid. Violence associated with drugs tends to be that associated with the commercial aspects of its sale and distribution i.e. a desire to defend to establish market position
23   Adang- page 79

did indeed cluster around transport hubs including bus stops, train stations and tram stops: evidence that the problems were more pronounced in high crime communities was, however, not established.[24] In a Belgian context, her conceptualisation of generators and attractors may also help to shed some light on the incidence of violent crime committed in or around train bus and metro stations alongside the disproportionate incidence of public transport related crime in Brussels in relation to the city's population as highlighted earlier in this article.

In a 2006 study concerning situational crime prevention on public transport, Smith and Cornish (and their co-authors) make a number of interesting observations concerning violent crimes committed against both passengers and employees. They emphasise that there are two high risk periods for violence: late afternoon (when the combination of children leaving school and the onset of the afternoon rush hour can lead to overcrowding, disputes over behaviour and personal space and increased anonymity) and late evening (when low passenger densities can increase vulnerability for staff and passengers alike). With specific reference to violence against staff, they also highlight that certain employees suffer a disproportionate number of assaults and emphasise the importance of three contributory factors:-

- the times, routes or locations the employee works are associated with a high risk of assault
- the role of the employee is associated with a high level of assault
- the behaviour of the employee may contribute to the assaults.

In addition, the authors point out that disputes over fares have been found to be important precursors for employee assaults on public transport with those at greatest risk being members of staff required to challenge members of the public. This observation appears to chime with the findings of our dossier analysis in Antwerp and with the notion that 'expressive' violence can sometimes be triggered in situations where supervision and oversight of potential offenders is high. Smith and Cornish also note that employees under stress from high workload will have less patience for passengers and may be less likely or able to diffuse potentially violent situations.[25]

Interestingly, in spite of the apparent importance of these assertions, we were not able to source any further studies which could cast light on the dynamics of problematic or violent incidents during ticket checks (as opposed to studies which attempted to establish the incidence and frequency of such incidents or which concentrated on related topics e.g. reducing fare avoidance).

In a discursive article concerning violence and disorder on the Dutch public transport system, Hauber (2001) highlights a number of factors which may have indirectly contributed to a reported rise in the number of violent incidents during the 1990's and a parallel increase in their severity. Emphasising that the increasingly commercial climate in which public transport providers must operate has led to a reduction in personnel on trams and trains in the Netherlands, he asserts that personal oversight has been largely replaced with technical measures. One practical consequence of this development is that the remaining train guards are sometimes less inclined to check

---

24 Burrell, A. (2007). Violence On And Around Public Transport, University of Central London
25 Smith, Martha J & Cornish, Derek B (eds) (2006). Secure & Tranquil Travel – Preventing Crime & Disorder on Public Transport, Jill Dando Institute of Crime Science/Univesity College London, pages 77 – 106.

tickets out of concern for their own safety leaving carriages less policed and consequently prone to more crime. In contrast, he points out that there were considerably fewer incidents on trams with conductors than on those without and, that feelings of safety and security for both staff and passengers were considerably enhanced as well. The absence of platform controls, access restrictions and ticket checks before boarding a train is also highlighted in that this can increase resentment from passengers who are subsequently asked to produce their ticket on the train itself. Open access to trains can, of course, allow passengers to board without first obtaining a ticket creating a potential conflict situation which might arguably be avoided in circumstances where a ticket check was obligatory before getting on the train.[26] This finding was also highlighted by LaVigne (1997) in her research into situational crime prevention initiatives on the metro system in Washington DC.[27]

Finally, a major piece of research undertaken by the charity Crime Concern on behalf of the United Kingdom government's Department of Transport (1997) focused on a number of broad themes concerning young people and crime on public transport. Perhaps unsurprisingly, the study concluded that the types and patterns of crime which occur on public transport reflect the opportunities presented by the mode of travel and the location of the route. The majority of assaults on public transport networks took place in metropolitan areas and occurred later in the day: approximately half of all assaults involved young people. Interestingly, the report also highlighted that whilst most young people felt safe using public transport during daylight hours, in common with adults their perceptions of personal security changed after dark with girls and young women in particular (but not exclusively) expressing concerns about their safety.[28]

6   Criminological theory

The diversity of violent behaviour in the public transport arena is reflected in the range of available explanations for violence which range from the 'socio-biological through psycho-analytical and psychological to sub-cultural and other sociological theories which focus on (on issues such as) hegemonic masculinity.'[29] Interesting and valuable though many of these works may be, much of their focus lies in attempting to identify the underlying background, developmental and structural causes of violence. They are, so to speak, concerned primarily with distal factors which may lead to violent offending in the future rather than the more immediate proximal factors which precede the commission of an offence in the here and now. As such, they cast relatively little light on the two of the key problematics on which our research was focused: to ascertain why offenders committed a particular crime against a particular person in particular circumstances and, with this information, to establish what if anything could be done to reduce the frequency of such incidents or prevent them from hap-

---

26   Hauber, A. R. (2001). "Openbaar Vervoer – Reizigers, Agressie En Onveiligheid." Justitiele Verkenningen(1), page 115

27   Lavigne, N. G. (1997). Visibility & Vigilance: Metro's Situational Approach To Preventing Subway Crime, NIJ Research In Brief.

28   (1999). Young People & Crime On Public Transport, Department For Transport UK, pages 13-15

29   Levy & Maguire, page 810

pening again – in other words, to consider what is realistically preventable in the short to medium term.

In the paragraphs which follow, we provide a synopsis of some key criminological theories which we feel can contribute to an understanding of our problematic. All the identified theories incorporate the premise that violent incidents are essentially the culmination of a dynamic, interactive and sometimes escalating process. They have also been chosen because they should allow us to make sense of a range of violent behaviours with differing origins, outcomes and underlying motivations committed on or around the Belgian public transport system.

The Routine Activity, Rational Choice, Opportunity And Lifestyle Exposure Theories

Historically, theories of crime have taken either a 'dispositional' stance, focusing on the individual offender or a 'sociological' approach which emphasised the social conditions associated with crime. The work of Cohen & Felson (1979) altered this paradigm somewhat with the development of the routine activity theory. They suggested that crime occurred when three particular elements were combined: a specific situation (a time and location), a target and the absence of effective guardians. The combination of these three elements provides the opportunity for successful offending. The three main components of the theory have evolved over time. Thus, for example, another construction drawn from the same theoretical background might focus on the exposure of victims to motivated offenders (proximity), potential yields as targets (reward) and accessibility (the absence of capable guardians). In this view, victims' degree of risk or exposure to criminal activity is influenced by their individual characteristics and lifestyles. These theories all depart from the premise that offenders seek to gain some advantage from their criminal behaviour and that this involves a degree of rationally based decision making whereby a range of alternative courses of action are contemplated. The theories are closely linked to much of situational crime prevention thinking by way of their efforts to understand an offender's initial decision to become involved in crime and, the process leading up to the commission of a specific crime – target selection, deterrence factors etc. It is in the second area that most of their practical application with regards to offenders has taken place.

Proponents assert that it is possible to significantly impact on the opportunities for crime and the offender's decision making by changing elements of the situation – by reducing target availability (in either a physical or personal manner) or increasing surveillance for example. In this sense, these theories may be of some relevance to the typologies of violent crime on public transport with a financial or material motivation which were highlighted earlier in this contribution. It is also suggested that the use of these theories in work with offenders can add considerably to the body of knowledge concerning situational deterrence factors and the way in which victims are selected. Critics argue, however, that the conception of offender motivation which is implicit in all four theories is overly simplistic. Offender motivation is either assumed to be constant or there is no explicit reference as to what motivates people to commit crime at all. This restricted focus means that many potential prevention opportunities are arguably missed. A further problem is that these theories are arguably more applicable to predatory crime committed for material gain rather than more 'expressive' acts of offending behaviour.[30] Even here, as we have seen, the selection of targets by offenders is not always an inherently rational process (Wright 2007). Thus, whilst increasing guardianship (by ensuring the presence of conductors on all buses and trains for example) is likely to have some impact on violent crime committed for financial gain, it is unlikely to eradicate the problem in its entirety. Such measures may, furthermore, serve to actually increase the incidence of violence committed against public transport personnel.

The Situational Action Theory

What should be apparent from the above discussion is that considering the problem of violence on public transport from a largely situational perspective is probably too restrictive in focus. Moreover, in view of the diverse forms of violent behaviour which occur on public transport, theories which concentrate only on violence with a notionally rational motivation will not suffice. We have already noted the need to consider a range of differing forms of violent behaviour alongside the difficulties in clearly delineating violent acts as being either rationally or instrumentally driven as opposed to more irrational and expressive. Equally, for some offenders 'individual characteristics and experiences may be the most important factor influencing their problematic behaviour; for others it may be the environment in which they operate.'[31] These differing perspectives are encompassed within Wikström's situational action theory which affords the potential to explore both individual and ecological factors alongside questions relating to an individual's development and the relative stability or change in the settings in which they participate.

The fundamental argument upon which the situational action theory is based is that peoples' acts (including criminal acts) are a consequence of how they see their options (how they react to their environment) and make their choices. Individual factors and environmental factors may be regarded as causes (or part causes) of crime to the extent that they can be shown to influence people's perceptions of alternatives and the process of choice relevant to their engagement in acts of crime.[32] The theory fuses together a number of complex concepts revolving around the interaction between an individual and the various settings in which they live their lives (friends, family, school, work etc.) Changes in an individual, their setting or the relationship between the two can have an influence on behaviour. The theory is, in other words, 'not about "kinds of individuals" or "kinds of settings" but about "kinds of individuals in kinds of settings".[33]

According to Wikström, at each stage of life the likelyhood of an individual engaging in a criminal act (and indeed the nature of these acts) may be viewed as an outcome of both external and internal factors. External factors relate to an individual's tempta-

---

30 McLaughlin, E & Muncie, J (Eds) (2006) The Sage Dictionary Of Criminology, Sage Publications pages 278-280, 339-340 & 365-366. See also Meier, R. F., Miethe, T.D. (1993). Understanding Theories of Criminal Victimisation. Crime & Justice – A Review of Research. Chicago, University of Chicago Press, pages 470-479

31 Wikström, P. O. & Treiber, K. (2008). Offending Behaviour Programmes, Youth Justice Board

32 Wikström, P. O. (2004) Crime As Alternative – Towards A Situational Action Theory Of Crime Causation in, Beyond Empricism: Institutions and Intention in the Study of Crime, New Brunswick, Transaction, page 7

33 Ibid – page 23

tion/provocation thresholds and their sensitivity to deterrence. Internal factors are an individual's current moral values and executive functions. Temptation is defined as the perceived option to satisfy a particular desire in a certain way whilst provocation can occur via a personal attack on an individual (or someone close to them), their property, security or self respect resulting in an unlawful response. An individual's morals and executive functions influence their propensity to commit crime. In essence, they determine what people find tempting and what they construe as provocative, alongside the tendency to see crime as a morally acceptable option and – ultimately – to choose that option. These tendencies can be mitigated by deterrance in any setting. Deterrance is defined as an inhibiting mechanism which involves the risk of monitoring or intervention and associated sanction. Importantly, it is conceived more widely than situational crime prevention or punitive interventions and includes the impact of, for example, personal relationships. Individuals vary in their sucseptibility to deterrance. Those with poor executive functions (i.e. weaker potential to exercise self-control) are likely to be less easily deterred than individuals with more developed executive functions. Changes in an individual's characteristics and experience will influence changes in their action through associated shifts in how they perceive alternatives and make choices in each of the settings in which they operate. The nature of settings themselves are also important in that some settings are likely to be more criminogenic than others. For Wikström, a criminogenic setting is one which is more likely to contribute to to an individual's perception of crime as an alternative and, to their choosing to act on such an option. Such settings are criminogenic to the extent that they create opportunity, cause friction and provide monitoring. Opportunity is construed as the presence of people, objects and events which are necessary for carrying out unlawful activities for either pleasure or gain. Friction refers to events that trigger adverse reactions to other people's behaviour (e.g. anger or irritation) and which increase the liklelyhood for some form of unlawful response. Monitoring in this context relates to the risk of intervention and detection if people carry out unlwaful acts. Conversely, some types of setting are more favourable to the development of the the kinds of values and emotions that suppport law abiding behaviour and to the development of executive functions that support the exercise of self control.

The situational action theory attempts to cast some light on precisely what causes a person to see crime as an alternative and what subsequently makes them act upon that option. Importantly, in the context of research into violence on public transport, the theory has a number of other advantages. Firstly, it allows us to consider both individual and ecological causes of particular crimes from both a proximal and distal perspective within an integrated theoretical framework. Moreover, its relative neutrality on the priority afforded to the causal components of crime – these being viewed as particular to an individual and their intercation with a specific setting – is particularly helpful when attempting to understand a phenomenon around which, as we have seen, relatively little previous research been has done. Finally, the theory's flexibility makes it potentially "fit for purpose" in attempting to understand the diverse typologies of violent crimes committed in the public transport arena including those with both a rational or material motivation and those in which no such motivation is apparent. For these reasons, we considered the situational action theory to be the most helpful of all the theoretical perspectives identified on which to base the empirical component of our interviews with perpetrators of violence on the Belgian public transport system. The results of these interviews will provide the basis for a subsequent article in the near future.

7   Crime prevention on public transport

Having examined several differing aspects of violent behaviour which are of relevance to the public transport arena and provided an overview of theories which may assist with their analysis, this contribution will conclude with some observations on the subject of crime prevention and reduction on public transport. In the paragraphs that follow, we provide an illustration of some of the initiatives taken to counteract violence on or around public transport in a number of countries both in Europe and beyond.

Mukherjee (1995) describes how in the Netherlands, the government responded to concerns about insecurity and aggression on public transport during the 1980's by authorising the public transport authorities to employ unemployed young people as VIC staff (Veiligheid, Informatie, Controle or Safety, Information and Control) in three major cities – Rotterdam, Amsterdam and the Hague. The VICs were deployed in different ways; some worked in groups of two to four, some were authorised to impose fines to defaulters, some implemented random ticket checks, and some staffed metro stations. The role of those staffing metro stations was to provide information and not to check passengers. Passengers caught without a valid ticket could either buy one from the driver or leave the train. In case of problems the VICs could rely on support from a special team or the police. The deployment and competencies of such private security staff raised some interesting boundary issues the exploration of which lies beyond the scope of this contribution.[34] The impact of the initiative was, however, positively evaluated with the numbers of passengers either witnessing or reporting that they had been the victim of assault or harassment both declining. These results were attributed to the visible presence of staff around the public transport system.[35]

In contrast, the French city of Nantes chose to respond to its problems of crime on public transport by establishing a dedicated unit from the local police (La Brigade de Surveillance des Transports Commun or Public Transport Surveillance Section – B.S.T.C.) which works in partnership with the city's public transport authority. Groups of police agents are assigned to travel by public transport: they can be supplemented by others in problem neighbourhoods or if particular problems occur. The public transport authority itself has established a security section comprised of some 48 agents. The work of these agents is primarily dissuasive in character their presence being viewed as a disincentive to those inclined to commit crime.[36]

A more diverse approach has been taken by the Flemish public transport provider De Lijn whereby a series of temporary or permanent prevention measures can be deployed in response to problem incidents or on a wider neighbourhood basis.

34   See for example Cachet, L and Ponsaers, P (2008). Security Policy Questions, in Governance of Security in the Netherlands and Belgium, Cachet, L, De Kimpe, S, Ponsaers, P & Ringeling, A (Eds), Den Haag, Boom Juridische Uitgevers.
35   Mukherjee, S. (1995). Reducing crime on public transport in The Netherlands, The promise of crime prevention: leading crime prevention programs <http://www.aic.gov.au/publications/rpp/01/index.html> Research and Public Policy Series. P. J. Grabosky, M. Canberra, Australian Institute of Criminology.
36   Antoine, Christophe et Jean-Hubert, La violence dans les Transports en commun , http://membres. lycos.fr/tpeeden/acceuil.htm

Alongside technical measures such as the installation of CCTV cameras on its ve-hicles, De Lijn also deploys dedicated personnel to enhance oversight and provide a dissuasive presence on buses, trams and bus/tram stops. Such initiatives are often undertaken in partnership with the local police. The organisation also works along-side local government, schools, youth organisations and street workers on a range of prevention projects.[37]

On a more global scale, a 2007 study examining violent crime and nightlife, the World Health organisation's Violence Prevention Alliance highlighted that in many parts of the world, transport options are often limited late at night when nightlife areas are at their busiest. Consequently, crowds often gather in the streets around bars and nightclubs increasing the potential for violent encounters. Long waits, frustration and competition for limited transport facilities can make bus stops and taxi ranks hotspots for violence. Ancillary problems can also occur as a result with people adopting risky methods to get home (accepting lifts from strangers, driving under the influence of alcohol etc.) Transport workers have, furthermore, been identified as being at risk of violence in several countries.

The report found some evidence to suggest that improvements to late night trans-port facilities can contribute to a reduction in levels of assault. The value of late night bus services is emphasised with the important provisos that such services need to protect both customer and driver safety and, that consideration must also be given to where customers are deposited to prevent any redistribution of violence elsewhere. The value of security and marshalling staff to assist with queuing arrangements is highlighted as is the need to protect transport staff using measures such as security staff and police on vehicles, CCTV, radio links between drivers and the police and the modification of vehicles to enhance the physical protection of drivers. Importantly, however, the report also cautions that transport improvements will not in themselves eradicate late night violence entirely. Consequently, transport measures must be seen as one component in a wider strategy to reduce the risk of violence.[38]

As part of a study evaluating a crime prevention initiative on bus routes in Liverpool, United Kingdom, Newton, Johnson and Bowers (2004) describe how many crime re-duction strategies on public transport have focused on attempting to influence one or more of the three routine activity theory elements: by improving surveillance, for ex-ample, and by helping to ensure that sufficient 'guardians' are present at specific times of the day to make it more difficult for a motivated offender to target a victim. They cite a range of good practice examples drawn from other research projects including improving visibility or lighting, increasing staff presence, allowing police in uniform to travel on buses for free, the use of CCTV, emergency help points, cleaning and speedy maintenance and the use of transport wardens. Interestingly, they also suggest that the most successful schemes tend to be the ones which are multi-agency and that adopt multi-tactical approaches.[39]

A practical example of such multi-agency work is provided by Burrell (2007) in her research into violence on public transport focused (in the main) around another ma-jor British conurbation, Birmingham. She describes how the police and bus operators

---

37   http://www.delijn.be/over/veiligheid/veilig_op_weg_projecten.htm?ComponentId=6065&SourcePag eId=6028
38   Youth Violence Alcohol & Nightlife – Late Night Transport.
39   Newton, Johnson & Bowers

conduct high visibility ticket checks in specific locations where a variety of bus routes converge. Passengers travelling without a ticket are asked to leave the bus, can be administratively fined and occasionally subject to a police search which have resulted in the recovery of weapons or other suspicious or illegal items. This initiative is positively assessed from the perspective of the Police but no evidence is actually provided concerning its utility as a crime reduction initiative. Burrell also describes how, following a rise in fare evasion and anti-social behaviour, the metro train operator decided to employ conductors and found that there have been fewer problems on the system since their introduction. Interestingly, the conductors have also been issued with badges which have the ability to audio record speech at the touch of a button. In the event of a dispute, the recording is activated and can be used as evidence in the event of an operator pressing charges against a passenger. Finally, the need to incorporate security measures into the design of all new bus and train stations is highlighted.[40]

The manner in which design, management and maintenance arrangements can impact on crime on public transport is considered by LaVigne (1997) who analysed situational crime prevention initiatives on the metro system of Washington DC in the United States. Crime rates on the system are stable and significantly lower than those in a number of comparable American cities. The author asserts that the open design of stations giving clear views for passengers reduces the potential for criminals to hide in dark corners or passageways. Stations have also been designed without toilets or excess seating thereby reducing the potential for loitering. Access to platforms is restricted to those having a valid travel ticket which must also be used to exit the system. Passengers can buy multiple-use farecards in any dollar amount thereby reducing the amount of cash which is physically available to pickpockets. Stations are continuously staffed during opening hours and are fitted with CCTV cameras. Such measures are complemented by strict enforcement of "quality of life" violations on trains (no smoking or eating) by specifically dedicated transit police officers and the prompt reporting and repair of all vandalism and graffiti. Information regarding 'house rules' for the metro are clearly displayed in stations and on trains. Those who violate these rules risk an element of public humiliation as station masters broadcast public reprimands of rule breakers.

Whilst recognising that precise attribution of crime rates to these preventative measures is not straightforward, the author believes that on balance, the Washington metro's design characteristics and operating policies have contributed to its low crime rates. An important exception to these successes is, however, assault, which, she suggests may not be as situationally influenced as other crime types.[41]

A number of other studies have, however, raised questions concerning the efficacy of some situational crime prevention measures such as CCTV. In a systematic review designed to assess the available research evidence on the impact of CCTV on crime, Welsh and Farrington (2008) warn that the significant investment in CCTV infrastructure in many countries has not been matched by sufficient rigour in evaluation. They identified only four specific studies concerning the use of CCTV in a public transport context all of which related to underground railway and metro systems. The studies provided conflicting evidence of success: in two cases a reduction in crime was dem-

---

40 Burrell
41 Lavigne

onstrated, in one case crime increased and in the other no discernable impact was noted. Furthermore, in both instances where reductions in crime were noted, the use of other crime reduction/prevention initiatives alongside CCTV meant that it was very difficult to attribute these outcomes to the introduction of CCTV alone. Overall, the review concluded that CCTV has a modest but significant effect on crime. It appears, however, to be more effective when targeted at vehicle crime in car parks than in other public or semi-public spaces (public housing, city/town centres or public transport systems).[42]

The impact of CCTV as a deterrent is questioned in article by Sivarajasingham, Shepherd and Matthews (2003) who concluded that many offenders had insufficient knowledge of the factors that increased their chances of detection when committing an offence either in general or, because of temporary cognitive impairment arising from stress, the use of alcohol or drugs etc. The authors emphasise however that CCTV can play an important role in reducing harm arising from incidents of violence in that intervention from law enforcement or other authorities often comes about more rapidly than would otherwise have be the case.[43]

Issues relating to deterrence, certainty and punishment in the context of the Zurich public transport system are the subject of an interesting article by Killias, Scheidegger and Nordenson (2009). They emphasise that the deterrent effect of punishment is based not only on the severity of the sanction itself but also on issues relating to consistency of application: the probability of being caught and that any eventual sanction is actually applied. This assertion is based upon research conducted on the Zurich public transport system from 2003 onwards. The authors describe how, following the abolition of ticket checks on suburban trains in 1993, an increasing proportion of users started to evade payment of fares (approximately 4% of passengers were regularly found without a ticket). This rate remained stable despite different periodic attempts to address the problem including patrols by plain clothes inspectors, increasing fines for repeat offenders, identity checks and ultimately criminal prosecutions. In a move designed to address fear of crime amongst passengers rather than fare evasion as such, the Zurich public transport authorities decided to re-introduce systematic ticket checks for passengers during evening hours. As a result of this initiative, approximately one passenger in three was subjected to a ticket check after 9.00pm. The percentage of passengers detected without having paid a fare fell to 1% whilst – interestingly – there was also a decrease in passengers travelling without tickets during daytime hours: a classic example of 'diffusion of benefits.' The authors also highlight that beyond a certain point, increasing the number of ticket checks will not actually result in any further reduction in passengers travelling without a valid ticket. This is because passengers who are still not discouraged by dramatically increasing levels of control may have social or personality characteristics which are not amenable to policy changes of this nature.[44]

---

42   Welsh, Brandon C & Farrington, David P, (2008). Effects of Closed Circuit Television Surveillance on Crime, Campbell Systematic Reviews, 2008, page 14

43   Sivarajasingham, V, Shepherd, J P & Matthews K (2003). Effect of urban closed circuit television on assault injury and violence detection, Injury Prevention, 9, 312-316

44   Killias, M, Scheidegger, D & Nordenson P (2009). The Effects of Increasing the Certainty of Punishment – A Field Experiment on Public Transportation, European Journal Of Criminology, Volume 6 (5), pages 387 – 400.

The above examples appear to tentatively illustrate that it is possible for public transport operators, police and security service providers together with public authorities to design and implement crime prevention or reduction initiatives which can have some impact on crime levels in and around public transport albeit that great care must be taken in interpreting changes in crime levels and the particular factors which may have brought them about. In that most of these initiatives focus on reducing the opportunities and increasing the difficulties and risks associated with offending – via both personal and technological interventions – they can be broadly categorised as falling under the situational crime prevention umbrella. The examples also raise questions, however, concerning the limits of situational crime prevention and whether such initiatives alone will be sufficient to prevent the commission of violent crime on public transport. Given the multiplicity of offence types, configurations of perpetrators and victims and potential causatory factors highlighted earlier in this article, this should probably not surprise us.

8   Conclusion

This contribution has considered a number of questions concerning violence on public transport. In attempting to establish the extent, nature and trends of violent behaviour on the Belgian public transport network, we have seen that the use of official crime statistics alone is of limited utility in that data collection by the police concerning the phenomenon is somewhat incomplete, time series comparative information is unavailable and, in all probability, many crimes remain unreported to the Police or other authorities. Official statistics do, however, provide some indication that the problem is significant enough to warrant further attention. The incidence of violence in and around the public transportation system in Brussels appears, furthermore, disproportionately high in relation to the city's population. Explanations for this discrepancy could form a useful focus for additional research in this area. It is also noteworthy that Belgium's four public transport authorities do not operate a uniform recording system for violent incidents which occur within their jurisdiction. Moves to redress this situation would allow for a more complete picture of the phenomenon to emerge.

Information gleaned from differing sources – legal dossiers, surveys undertaken by public transport providers themselves and academic research – indicates that violent behaviour on public transport in Belgium involves a diverse range of behaviours with differing motivational drivers committed by varying configurations of actors. Such diversity is not confined to the typology of incidents alone. The authors' own analysis of legal dossiers in Antwerp and Brussels appears to indicate that differing forms of violent crime, configurations of actors and offender motivation may be apparent in Belgium's two biggest cities. This too, could provide a useful focus for further research.

The breadth and complexity of violent incidents on public transport effectively entails that differing etiological explanations are required. Our literature review examined a number of criminological theories which attempt to explain violent behaviour concentrating, in particular, on theories with a primarily (but not exclusively) proximal focus. In light of the diverse typologies of violence on public transport, we conclude that the flexibility afforded by Wikström's situational action theory is particularly appropriate for those wishing to gain a better understanding of the problematic.

The findings on incidence and typology above have been complemented by a review of international literature illustrating the range of both personal and situational causal factors for violent behaviour and the initiatives undertaken by various public transport providers to counter it. In so doing, we have highlighted a number of examples where situational crime prevention approaches appear to have had a positive impact on violent crime in the public transport arena. However in light of the diverse typologies involved, such initiatives should not necessarily be seen as a panacea in that they will be unlikely to deter all types of offender and may, in certain situations, actually create the conditions which can provoke violent incidents between passengers and staff. There remains a need for robust evaluation of both crime prevention initiatives and further analysis of all stages of the violent crime commission sequence from the offender's perspective. In this way, the shortcomings of overly generic crime prevention measures which can operate at too high an aggregate level to address the actual issues involved in crime commission may be overcome (Cornish 1994).

Valuable though existing crime prevention initiatives may be, too little is known about the dynamics of violent crime in general and its commission in the arena of public transport in particular for us to be confident that a fully stocked tool box of solutions is currently available. The development of a more specific knowledge base in this area should, therefore, be considered a priority.

9   Bibliography

Adang, O. M. G. (2000). "Jonge Mannen In Groepen – Een Geweldige Combinatie? "*Justitiële Verkenningen (1)*, Den Haag

Bandura, A. (1986) *Social Foundations Of Thought & Action: A Social Cognitive Theory*, Englewood Cliffs, NJ, Prentice Hall

Bradley, H. (1996). *Defining Violence*, Aldershot, Avebury

Brown, L. (2008). *Violent Crime: Some Basic Facts & Implications For Social Work Practice*, Edinburgh, Criminal Justice Social Work Development Centre For Scotland.

Burrell, A. (2007). *Violence On And Around Public Transport*, London, University of Central London

Cachet, L and Ponsaers, P (2008). Security Policy Questions, in *Governance of Security in the Netherlands and Belgium*, Cachet, L, De Kimpe, S, Ponsaers, P & Ringeling, A (Eds), Den Haag, Boom Juridische Uitgevers

Campbell, B (1993). *Goliath: Britain's Dangerous Places*, London, Methuen Publishing

Celano, A. (2006). "Using Words To Diffuse Violence " *Security Management* Vol 50, Alexandria

Cohen, L.E. and Felson, M. (1979) "Social change and crime trends; a routine activity approach", *American Sociological Review*, 44, Nashville

Cornish, D. (1994). The Procedural Analysis Of Offending And Its Relevance For Situational Prevention in, *Crime Prevention Studies Volume 3*, Clark R. (Ed), New York, Criminal Justice Press, 1994

Cozens, P, Neale, R, Hillier, D, & Whitaker, S (2004). Tackling Crime and Fear of Crime While Waiting at Britain's Railway Stations, *Journal of Public Transportation*, volume 7, no. 4, Tampa, University of South Florida

De Lijn (2008) *Veiligheidsmonitor 2008*

Department For Transport (1999). *Young People & Crime On Public Transport*

De Wree, E., Vermeulen, G. en Christiaens, J., 'Aanpak op het juiste spoor: (Strafbare) overlast door jongerengroepen in het kader van openbaar vervoer', Antwerpen, *Panopticon*, 2006, nr. 6

De Wree, E., Vermeulen, G., Christiaens, J. (2006). *Strafbare Overlast Door Jongerengroepen In Het Kader Van Openbaar Vervoer*, Anwterpen/Appeldoorn, MAKLU

Dobash, R.E, Dobash, R.D, Cavanagh, K and Lewis, R (2001) *Homicide In Britain*, Reserach Bulletin No.1, Manchester, University of Manchester Farrington, D. P. (2002). Developmental Criminology & Risk Focused Prevention, *Oxford Handbook of Criminology*. eds. M. Maguire, Morgan, R., Reiner, R., Oxford, Oxford University Press

Farrington, D. (1989) Early Predictors Of Adolescent Aggression & Adult Violence, *Violence & Victims* 4, New York, Springer Publications

Farrington, D.P. (1992), Explaining the Beginning and Ending of Antisocial Behavior from Birth to Adulthood in *Facts, Frameworks and Forecasts: Advances in Criminological Theory Volume 3*, McCord, J (ed), New Brunswick, Transaction

Federale Politie – CGOP Beleidsgegevens (2008) *Politiële Criminaliteitsstatistieken België 2000/2007*

Frieze, L and Browne, A (1989) "Violence in marriage" in Ohlin, L and Tonry, M (Eds), *Family Violence*, Chicago, University of Chicago Press

Gadon, L., Johnstone, L., Cooke, D. (2006). "Situational Variables And Institutional Violence: A Systematic Review Of The Literature." *Clinical Psychology Review* 26, London

Gane, S. R. (2007). "Avoiding Violent Outcomes." *Security Management* 2.

Gresswell, D and Hollin, C. (1994) Multiple Murder: a review, *British Journal Of Criminology* 34: 1-29, Oxford

Gold, R (1958), 'Roles in sociological field observation', *Social Forces*, vol. 36, no 3, Chapel Hill

Hare, R. (2001) Psychopathy and risk for recidivism and violence in Gray, N, Laing, J, and Noaks, L. (Eds), *Criminal Justice, Mental Health & The Politics Of Risk*, 27-48, London, Cavendish

Hauber, A. R. (2001). "Openbaar Vervoer – Reizigers, Agressie En Onveiligheid." *Justitiele Verkenningen*(1) Den Haag

Hirschi, T. (1969) *Causes of Delinquency*, Berkley, CA, University of California Press

Hissel, S., Gekkers, S. (2008). Evaluatie Cameratoezicht Op Openbare Plaatsen, *Reiobeleidsonderzoek*. Amsterdam

Homan, R. (1978), 'Interpersonal communications in Pentecostal meetings', *Sociological Review*, vol. 26, no.3, Hoboken

Humphreys, L.(1970) *'Tearoom Trade'* London, Duckworth Indermaur, D. (1999). "Situational Prevention of Crime – Theory & Practice In Australia." *Studies on Crime & Crime Prevention* 8(1), New York

Jones, S (2000). *Understanding Violent Crime*, Buckingham, Open University Press,

Junger-Tas, J., Terlouw, G., Klein, M. W. (ed) (1994). *Delinquent Behavior Among Young People In The Western World – First Results Of The International Self Report Delinquency Study*, Dordrecht, RDC Ministry Of Justice/Kugler Publications

Killias, M, Scheidegger, D & Nordenson P (2009). The Effects of Increasing the Certainty of Punishment – A Field Experiment on Public Transportation, *European Journal Of Criminology*, Oxford, Volume 6 (5)

Krug, E.G, Dahlberg, L.L, Mercy, J.A, Zwi, A.B, Lozano, R (2002) *2002 World Report On Health*, Geneva, World Health Organisation

Khondaker, M. I. (2007). "Juvenile Deviant Behavior In An Immigrant Bangladeshi Community: Exploring the Nature and Contributing Factors " *International Journal Of Criminal Justice Sciences* 2(1), Abishekapatt

Kyvsgaard, B. (2003). *The Criminal Career: The Danish Longitudinal Study*, Cambridge, Cambridge University Press

Lavigne, N. G. (1997). *Visibility & Vigilance: Metro's Situational Approach To Preventing Subway Crime*, Washington, NIJ Research In Brief

Levi, M., Maguire, M. (2002). Violent Crime. *Oxford Handbook of Criminology*, M.Maguire, Morgan, R., Reiner, R., Oxford, Oxford University Press

Loeber, R., Farrington, D.P., Ed. (1998). *Serious & Violent Juvenile Offenders – Risk Factors & Successful Interventions*, Thousand Oaks, Sage Publications

Longmore-Etheridge, A. (2007). "Nurses On Guard." *Security Management*, Alexandria

Maesschalck, J & Ringeling, A, (2008). What goes up must come down? The career of 'safety and security' as a policy issue in *Governance of Security in the Netherlands and Belgium*, Cachet, L, De Kimpe, S, Ponsaers, P & Ringeling, A (Eds), Den Haag, Boom Juridische Uitgevers

Mattinson, J, (2001). *Stranger And Aquaintance Violence: Practice Messages From The British Crime Survey*, Briefing Note 7/01, London: Home Office

McCafferty, K. (2007). "Show No Fear." *Security Management*(2), Alexandria

McLaughlin, E & Muncie, J. (Eds) (2006) *The Sage Dictionary Of Criminology*, London, Sage Publications

McNeill, F. (2002). *Beyond 'What Works': How and Why Do People Stop Offending?*, Edinburgh, Criminal Justice Social Work Development Centre for Scotland

Meier, R. F., Miethe, T.D. (1993). Understanding Theories of Criminal Victimisation. *Crime & Justice – A Review of Research*. Chicago, University of Chicago Press

Mukherjee, S (1995). *Reducing Crime On Public Transport In The Netherlands: The Promise Of Crime Prevention: Leading Crime Prevention Programmes*, Research & Public Policy Series, eds. Grabosky P.J., Canberrra, M, Canberra, Australian Institute Of Criminology

Newton, A.D, Johnson, S.D, Bowers, K.J (2004). Crime on bus routes: an evaluation of a safer travel initiative, *Policing: An International Journal of Policing Strategies & Management*, 27,3, Bingley

Olewus, D. (1987) "Testosterone and adrenaline: aggression and anti-social behaviour in normal adolescent males" in Mednick, S, Moffit, T, and Stack, S (Eds), *The Causes Of Crime: New Biological Approaches*, New York, Cambridge University Press

Pardoel, K. V., K. (2003). *Handreiking Cameratoezicht In En Rond Het Openbaar Vervoer*. Tilburg, IVA

Pauwels, L. (2007). *Buurtinvloeden en Jeugddelinquentie*, Den Haag, Boom Juridische Uitgevers

Pease, K. (2002). Crime Reduction. *Oxford Handbook of Criminology*. M. Maguire, Morgan, R., Reiner, R., Oxford, Oxford University Press

Pease, K. (2005) No Through Road – Closing Pathways To Crime, in *Crime Reduction The Law*, Moss, K. S., M., Ed. Oxford, Routledge

Pogrebin, M., Ed. (2007). *A View Of The Offender's World About Criminals*, New York, Sage Publications

Pogrebin, M, Stretesky, P, Prabha Unnithan, N, Venor, G (2006). Retrospective Accounts of Violent Events by Gun Offenders, *Deviant Behavior*, Volume 27, Number 4, Lafayette, Routledge

Reiss, A.J, Roth, A (ed), *Understanding & Preventing Violence*, Panel on the Understanding and Control of Violent Behaviour, Washington DC, National Research Council, 1993

Runyan, C. W. (1998). "Using The Haddon Matrix: Introducing The Third Dimension" *Injury Prevention*(4), London

Sivarajasingham, V, Shepherd, J P & Matthews K (2003). Effect of urban closed circuit television on assault injury and violence detection, *Injury Prevention*, 9, London

Smith, Martha J & Cornish, Derek B (eds) (2006). *Secure & Tranquil Travel – Preventing Crime & Disorder on Public Transport*, Jill Dando Institute of Crime Science/Univesity College London

Summerell, J, Shorrocks, T & Mitchell, K (2004). *Reducing Assaults On Railway Staff – A Report For Railway Safety*, London, Rail & Safety Standards Board

Toch, H (1992). *Violent Men – An Enquiry Into The Psychology Of Violence*, Washington, American Psychological Association

Trickett, A., Osborn, D., Seymour, J., Pease, K. (1992). "What Is Different About High Crime Areas?" Oxford, *British Journal Of Criminology* 32(1)

Union Internationale des Transports Publics.(2007). *Tackling Social Exclusion – The Role of Public Transport*. Brussels

Van Den Bogaerde, E, Van Den Steen, I, Klinckhamers, P & Vandendriessche, M (2007). *Veiligheidsmonitor 2006*, Brussel, Federale Politie – Directie van de nationale gegevensbank

Verwee, I., Ponsaers, P. en Enhuis E. (2007). *Inbreken Is Mijn Vak,,* Den Haag, Boom Juridische Uitgevers

Verwee, I., Ponsaers, P. en Enhuis E. (2007) *Onderzoeksrapport – Woninginbraken: diefstalpraktijk & preventiebeleid*, Gent, SVA/Universiteit Gent

The Wave Trust (2005). *Violence & What To Do About It,* London

Weaver, B., McNeill, F. (2006). *Giving Up Crime: Directions For Policy*, Glasgow, Scottish Centre For Crime & Criminal Justice Research

Welsh, Brandon C & Farrington, David P, (2008). *Effects of Closed Circuit Television Surveillance on Crime*, Oslo, Campbell Systematic Reviews, 2008

W. H. O. G. C. F. V. Prevention. (2007). *Youth Violence Alcohol & Nightlife* – Late Night Transport. Liverpool, John Moores University

Whyte, B. (2001). *Effective Intervention for Serious and Violent Young Offenders*, Edinburgh, Criminal Justice Social Work Centre For Scotland

Wikström, P. O., Treiber, K. (2008). *Offending Behaviour Programmes*, London, Youth Justice Board, United Kingdom

Wikström, P. O. (2005). The Social Origins Of Pathways In Crime: Towards a Developmental Ecological Action Theory Of Crime – Involvement & Its Changes. *Integrated Developmental & Life Course Theories of Offending*. D. P. Farrington Ed, New Brunswick, Transaction Publishers

Wikström, P. O. (2004) Crime As Altrnative – Towards A Situational Action Theory Of Crime Causation in, *Beyond Empricism: Institutions and Intention in the Study of Crime*, New Brunswick, Transaction Publishers

World Health Organisation Europe. (2007). *Breaking The Cycle: Public Health Perspectives On Interpersonal Violence In The Russian Federation,* Geneva

Myths and Reality in the History of Restorative Justice

Nikolaos Stamatakis
Tom Vander Beken

Initially, it is important to recognise that the advocates of restorative justice do not all have precisely the same expectations in mind. In the last decades several definitions of restorative justice have been constructed by practitioners and theoreticians couched within several parameters. Definitions that are trying to captivate the basic concept of the 'new wave' mediation, often underlying not the *Law* as the basis for a decision imposed by a judge, but the *Process* itself. Today, it is an accepted altruism to claim that restorative justice is not a single academic theory of crime or justice. It is an 'umbrella concept', which covers beneath a variety of practices with no universally established definition. The same concept contains a number of elements that we need to unpack in order to decide which activities could be seen as truly restorative and which less (or not at all). As Shapland (2003) has commented, it is almost impossible to draw solid boundaries round what would or would not be seen as restorative justice.[1]

As it is mentioned above, the (re)discovery of the historical roots of restorative justice might be able to provide credible answers to contemporary dilemmas concerning, its genesis and prospective development. Thus, bringing all those facts back into focus, the present article is devoted to restorative justice and it is divided into two sub-sections: the *first* sub-section provides a theoretical overview of the main concepts of restorative justice as they are conceived and applied today, sketching the contours of a restorative and transformative approach to having offenders reconciling with their victims, their families and finally with their communities. The *second* sub-section provides a historical overview of restorative justice. Starting from the non-state or acephalous societies and with an intermediate stop at the early-state ones, we are reaching the modern societies seeking for traces of restorative justice. In this sub-section it is also examined the ongoing contribution of restorative justice to conflict resolution looking for the presence of historical roots and signs in the restorative justice movement. Following this historical sequence, the accuracy of the early and contemporary signs of restorative justice is checked based on historical and anthropological facts in order to avoid overstatements or give false impressions.

1   'Breaking Down' the Notion of Restorative Justice

Repentance, forgiveness, reconciliation, healing – these all find a place under the same roof that is constructed by restorative justice and aims to shelter all these inter-related factors that have been neglected by the adversary model of justice. For instance, the

---

1   Shapland, J. (2003). Restorative Justice and Criminal Justice: just responses to crime? In A. von Hirsch, J. Roberts, A.E. Bottoms, K. Roach & M. Schiff (Eds.), Restorative Justice and Criminal Justice: competing or reconcilable paradigms?. Oxford: Hart Publishing, pp. 195-217

opposing parties are discouraged from meeting before the court, the offenders have no opportunities to apologise and victims never get a chance to express their feelings and seek answers to their questions. The way that the process is designed supports the denial rather than the acceptance of responsibility by the offender.[2] If not rejected, accountability is impersonal and counter-productive to any process of healing and restoration; such a process would require the offenders to deal with their sense of guilt and face their victims and the consequences of their unlawful actions. Therefore, restorative justice could be broadly pictured as healing or peace-making justice.

The Dominant 'Schools' of Restorative Justice

The process of achieving healing, peace and finally, restoration is dependent upon the various 'schools' of restorative justice. These are: I) The **Maximalist** (or fully-fledged) **Model** admits the possibility of the use of *coercion* in restorative interventions, incorporating 'court-imposed', and thus coercive, restorative sanctions to the offender. [3] Bazemore's and Walgrave's maximalist approach is condensed to the argument that the handling of conflict through voluntary settlements between the victim and the offender is not appropriate for all situations. Furthermore, they claim that the value of mediation programmes is limited due to their voluntary nature and the inability of the parties to reach an agreement.[4] II) Opposite to the Maximalist model lays the **Diversionist** (and **Purist**) **Model** of restorative justice. McCold, as one of its main supporters, advocates that any restorative intervention should aim towards the reparation of harm, as the needs of the wider society for constructive responses provide limited role for coercive formal responses.[5] The Purist model is mainly based on a *voluntary* cooperative approach, which claims to be holistic in response to stakeholders' needs.[6] The Diversionist model also tries to withdraw as many cases as possible from the criminal justice system as operated by the State and propose the development of programmes parallel to the traditional penal justice system.[7] III) Another trend diverse to both the above models that coexists within the restorative justice movement is the **Republican Theory**. According to this theory, the fundamental function of a criminal justice system is to safeguard and promote the rights of all citizens, clearly dividing the 'functions' of the State from those of the community. Since here restorative justice flows from the republican theory, the role of the government is to preserve the public order, while the role of the community is to built and maintain peace. Thus, the goal of criminal justice should be to maximise dominion, which is freedom, holistically

---

2   McElrea, FWM. (2001). A Christian Approach to Conflict Resolution. A contribution to the seminar What does the Lord require of Christians in conflict? Australasian Christian Legal Convention. Melbourne. 1-4 February

3   Bazemore, G. & Walgrave, L. (1999). Restorative Juvenile Justice: in search of fundamentals and an outline for systemic reform. In G. Bazemore & L. Walgrave (Eds.). Restorative Juvenile Justice: Repairing the harm by youth crime. Monsey NY: Criminal Justice Press, p. 48

4   Walgrave, L. (2003). Repositioning Restorative Justice. Cullompton: Willan publishing, p. 47

5   Walgrave, L. (2000). How Pure Can a Maximalist Approach to Restorative Justice Remain? Or Can a Purist Model of Restorative Justice Become Maximalist?. Contemporary Justice Review, 3, 418

6   McCold, P. (1999). Toward a Holistic Vision of Restorative Juvenile Justice: a reply to Walgrave. Paper presented at the 4th International Conference on Restorative Justice for Juveniles. Leuven, p. 32

7   McCold, P. (2000). Toward a Mid-Range Theory of Restorative Criminal Justice: a reply to the Maximalist model. Contemporary Justice Review, 3, 357 – 414

conceived.[8] Here, a Republican conception of freedom includes concepts such as reprobation, checking of state power over the individual and reintegration of victims and offenders.

Nowadays, it is manifest that each theoretical approach or 'school' of restorative justice is reflected upon the various definitions. But, what makes a restorative justice definition better than another? And, what concepts or 'ingredients' this definition should entail? In general terms, the philosophy of restorative justice is mainly based on community healing, while its focus is on the human beings closely affected by the crime: on offender accountability, problem solving, and creating an equal voice for offenders and victims. Restorative justice is a theory of justice that is proactive rather than reactive. It tries to deconstruct the idea of retribution by recognising the need for community healing; so, it evaluates the hurts in terms of the values of the economic and social justice perspectives that community must adopt in order to obtain the best possibility of preventing future crime. From the community's perspective, restorative justice is a claim on it (community) to frame the issues of justice differently; to undergo a radical shift of mind and to own a restorative vision.

In other words, restorative justice is a notion that reflects an alternative way of thinking about and responding to crime. Moreover, it is an analytical framework for assessing the socio-cultural factors feeding disputes and fears of crime. It involves individuals who are not detached from the incident and the participation of the community is not an abstract process. Restorative justice may also be well described as an approach to remedying crime in which it is understood that all things are interrelated and that crime disrupts the harmony which existed prior to its occurrence (or at least which it is felt should exist).[9] In this sense, restorative justice could be viewed more as a humanitarian approach, and less as a philosophical system based on strictly defined principles and norms.

Discovering the Main 'Ingredients' of a Widely Accepted Restorative Justice

As a humanitarian approach – and so focused primarily on people rather than facts and deeds – restorative justice could never be positioned along with other forms of criminal justice. Restorative justice occupies a place somewhere between penal theory and carefully designed practices, which reflect a way of thinking, responding and, in a long-term, controlling crime. The exact place of restorative justice should not be determined by whether restorative justice is conceived as an end in itself, irrespective of the outcome (if any) or as a means to some other end.[10] In both those cases restorative justice could bring desirable outcomes as long as it remains 'restorative' and expresses particular restorative values, which derive their potency from the fact that they reflect moral and legal truth.

Consequently, restorative justice could be conceived as a 'world notion' providing that it shares by default three fundamental principles: I) **Accountability**: there is

---

8   Braithwaite, J. & Pettit, P. (1990). Not Just Desserts: a Republican theory of criminal justice. Oxford: Claredon Press

9   Supreme Court of Canada in R v. Gladue (1999) 133 C.C.C. (3d) 385 at paragraphs 71 & 72; and Pickard, P., Goldman, P., Cairns-Way, R. & Mohr, R.M. (2002). Dimensions of Criminal Law. (3rd ed.). London: Pearson, p. 151

10   Dignan, J. (2005). Understanding Victims and Restorative Justice. Buckingham: Open University Press, p. 5

no restorative practice for those who 'didn't do it'. Restorative justice is about creating processes that allow offenders to take responsibility of the harm created by their actions, trying to balance the accountability of the offender with the need of the affected by the offence. In addition, it is an opportunity for community to see its role in contributing to the crime. In all cases, accountability involves not only accepting responsibility for the crime, but also accepting responsibility for addressing the harms and needs arising from it. Hence, accountability *is* an important feature of restorative justice representing the opportunity to denounce the criminal act and reinforce social rules and laws. II) **Inclusion**: restorative justice is driven by the involvement of all the parties affected by crime in the determination of punishment. These are more other identified as the victim, the offender, their individual support people (family, friends) and the community. This requires elevating the roles of those traditionally excluded from the process, as well as the involvement of criminal justice professionals. III) **Transformation**: it has become obvious that, at a practical level, the restorative justice process itself can often transform the relationship between the community and the criminal justice as a whole transformation implies the forward-looking aspects of restorative justice. The potential outcomes of restorative justice interventions typically include healing, reparation of harms and moreover, restoration of positive relationships. These goals apply equally to all parties involved, but are not always achieved or within the scope of all circumstances. Restorative justice interventions foster movement towards these aims via communication and interaction between each individual participant impacted by the crime.

Restorative justice processes take into consideration and value the full breadth of each individual participant in relation to the larger context in which they function. This includes appreciation of the psychological, mental, emotional, spiritual and social context surrounding each person. Regarding the spiritual component, it is worth mentioning that for many participants a restorative experience can connect deeply to belief systems. At the same time, criminal behaviour arouses strong emotional reactions within those who are offended and the inevitable consequence is conflict, hostility and alienation. The criminal justice system endeavours to channel those volatile reactions to prevent further injury as it establishes guilt and comes to an appropriate penalty. But it does little to promote constructive resolution of the destructive interpersonal chain reaction that criminal behaviour triggers. In fact, the criminal justice system seems to work in a pattern that compounds the sense of alienation between the victim and the offender. In contrast, restoration of wholeness to community (including the victim and the offender) is the ultimate goal of restorative justice. Such holism can be accomplished in relation to a willingness on the part of both parties to work towards reconciliation and peace making. Restoration is costly for both parties; even more costly personally and socially than to remain alienated.

Restorative justice could become an immediate response to isolation imposed by the current way of doing justice on the condition that some basic 'ingredients' which could apply to any type of offence are included. First of all, *safety* should be the primary consideration of the community. All the decisions regarding crimes should be based on a safety axis. Afterwards, the offender must be made accountable for his/her criminal behaviour. As long as the offender's *responsibility* is secured, victims and the community (seen as the secondary victim) that have been harmed require *restoration*. The maintenance of the sequence between safety-responsibility-restoration is dependent upon the existence of a continuum of services and/or treatment options in a variety of settings. Most importantly, it is also dependent upon a coordinated systems approach and cooperation between public and private resources; something that seems the stiffest part.

Justice, when spelled with capital 'J', should be distinctive for its social, all-inclusive character and for the promotion of a common good taking advantage of all the available, but unfortunately modestly used, resources. Similarly, the proponents of restorative justice urge that society should seek for more comprehensive responses to crime that will encourage the reintegration of offenders; through or in parallel to legal stream.

Restorative Justice between Law and Common Good

Where does law come from? In the Western world there is a general understanding that law and justice are not synonymous terms.[11] The Law does not stand-alone as it should establish what Justice is. In a secular society, for law and justice to find a common ground, they have both to abide to the principles of 'common good' and peace-making. Law should not be merely a civil generalization irrelevant to daily life. It should actually preserve in words a number of moral values that should underlie the rules of social harmony and order. In the same vein, the main 'ingredients' of restorative justice are not laws or rules imposing arbitrary regulations irrelevant to peoples' needs. The principles of restorative justice could be seen as groundwork for true justice and equity based on universal principles to be implemented in society. These principles are not retributive but restorative and forgiving, presupposing the possibility of an ongoing relationship between individuals. However, ruptures do happen and in such cases the emphasis is on restitution and restoration – as a way of setting things and relationships right – not on vengeance and retributive punishments.

A major component of restorative justice is about reintegration and inclusion paying special attention to those who have been off the law-beaten track. Thus, a core theme of the ministry of the restorative justice ethos is that anyone is welcomed back into the community trying to eliminate anything akin to second-class citizenship. In the final analysis, Restorative Justice is about redemption. This ongoing connection provides a multi-dimensional, people-centred view of justice. Restorative justice best corresponds to the humanitarian vision of Law viewing crime as damaging to relationships, not merely as abstract law-breaking. The rationale behind this approach to punishment and the view that all affected parties should be involved in the criminal justice process goes to the very heart of the restorative justice philosophy.

This brings the assumption that crimes or violations are committed against real individuals, rather than against the state. Violations[12] should always require recompense for the harm inflicted to the victim. Restorative justice, therefore, advocates restitution to the victim by the offender rather than retribution by the state against the offender. Instead of continuing the 'revolving door' effect on persistent offenders and escalating

---

11   Allard, P. & Wayne, N. (2003). Christianity: the Rediscovery of Restorative Justice. In G.A. Johnstone (Eds.). Restorative Justice Reader. Devon: Willan publishing, p. 158
12   Mainly against life that God has created, but also other violations like injury (Exodus, 21: 18-23) or theft (Exodus, 22: 1-4)

the cycle of violence, it tries to restore relationships and stop the violence.[13] But, what exactly is justice in the sphere of restoration? Definitely, RJ's perception of justice is based on peoples' justice and not on the 'dry', emotionless imposition of punishment of the wrongdoer; it disfavours the literal[14] proportional concept of 'eye for eye and tooth for tooth'[15] (Exodus, 21: 23-24) and placed great importance on the concept of forgiveness – either through the (partial) restitution of the damage or through a single but genuine apology. Singh (2001) finds an emphasis on the virtues of mercy, forgiveness and compassion; these provide foundations for healing and reconciliation.[16] Especially the word 'forgiveness' is omnipresent in the restorative justice doctrine. Drawing, at the same time, a clear line of distinction between 'forgive' and 'forget', forgiveness does not necessitate forgetting actions of those who hurt us through an imposed amnesia. Restorative justice tries to endorse the value of forgiveness, which entails liberation from the past and forms a precondition for reconciliation.[17] Both now forgiveness and reconciliation are important aspects of restoration[18], and restoration is the very definition of justice.

I also recognise that restorative justice is much more than a set of values and an inclusive process. Restorative justice is something deeper that explains both its power and cross-cultural applicability. It boasts an ideology that centres on forgiveness, repentance and reconciliation; values that lie at the heart of Common Good. In contrast, the word retribution as punishment has no place in the restorative justice vocabulary. Apparently, these values are not exclusive to restorative justice as elements of inclusion and community cohesion could be found in other criminal justice systems too; nevertheless, in a scattered, fragmented and hence form placed low on the priority list of a retributive justice system. However, restorative justice is the only approach that regards these values as top priorities placing the individual in an integrated relationship of solidarity with the community. The individual is 'part and parcel of society [...], which is not a system, but the highest form of integral and integrated collectivity'.[19] Within this view of the intimate relationship between community and individual,

---

13   Hutchison, P. & Wray, H. What is Restorative Justice? Available from http://gbgm-umc.org/nwo/99ja/what.html (08/06/2009)

14   The term lex talionis in the Torah does not always refer to literal eye-for-an-eye codes of justice. The Talmud (a record of rabbinic learning pertaining to Jewish law, ethics and history) states that punishments serve to remove dangerous elements from society and to deter potential criminals from violating the law within their context and epoch.

15   Under the British Common Law, successful plaintiffs were entitled to repayment equal to their loss (in monetary terms); and in the modern tort law system, this has been extended to translate non-economic losses into money as well. However, the strict proportional concept of 'an eye for an eye' (interpreted either literally or in monetary terms) is opposed to both Christianity and Restorative Justice, as it favours retribution. A 'monetary' interpretation of this lex talionis term can be found in the early Anglo-Saxon legal code, which had substituted payment of wergild (or reparational payment) for direct retribution: a particular person's life had a fixed value, derived from his social position; so, any homicide was compensated by paying the appropriate wergild, regardless of intent.

16   Singh, P. (2001). Sikhism and Restorative Justice: Theory and Practice. In M.L. Hadley (Ed.). The Spiritual Roots of Restorative Justice. Albany, NY: SUNY Press, p. 200

17   Allender, D. (2000). Forgive and Forget and Other Myths of Forgiveness. In L.B. Lampman & M.D. Shattuck (Eds.). God and the Victim. Grand Rapids, MI: Eerdmans Publishing Company, p. 213

18   Hadley, M.L. (2001). The Spiritual Roots of Restorative Justice. Albany, NY: State University of New York Press, p. 48

19   Ammar, N.H. (2001). Restorative Justice in Islam: Theory and Practice. In M.L. Hadley (Eds.). The Spiritual Roots of Restorative Justice. Albany, NY: SUNY Press, p. 166

crime seems to fit in both private and public spheres: as a repeal of one's responsibility towards the Law and as a rupture of community's harmony and unity.

Focusing more on mediation, restorative justice could be considered as a close cousin to dispute or conflict resolution. It expresses other humanistic values or facilitates their expression. One of these is definitely the value of reconciliation, which urges people to shape their lives in positive relation to one another. As mentioned above, restorative justice practices (especially mediations and conferences) encourage the expression of remorse and the willingness to change – to become better. They also reject high formalism and legalism in favour of the personal encounter and engagement of those directly engaged. Zehr (1998) finds some support for restorative justice in the Old Testament concept of 'shalom', which could be broadly translated as a multileveled 'peace' between God and humanity, as well as between people in conflict[20]: 'Administer true justice; show mercy and compassion to one another' (Zechariah, 7: 9). In a similar way, restorative justice upholds the possible benefits of peace-making, restoration and reconciliation that can only be gained via *direct* contact with each other and not by legal authorities or by just following orders.

The key element in all these approaches in the desire to heal victim and offender wounds in the process of justice. Sande (1997), although not using the term restorative justice, encapsulates the concept by saying: to some, conflict is a hazard that threatens to sweep them off their feet and leave them bruised and hurting. To others, it is an obstacle that they should conquer quickly and firmly. But a few people have learned that conflict is an opportunity to solve common problems in a way that offers benefits to those involved.[21] However, building cultures of reconciliation implies a process. Unfortunately, social conflict is inherent in human relations, and in order to learn how to deal with practices and attitudes that contribute to conflict rather than mutuality, we should recognize the origins of conflict. The Scriptures refer to humans as a 'body' connoting a structure of stability but also one with flexibility.[22] So, communities, built using the same 'body, provide support, encouragement and space for those involved in restorative processes encouraging opportunities for reparation and peacemaking so that offenders and victims find healing in a society of hope.

To sum up, restorative justice *is* addressing the needs of the victims and of offenders in such a way that they and the community may be healed. However, restorative justice is *not* yet a widely accepted paradigm, even in religious circles. It is a concept in progress that needs further definition and communication. Nevertheless, even not a finished concept, it is not untried. There is a number of restorative initiatives already been undertaken around the world (e.g. victim-offender mediations, family group conferences, circles of support and accountability), but unfortunately, they are often underfunded and (hence) not broadly known.

20   Zehr, H. (1998). Restoring Justice. In L.B. Lampman & M.D. Shattuck (Eds.). God and the Victim: Theological Reflections on Evil, Victimization, Justice and Forgiveness. Grand Rapids, MI: W.B. Eerdmans publishing, pp. 131-159
21   Sande, K. (1997). The Peacemaker: a Biblical Guide to Resolving Personal Conflict. (2nd ed.). Grand Rapids, MI: Baker Academic, p. 17
22   Van Ness, D.W. (2002). The Role of Church in Criminal Justice Reform. Paper presented in the 'Justice that Restores' Forum. Orlando, FL: 14-16 March, p. 6

2   Digging to the Historical Roots of Restorative Justice

As restorative justice moves constantly forward, there is a tempting desire to glance back to history. At the same time, one thing that is prominent in the restorative justice literature is that its history is given very little consideration. The historical roots of restorative justice and the exact date of its founding remain an often debatable enigma. Some theorists, like Sarre (1999), Sylvester (2002) and Bottoms (2003), admit that the restorative justice movement is of recent origin or a novel approach to criminal justice that really only achieved a sustainable viability in the early[23] or mid-1990s[24] and that authors are using history to legitimise restorative justice in present[25]; while others, like Weitekamp (1999) and Braithwaite (1999), proclaim that it has been tracked since humans began forming communities[26] and that it had been the dominant model of criminal justice for all of the world's people[27]. Nevertheless, both conflicting assumptions appear to be correct and scientifically justified.

Restorative justice narratives fall somewhere between myth and history. Hence, the purpose of the historical analysis given below is to shade light on the 'dark' points behind this partition offering a constructive criticism and accurate interpretations of previous cultures' approaches to crime without favouring one theoretical side against the other. In the present article, history does not start at the point when the term 'restorative justice' became popular; it goes further back in time, trying to discover possible links between modern restorative practices and historical or anthropological precursors, such as restitution and indigenous justice.

Pre-State or Acephalous Societies

One cannot clearly distinguish the historical track of restorative justice or understand its implications without first familiarizing oneself with the restitution literature that predates it. It is important to note a critical parallel between justice of the past and Restorative Justice. Many of the key values of restorative justice are deeply rooted in the tradition of indigenous cultures and can be best appreciated when they are seen through this prism. Starting from the acephalous, pre- or non-state societies, restitution was a viable criminal justice response. At the same time, their frequency was fluctuating and their dominance was subject to various socio-economic factors.

For certain restorative justice theorists, restitution was regarded as the normal way of handling 'crime'[28] and, what we know today as Restorative Justice, it was representing a return of the simple common wisdom of viewing conflict as an opportunity for a community to learn and grow. Braithwaite (1998) makes a generic claim that "re-

---

23   Sarre, R. (1999). Restorative Justice: Translating the Theory into Practice. Australia Law Review, 1, 13

24   Sylvester, D.J. (2003). Myth in Restorative Justice History, Utah Law Review, 1, 471-522

25   Bottoms, A. (2003). Some Sociological Reflections on Restorative Justice. In A. von Hirsch, J. Roberts, A. Bottoms, K. Roach & M. Schiff (Eds.). Restorative Justice and Criminal Justice: Competing or Reconciling Paradigms?. Portland, OR: Hart publishing, p. 88

26   Weitekamp, E.G.M. (1999). The History of Restorative Justice. In G. Bazemore & L. Walgrave (Eds.). Restorative Juvenile Justice: Repairing the Harm of Youth Offending. Monsey, NY: Criminal Justice Press, pp. 75-102

27   Braithwaite, J. (1999). A Future where Punishment is Marginalised: Realistic or Utopian?. UCLA Law Review, 46, 1725-1750

28   Van Ness, D. (1993). New Wine and Old Wineskins: four challenge of restorative justice. Criminal Law Forum, 4, 252-257

storative justice had been the dominant model of Justice throughout most of human history for all the world's peoples".[29] More specifically, Weitekamp (1999) claims that in acephalous (non-state) societies negotiated restitution was more common than retribution for all crimes.[30] He emphasizes the concept of collective responsibility by stating that non-state societies formed a kin-based social organisation that sought not to punish the culprits but to seek resolution and regain balance facilitating a quick return to daily life.[31] Their motivation was to disfavour retributive and retaliatory reactions to crime (or to any behaviour that was then deemed unlawful and dishonest) by doing something for the victim through restitution.

In contrast, Nader and Combs-Schilling (1977) have emphasized less these claims explaining that restitution was only one among many sanctions operating in a larger social control system of such societies.[32] The same society that was using restitution as a strategy could also use retaliation.[33] In other words, the adopted theory of restitution to the injured was indeed a liberal and merciful response to crime, but we should not overlook the ability of pre-state societies to impose hierarchical penalties or the devastating consequences of the offender's inability to pay the imposed debt.

As far as the evidence allow, restorative justice writings use very often anthropological works to sustain the above views; sometimes scratching only the surface of the relevant literature and falling into slight overstatements and contradictions. For example, restorative justice proponents argue that within the *Eskimo* society, retaliation was very rarely used and that, as a victim-centred society, it was responding to crime with restitution.[34] At the same time, Adamson-Hoebel's famous work 'The Law of Primitive Man' guides us to a different direction. He concludes that Eskimo's criminal justice system was permitting retaliatory killings (for specific offences, such as rape, homicide and insult) and he admits that homicidal dispute, though present and prevalent, was made less frequent (not 'only rarely') than restitution.[35] Often, there was no further need for a community response to crime as some Eskimo villages were financially able to maintain the victim's family without creating a burden to the village.

Comparable occurrences could be found in the laws and institutions of the Māori nation in New Zealand. Presenting the New Zealand sanctioning structure as developed in a 'perfect society', Pratt (1992) noted that the Māori legal system in pre-contact times accentuated a restorative nature.[36] Today, Consedine (2003) describes how Māori justice operates alongside the dominant English-derived system in New Zealand, stating that the Māori integrated system of justice emphasizes aiding, forgiving and heal-

29 Braithwaite, J. (1998). Restorative Justice. In M. Torny (Eds.). The Handbook of Crime and Punishment. Oxford: Oxford University Press, p. 323
30 Weitekamp, E.G.M. (2003). The History of Restorative Justice. In G.A. Johnstone (Eds.). Restorative Justice Reader. Devon: Willan publishing, p. 111
31 Weitekamp, E.G.M. (1999). The History of Restorative Justice. In G. Bazemore & L. Walgrave (Eds.). Restorative Juvenile Justice: Repairing the Harm of Youth Offending. Monsey, NY: Criminal Justice Press, p. 76
32 Nader, L. & Combs-Schilling, E. (1977). Restitution in Cross-Cultural Perspective. In J. Hudson & B. Galaway (Eds.). Beyond Restitution – Creative Restitution. Lexington, MA: Lexington Books, p. 32
33 Nader, L. & Combs-Schilling, E. (1977). Idem, p. 36
34 Weitekamp, E.G.M. (1999). supra note 25
35 Adamson-Hoebel, A. (1961). The Law of Primitive Man. Cambridge, MA: Harvard University Press, p. 99
36 Pratt, J. (1992). Punishment in a Perfect Society: the New Zealand Penal System 1840-1939. Wellington: Victoria University Press, p. 35

ing, as opposed to retribution.[37] In contemporary Māori justice people are less anxious to find truth and justice than to abate conflict and promote harmony.[38] The offenders are encouraged to experience shame and the victim gets the chance to express his/her feelings (such as pain, anger etc.) to the perpetrator to achieve forgiveness.

It is important to mention that the contemporary notion of Māori justice has been adjusted to the customs of values of the contemporary era, leaving behind certain non-restorative characteristics. Bearing in mind that the committing of a *hara* (crime) would damage and upset the *hapu* (social balance), an integral component of the Māori process was to redress the harm inflicted to victims. The form as well as the level of compensation (*utu*) was depending on the degree of the offence and sometimes it was accompanied by mediation. At the same time, for serious offences against *hapu*, the death penalty was demanded; and the victim's right to redress could also be passed from one generation to another until a resolution was being negotiated and the aggrieved party was compensated.[39] This means that the scale of restitution or remuneration was clearly determined by the victim and his or her status and it was slightly concerned with the offender or the larger community; as restorative justice argues. In other words, the restitution was not a fruit of a compromise that satisfied both parties, but it was rather imposed than negotiated.

The Māori law thus embodied ideals that sometimes were mutually contradicted and not necessarily restorative. However, the same ideals did provide direction to individuals and strength to local communities. *Utu* or compensation and satisfaction were placed in the heart of Māori justice; in parallel, *muru* – a formalised concept of retributive compensation – was another central feature aimed to restore the lost balance. *Muru* was a means to achieve *utu* that parties agreed to – but not all cases were ended in reconciliation.

In both the above Indigenous groups, as well as other indigenous ethnic groups around the world – such as the Yurok Indians in Northern California, the Ifugao in the Philippines or the Nuer tribe in Sudan – even where restitution may have predominated, it is clear that it contained elements non-restorative in nature. Avoiding, though, oversimplifications and broad generalisations, it would be worthwhile to mention that the above ambiguities could not of course be extended to all pre-state societies. In contrast to all the above acephalous societies, the **Navajo** Indian nation in North America (U.S.A. and Canada) separates the action from the person and perceives people not simply as individuals but everyone owes special obligations to others. The basic concepts of Indian peacemaking justice are reciprocity and solidarity as opposed to hierarchy.[40] Within the horizontal system of justice, no person is above the other. Navajos do not think of equality as treating people as equal *before* the law, but they are equal *in* the law.[41]

---

37   Consedine, J. (2003). The Maori Restorative Justice. In G.A. Johnstone (Eds.). Restorative Justice Reader. Devon: Willan publishing, pp. 152-157

38   Rudolph, L. & Rudolph, S.H. (1967, p. 258) quoted by Danzig, R. (1973). Toward the Creation of a Complementary, Decentralized System of Criminal Justice. *Stanford Law Review*, 26, 1-54

39   Joseph, R. (1999). Māori Customary Laws and Institutions for Aotearoa / New Zealand. (Draft). Hamilton: University of Waikato – Te Matahauraki Research Institute, pp. 4-5 & 16-17

40   Zion, J.W. (1998). The Dynamics of Navajo Peacemaking. Journal of Contemporary Criminal Justice, 14, 58-74

41   Yazzie, R. (1994). Life Comes from It. The Ecology of Justice, 38, 29

The Navajos believe that crime happens when an offender has failed to remember that all human beings are to live in a peaceful, harmonious relationship. In Navajo justice, healing is achieved via the very process of decision-making where peacemaking comes as a remedy for the 'we can't do anything' excuse. In general, excuses do not prevail in a process that fosters full discussion to solve problems.[42] Navajos recognise both a form of traditional case law and a corpus of legal principles[43] and most cases in peacemaking, which is based on family and societal relationships, are brought directly by the affected parties. In Navajo legal doctrine, a person is not framed in the isolated context of individual rights, but only as a member within a community. Traditional Native American spirituality emphasizes that ultimate justice is up to the Creator, and human justice is to focus on restoration of harmony rather than on revenge or punishment.

The Navajo peacemaking process helps an offender realize that what he or she has done is incorrect and urges each member of the community owes to treat everyone as if that person was his/her relative. The Navajo process brings the victim and the offender together to talk each other. The main purpose of this process is to create bonds between the two parts, to connect or bind each other as a community. Flies-Away (2004) rightly mentions that 'people do the worst things when they have no ties to people' (Flies-Away).[44]

Central to Navajo justice is the concept of *"k'e"* which could be translated as unity and cohesion: what you do, either good or bad, there are consequences on me and *vice versa*. According to Yazzie (2004), *k'e* means to restore one's dignity, to restore one's worthiness[45]; this illustrates that Indians do not resolve matters by rules, but by examining relationships. To the Navajo way of thinking, justice is related to healing. Compensating a victim in accordance with the victim's feelings and the offender's ability to pay is more important than using a precise measure of actual damages. Subsequently, the Navajo justice abandons guilt and adequate compensation in favour of achieving well-being for everyone.

The concept of peacemaking is found in the cultures of many indigenous nations[46] and its usage has spread o include other forms of restorative justice, not necessarily indigenous. All the same, it has been stated that Navajo nation peacemaking is not a form of restorative justice that can be adopted in a non-Indian environment.[47] In response to this, my argument is that the cultural and spiritual specifics of the Navajo group may be unique and non-directly transferable; however, their values beyond the Navajo rituals – like harmony and cohesion – could be (or are already) incorporated into any restorative process developed into a non-Indian context. The Navajo peace-

42   Yazzie, R. & Zion, J.W. (1996). Navajo Restorative Justice: the law of equality and justice. In B. Galaway & J. Hudson (Eds.). Restorative Justice: International Perspectives. Monsey, NY: Criminal Justice Press, p. 146

43   Idem, Yazzie & Zion, 1996, p. 147

44   Flies-Away, J. In L. Mirsky (2004). Restorative Justice Practices of Native American, First Nation and other Indigenous People of North America: Part One. Restorative Practices e-Forum. Available at: Available at: http://www.restorativejustice.org/articlesdb/articles/4230 (06/06/2009), p. 6

45   Yazzie, R. In L. Mirsky, (2004). Idem, p. 3

46   Stuart, B. (1997). Building Community Justice Partnerships: Community Peacemaking Circles. Ottawa, ON: Department of Justice of Canada – Aboriginal Justice Directorate

47   Goldberg, C.E. (1997). Overextended Borrowing: Tribal Peacemaking Applied to Non-Indian Disputes. Washington Law Review, 72, 1019

making and (original) dispute resolution[48] are not mediation and they are not alternative dispute resolution. It is, though, restorative justice but not exactly the same kind. The main diversification point lies on the role of the peacemaker who, in the Navajo case, is not a neutral mediator. He is a knowledgeable person who tries to actively involve support group members in the dispute (not just as participants) in order to achieve a consensus, as the process cannot end without it.[49]

Early-State Societies

Moving from pre-state to early-state societies, scholars have drawn differing conclusions: on one hand, the emergence of State and State punishment is seen as derived from a 'Hobbesian jungle'[50] or a 'Leviathan' depicting State as a Biblical monster[51]; while, on the other hand, early-state society responses were restorative in nature and the purpose of their justice process was to restore a community that had been sundered by crime.[52]

In support to the later conclusions Johnstone (2003) claims that in Europe between the 6th and 7th centuries, restorative justice was more the norm and state prosecution and punishment the exception.[53] The 'common conscience' or 'folklaw' of Europe developed prior to the emergence of the Western legal tradition – its development began to take shape in the late 11th and 12th centuries.[54] Braithwaite (2002) also mentions that restorative justice traditions could be found in ancient Arab, Greek and Roman civilizations, in which a restorative approach was accepted even to homicide.[55] Other restorative justice theorists[56] refer to various legal codes written in early-state societies; especially those of ancient Mesopotamia. Focusing on the famous codes of Ur-Nammu (c. 2100-2050 B.C.) and of Hammurabi (c. 1760 B.C.), it is claimed that restitution was used even for violent crimes. Moreover, the Babylonian Code of Hammurabi is the only source in the historical literature where the concept of restitution is restricted to property crimes[57] and victims were entitled to receive payment according to when and

---

48   The term 'original dispute resolution' has been added by both Zion and Yazzie (1997) referring to its origins in the distant past.*
     * Zion, J.W. & Yazzie, R. (1997). Indigenous Law in North America: the Wake of Conquest. Boston Law Review, 20, 55-56
49   Nielsen, M.O. (2005). Navajo Nation Courts and Peacemaking: Restorative Justice Issues. In M.O. Nielsen & J.W. Zion (Eds.). Navajo Nation Peacemaking: Living Traditional Justice. Tucson, AZ: University of Arizona Press, pp. 143-155
50   Hobbes, T. (1651/1991). Leviathan. Cambridge: Cambridge University Press
51   The Hebrew word 'Leviathan' is referred to in the Old Testament (Psalm 74:13-14; Job 41; Isaiah 27:1) and its Christian interpretation is considered to be a semitic mythological (sea) beast associated with evil.
52   Van Ness, D.W. & Strong, K.H. (1997). Restoring Justice. Cincinnati, OH: Anderson publishing, p. 9
53   Johnstone, G. (2003). Introduction – A Restorative Justice Reader. Devon: Willan publishing, p. 101
54   Berman, H.J. (1997). Law and Revolution: the Formation of the Western Legal Tradition. Cambridge: Cambridge University Press
55   Braithwaite, J. (2002). Restorative Justice and Responsive Regulation. New York : Oxford University Press, p. 3
56   See Weitekamp, E.G.M. (1999) & Van Ness, D.W. and Strong, K.H. (1997)
57   Weitekamp, E.G.M. (1999). The History of Restorative Justice. In G. Bazemore & L. Walgrave (Eds.). Restorative Juvenile Justice: Repairing the Harm of Youth Offending. Monsey, NY: Criminal Justice Press, p. 83

---

under what circumstances certain (property) offences were committed.[58] Edelhertz (1975), after reviewing some anecdotal sources from the Code of Hammurabi, concludes that "primitive penal law was largely a law of torts [...] that recognized the private and individual nature of the wrong, but sought to redress it through economic means".[59]

Similarly, the same theorists have supported the claim the early legal systems forming the foundation of Western law emphasized the need for offenders and their families to settle with victims and their families. In particular, it has been noted that the Twelve Tables of Roman Law (449 B.C.) included provisions for (disproportionate) restitution, as they required offenders to pay double or even triple the value of the inflicted damage to victims for certain crimes. Also in post-Roman European laws, the Laws of Ethelbert of Kent (602-603) are noted as an example of promoting restitution in an effort to avoid further crime and retaliatory responses. To achieve so, the Laws of Ethelbert of Kent adopted a community-based justice approach stating that the offender's clan was responsible for crimes committed by its members because every crime was against not the individual but the whole family or clan.[60]

However, historians who have studied these texts have found that, for example the Code of Ur-Nammu – the oldest of the ancient Mesopotamian law collections (2094-2047 B.C.) – and the Twelve Tables of Roman Law included provisions for imprisonment and even death for a number of crimes along with a compensatory scheme for dealing with criminal conduct.[61] Especially about the Twelve Tables, Drapkin (1989) stresses the dominant place of capital punishment concluding that the Roman Law was based on revenge.[62] Concerning the code of Ur-Nammu, it had been protecting the interests of the upper-class citizens imposing sharply differentiated penalties and treatment for upper-class and lower-class (slaves) offenders. Similar claims have been made for the Code of Hammurabi which embraced a talionic notion of criminal justice in order to enforce rigid norms of class, gender and other status distinctions.[63]

Thus, many of the restitutionary structures in early societies were at the same time punitive in nature, requiring the offenders to make payments far above that of the value lost to the victim (*see* the Drakonian fine structure) resulting to slavery or servitude of the offender.[64] Although restitution was a possible penalty, extreme penalties were also available, which few if any offenders were able to pay the required amount. Given this financial difficulty of offenders, they were often left at the mercy of victims. For instance, among certain Germanic tribes (e.g. the Visigoths) the high level of payment demanded for penalties was to shame offenders by assuring that they could not

58 Karmen, A. (1990). *Crime Victims: An introduction to victimology* (2nd ed.). Belmont, CA: Wadsworth, p. 279

59 Edelhertz, H. (1975). Restitutive Justice: a General Survey and Analysis. Seattle, WA: Battelle Human Affairs Research Centres, p. 4

60 Weitekamp, E.G.M. (1999). Idem, p. 85

61 VerSteeg, R. (2002). Law in the Ancient World. Durham, NC: Carolina University Press, p. 286

62 Drapkin, I. (1989). Crime and Punishment in the Ancient World. Lexington, MA: Thorsten Sellin, p. 232

63 Sylvester, D.J. (2003). Myth in Restorative Justice History. Utah Law Review, 1, 514

64 Lindgren, J. (1996). Why the Ancient May Not Have Needed a System of Criminal Law. Boston University Law Review, 76, 40-43

afford to make the appropriate compensation payments[65] rather than by serving as a deterrent and a process of victim's reintegration ('reintegrative shaming').

Consequently, even in those cases where restitution was employed, it did not appear to be restorative in nature. All criminal justice responses in the early-state societies were as equally punitive and retributive as they were forgiving and restoring. The ability of victims and their families to engage in the negotiations for compensation is a beneficial process for victims; however, the aforementioned isolated examples of restitutionary processes should not be characterised as inherently or purely victim-oriented. A basic argument why restitution favoured offenders and communities instead of victims is based on each society's motivations to strengthen community authority, to prevent the socially disintegrating effects of privately wrought restitution and moreover to eliminate the fear of vengeance by wrongdoers.[66] The problem was not crime as such, but the inability of victims to suppress their desire for vengeance. As Edelhertz (1975) concludes "a closer look at the history of restitutive justice reveals a penalty system far more concerned with the offender than with the victim".[67]

Later-State Societies

Although the restorative justice movement is of recent origin, the precise chronicle of its founding still remains vague. Restorative justice theorists have claimed not only that reparative approaches to crime have modern utility, but also that they are ore traditional than our modern approaches.[68] In parallel, some other scholars have indicated that the history restorative justice cites lacks accuracy[69] and that restorative justice adherents have been also criticized for creating a 'mythical' past for utilitarian purposes.[70]

With the rise of the later-state societies' prosecutorial and judicial functions, crime was regarded primarily as a public disorder.[71] Law considered a western concept[72] and importance was not attached to how truth could be revealed, but to how Justice could be brought. As a system of institutional reform, the restorative justice movement is generally dated in the early 1970s with the creation of the Minnesota Restitution Centre and Kitchener, Ontario's development of victim offender reconciliation programmes. However, theorists retrace a history for restorative justice and they place it among early fundamental theories of Beccaria, Bentham, Ferri, Garofalo etc., using criminological material. Weitekamp (1999) for instance claims that Bentham laid the groundwork for advocates of restorative justice and stressed the necessity of taking

---

65 Schafer, S. (1970). Compensation and Restitution to Victims of Crime. (2nd ed.). Montclair, NJ: Patterson Smith, p. 5
66 Edelhertz, H. (1975). Restitutive Justice: a General Survey and Analysis. Seattle, WA: Battelle Human Affairs Research Centres, p. 12
67 Idem, p. 16
68 Braithwaite, J. (1999). Restorative Justice: Assessing Optimistic and Pessimistic Accounts. (ed. M. Tonry). Crime and Justice, 25, 1-2
69 Daly, K. (2002). Restorative Justice: the Real Story. Punishment and Society, 4, 55-56
70 Idem
71 Karmen, A. (1990). *Crime Victims: An introduction to victimology.* (2nd ed). Belmont, CA: Wadsworth, pp. 279-280
72 Gulliver, P.H. (1969). Dispute Settlement without Courts: the Ndendenli of Southern Tanzania. In L. Nader (Ed.). Law in Culture and Society. Chicago, IL: Aldine, p. 12

care of the crime victim by means of restorative justice.[73] I would argue here that these claims do not accurately represent the work of Bentham. Bentham did promote restitution to victims but this constitutes only a segment of his contribution in the criminological field. His name is synonymous with the 'Panopticon' style of prison that he eagerly promoted; thus to boldly consider Bentham as an advocate or promoter of restorative justice seems rather imprecise.

On the whole, restorative justice as a social movement really only achieved a sustainable viability in the mid-1990s[74] and it was based on intensive studies on the history of restitution and compensation. One of the earliest studies was Schafer's '*Compensation and Restitution to Victims of Crime*' published in 1970. Through his work, he tried to present and explain restitution's historical lineage. Schafer (1970) claimed that for a long period of time restitution was almost inseparably attached to the institution of punishment.[75]

In relation to this argument, Edelhertz's publication '*Restitutive Justice*' (1975) – which constituted a comparative review of restitutionary process around the world – recognised that restitution and compensation programmes have long historical antecedents.[76] By claiming that restitution as an approach for dealing with criminal behaviour was 'by no means new', Edelhertz (1975) also recognised that later-state societies'[77] concern about restitution was directly connected with the protection of offender's and offender's social groups, not benefits to victims.[78] In fact, it was more the victim's behaviour that was being called into question than the offender's.[79]

A direct refutation of Schafer's and Edelhertz's statements was offered by Nader and Combs-Schilling who came to the conclusion that restitution is separate from punishment and that it represents different motivations and purposes.[80] They argued that restitution *was* victim-oriented as the appropriate level was usually determined according to the type of harm and the status and the victim.[81] For instance, in Britain during the 9th century, perpetrators were required to restore peace by making payments to the victim (and the victim's family). The main purpose of institutionalized restitution was to prevent retaliatory responses for wrongdoing, providing a more standardized mean of reparation. Although not yet a fully-fledged restorative justice approach, this redirection of restitution history motivated the more formal restorative justice movement to emerge. From all the above isolated initiatives on both theoretical and practical level, restorative justice has grown to become an international phenomenon being practiced into formal institutions.

---

73   Weitekamp, E.G.M. (1999). The History of Restorative Justice. In G. Bazemore & L. Walgrave (Eds.). Restorative Juvenile Justice: Repairing the Harm of Youth Offending, Monsey, NY: Criminal Justice Press, p. 90

74   Braithwaite, J. (1999). A Future where Punishment is Marginalised: Realistic or Utopian?. UCLA Law Review, 46, 1725-1750

75   Schafer, S. (1970). Compensation and Restitution to Victims of Crime. (2nd ed.). Montclair, NJ: Patterson Smith, pp. 3-12

76   Edelhertz, H. (1975). Restitutive Justice: a General Survey and Analysis. Seattle, WA: Battelle Human Affairs Research Centres, p. 1-20

77   The term 'later-state societies' refers to all those societies that followed the early-state ones.

78   Idem, p. 4

79   Idem, p. 13

80   Nader, L. & Combs-Schilling, E. (1977). Restitution in Cross-Cultural Perspective. In J. Hudston & B. Galaway (Eds.). Restitution in Criminal Justice. Lexington, MA: Lexington Books, p. 27

81   Idem, p. 32

3   Drawing Conclusions

All the above evidences derived from the anthropological and historical aspects of restorative justice appear competing or variously interpreted, albeit some common conclusions are reached: (a) restitution is inherently victim-oriented;[82] (b) restitution seeks to restore balance and community cohesion rather than to impose punishment;[83] and (c) the acephalous societies regarded and used restitution as the central remedy to criminal behaviour[84]; a priority that did not trespass into the early-state and contemporary societies. These elements proclaim a universal approach of societies to favour restitution and are echoing some values found in restorative justice, but with important differences. Restoration, being also present in non-state and early-state societies, constituted a form of conflict resolution that was bearing similar to restitution purposes, like: (a) to cover the victim's needs; and (b) to re-establish the broken relationship between the victim, the offender and the community as a whole. Along with these common priorities, restoration served a number of other purposes that were not covered by mere restitution, namely: (a) to rehabilitate the offender and prepare his/ her return to society avoiding stigmatization; (b) to familiarize the members of the society with the dominant values and norms; (c) to act as a deterrent for its members; and (d) to prevent recidivism.[85] These common features that aboriginal justice processes share should not be misconceived and broadly generalised to every aspects of justice and community's 'well-being'. They are processes that embrace both the notion of holistic peace and justice-making, but most importantly, the fundamental right of self-determination. There are many different Aboriginal people around the world and each has a unique and traditional understanding of what the experience of justice in their community was and still is.

As a general conclusion, restorative justice does bear deep historical roots in both pre-state and early-state societies. Nonetheless, the roots of the modern restorative justice movement are derived from many social, theological and philosophical movements. It is generally acknowledged that restorative practices strive to embody the values and principles that are akin to and informed by holistic peace and justice-making processes in many Aboriginal communities. Many Aboriginal cultures worldwide have generously offered their wisdom to this constantly growing notion of justice.

Hence, it would have been unfair and scientifically inaccurate to claim that restorative justice institutions were not important elements of past human societies – they were as important as retributive elements. A common misconception exists that restorative justice and Aboriginal justice are synonymous and that all indigenous people embrace existing restorative justice models and practices used by settlement communities. Although there are some overlaps and superficial similarities, Aboriginal justice processes are additionally grounded in nation-specific traditions with structural linkages of accountability and responsibility that differ considerably from its "restora-

---

82   Van Ness, D.W. (1990). Restorative Justice. In B. Galaway & B. Hudson (Eds.). Criminal Justice, Restitution and Reconciliation. Monsey, NY: Criminal Justice Press, p. 7
83   Van Ness, D.W. & Strong, K.H. (1997). Restoring Justice. Cincinnati, OH: Anderson publishing, p. 8
84   Weitekamp, E.G.M. (1999). The History of Restorative Justice. In G. Bazemore & L. Walgrave (Eds.). Restorative Juvenile Justice: Repairing the Harm of Youth Offending. Monsey, NY: Criminal Justice Press, pp. 78-79
85   Nader, L. & Combs-Schilling, E. (1977). Restitution in Cross-Cultural Perspective. In J. Hudston & B. Galaway (Eds.). Restitution in Criminal Justice. Lexington, MA: Lexington Books, pp. 34-35

tive" counterpart. As it is correctly put by Braithwaite (1996) '*I have yet to discover a culture which does not have some deep-seated restorative traditions. Nor is there a culture without retributive traditions*'.

Therefore, the (partial) 'indigenousness' of restorative justice alone should not allow its blind acceptance into criminal justice systems today. The long tradition of reparative schemes and reconciliation should always be recognised, but justice measures should not be implemented based solely on the fact that they have been practiced in indigenous societies. The economic and social factors in given disputes would need to be understood before conclusions could be reached. Deeper investigations into these factors may aid the restorative justice movement too. Although history matters, it is not the only thing that matters. Empirical evidence of restorative justice's applicability in various settings as well as the measurement of its effectiveness which will determine how restorative justice *should* be in the future are more important elements than historical evidence and how it 'used to be' in the past.

In conclusion, we might also have to draw a line on what is and what is not restorative justice in order to define both the 'age' and scope of restorative justice. Seeing restorative justice through the prism of restitution and compensation – a common technique to construct a past or a base for restorative justice – certainly gives us a sense of universality. Although, the concept of restitution is certainly one component of some restorative justice practices, there are surely important elements that differentiate restorative justice from mere compensation. All the aforementioned 'ingredients' of restorative justice are the basic elements towards a widely accepted 'recipe' for restorative justice, but not the only ones. Since there is no blueprint for restorative practices, the rest elements will be dependent upon each community's needs or aspirations, giving us a sense of locality and adaptation to new ideas, priorities and set of values in a society that constantly moves and changes.

4   Bibliography

Adamson-Hoebel, A. (1961). *The Law of Primitive Man*. Cambridge, MA: Harvard University Press

Allard, P. & Wayne, N. (2003). Christianity: the Rediscovery of Restorative Justice. In G.A. Johnstone (Eds.). *Restorative Justice Reader*. Devon: Willan publishing

Allender, D. (2000). Forgive and Forget and Other Myths of Forgiveness. In L.B. Lampman & M.D. Shattuck (Eds.). *God and the Victim*. Grand Rapids, MI: Eerdmans Publishing Company

Ammar, N.H. (2001). Restorative Justice in Islam: Theory and Practice. In M.L. Hadley (Eds.). *The Spiritual Roots of Restorative Justice*. Albany, NY: SUNY Press

Bazemore, G. & Walgrave, L. (1999). Restorative Juvenile Justice: in search of fundamentals and an outline for systemic reform. In G. Bazemore & L. Walgrave (Eds.). *Restorative Juvenile Justice: Repairing the harm by youth crime*. Monsey NY: Criminal Justice Press

Berman, H.J. (1997). *Law and Revolution: the Formation of the Western Legal Tradition.* Cambridge: Cambridge University Press

Bottoms, A. (2003). Some Sociological Reflections on Restorative Justice. In A. von Hirsch, J. Roberts, A. Bottoms, K. Roach & M. Schiff (Eds.). *Restorative Justice and Criminal Justice: Competing or Reconciling Paradigms?.* Portland, OR: Hart publishing

Braithwaite, J. (1998). Restorative Justice. In M. Torny (Eds.). *The Handbook of Crime and Punishment.* Oxford: Oxford University Press

Braithwaite, J. (1999). A Future where Punishment is Marginalised: Realistic or Utopian?. *UCLA Law Review, 46,* 1725-1750

Braithwaite, J. (1999). Restorative Justice: Assessing Optimistic and Pessimistic Accounts. (ed. M. Tonry). *Crime and Justice, 25,* 1-127

Braithwaite, J. (2002). *Restorative Justice and Responsive Regulation.* New York: Oxford University Press

Braithwaite, J. & Pettit, P. (1990). *Not Just Desserts: a Republican theory of criminal justice.* Oxford: Claredon Press

Consedine, J. (2003). The Maori Restorative Justice. In G.A. Johnstone (Eds.). *Restorative Justice Reader.* Devon: Willan publishing

Daly, K. (2002). Restorative Justice: the Real Story. *Punishment and Society, 4,* 55-79

Dignan, J. (2005). *Understanding Victims and Restorative Justice.* Buckingham: Open University Press

Drapkin, I. (1989). *Crime and Punishment in the Ancient World.* Lexington, MA: Thorsten Sellin

Edelhertz, H. (1975). *Restitutive Justice: a General Survey and Analysis.* Seattle, WA: Battelle Human Affairs Research Centres

Flies-Away, J. In L. Mirsky (2004). *Restorative Justice Practices of Native American, First Nation and other Indigenous People of North America: Part One.* Restorative Practices e-Forum. Available at: http://www.restorativejustice.org/articlesdb/articles/4230 (06/06/2009)

Goldberg, C.E. (1997). Overextended Borrowing: Tribal Peacemaking Applied to Non-Indian Disputes. *Washington Law Review, 72,* 1003-1019

Gulliver, P.H. (1969). Dispute Settlement without Courts: the Ndendenli of Southern Tanzania. In L. Nader (Ed.). Law in Culture and Society. Chicago, IL: Aldine

Hadley, M.L. (2001). *The Spiritual Roots of Restorative Justice.* Albany, NY: State University of New York Press

Hobbes, T. (1651/1991). *Leviathan.* Cambridge: Cambridge University Press

Hutchison, P. & Wray, H. *What is Restorative Justice?* Available from http://gbgm-umc.org/nwo/99ja/what.html (08/06/2009)

Johnstone, G. (2003). *Introduction – A Restorative Justice Reader*. Devon: Willan publishing Joseph, R. (1999). *Māori Customary Laws and Institutions for Aotearoa / New Zealand*. (Draft). Hamilton: University of Waikato – Te Matahauraki Research Institute

Karmen, A. (1990). *Crime Victims: An introduction to victimology*. (2nd ed). Belmont, CA: Wadsworth

Lindgren, J. (1996). Why the Ancient May Not Have Needed a System of Criminal Law. *Boston University Law Review, 76*, 40-43

McCold, P. (1999). *Toward a Holistic Vision of Restorative Juvenile Justice: a reply to Walgrave*. Paper presented at the 4th International Conference on Restorative Justice for Juveniles. Leuven

McCold, P. (2000). Toward a Mid-Range Theory of Restorative Criminal Justice: a reply to the Maximalist model. *Contemporary Justice Review, 3*, 357 – 414

McElrea, FWM. (2001). *A Christian Approach to Conflict Resolution*. A contribution to the seminar *What does the Lord require of Christians in conflict?* Australasian Christian Legal Convention. Melbourne. 1-4 February

Nader, L. & Combs-Schilling, E. (1977). Restitution in Cross-Cultural Perspective. In J. Hudson & B. Galaway (Eds.). *Beyond Restitution – Creative Restitution*. Lexington, MA: Lexington Books

Nielsen, M.O. (2005). Navajo Nation Courts and Peacemaking: Restorative Justice Issues. In M.O. Nielsen & J.W. Zion (Eds.). *Navajo Nation Peacemaking: Living Traditional Justice*. Tucson, AZ: University of Arizona Press

Pickard, P., Goldman, P., Cairns-Way, R. & Mohr, R.M. (2002). *Dimensions of Criminal Law*. (3rd ed.). London: Pearson

Pratt, J. (1992). *Punishment in a Perfect Society: the New Zealand Penal System 1840-1939*. Wellington: Victoria University Press

Rudolph, L. & Rudolph, S.H. (1967, p. 258) quoted by Danzig, R. (1973). Toward the Creation of a Complementary, Decentralized System of Criminal Justice. *Stanford Law Review, 26*, 1-54

Sande, K. (1997). *The Peacemaker: a Biblical Guide to Resolving Personal Conflict*. (2nd ed.). Grand Rapids, MI: Baker Academic

Sarre, R. (1999). Restorative Justice: Translating the Theory into Practice. *Australia Law Review, 1*, 11-25

Schafer, S. (1970). *Compensation and Restitution to Victims of Crime*. (2nd ed.). Montclair, NJ: Patterson Smith Schaff, P. (1997). *History of the Christian Church*. Oak Harbor, WA: Logos Research Systems

Shapland, J. (2003). Restorative Justice and Criminal Justice: just responses to crime? In A. von Hirsch, J. Roberts, A.E. Bottoms, K. Roach & M. Schiff (Eds.), *Restorative Justice and Criminal Justice: competing or reconcilable paradigms?*. Oxford: Hart Publishing

Singh, P. (2001). Sikhism and Restorative Justice: Theory and Practice. In M.L. Hadley (Ed.). *The Spiritual Roots of Restorative Justice*. Albany, NY: SUNY Press

Stuart, B. (1997). *Building Community Justice Partnerships: Community Peacemaking Circles*. Ottawa, ON: Department of Justice of Canada – Aboriginal Justice Directorate

Sylvester, D.J. (2003). Myth in Restorative Justice History, *Utah Law Review, 1*, 471-522

Van Ness, D.W. (1990). Restorative Justice. In B. Galaway & B. Hudson (Eds.). *Criminal Justice, Restitution and Reconciliation*. Monsey, NY: Criminal Justice Press

Van Ness, D. (1993). New Wine and Old Wineskins: four challenge of restorative justice. *Criminal Law Forum, 4*, 252-257

Van Ness, D.W. (2002). *The Role of Church in Criminal Justice Reform*. Paper presented in the 'Justice that Restores' Forum. Orlando, FL: 14-16 March

Van Ness, D.W. & Strong, K.H. (1997). *Restoring Justice*. Cincinnati, OH: Anderson publishing

VerSteeg, R. (2002). *Law in the Ancient World*. Durham, NC: Carolina University Press

Walgrave, L. (2003). Repositioning Restorative Justice. Cullompton: Willan publishing

Walgrave, L. (2000). How Pure Can a Maximalist Approach to Restorative Justice Remain? Or Can a Purist Model of Restorative Justice Become Maximalist?. *Contemporary Justice Review, 3*, 415 – 432

Weitekamp, E.G.M. (1999). The History of Restorative Justice. In G. Bazemore & L. Walgrave (Eds.). *Restorative Juvenile Justice: Repairing the Harm of Youth Offending*. Monsey, NY: Criminal Justice Press

Weitekamp, E.G.M. (2003). The History of Restorative Justice. In G.A. Johnstone (Eds.). *Restorative Justice Reader*. Devon: Willan publishing

Yazzie, R. (1994). Life Comes from It. *The Ecology of Justice, 38*, 29-38

Yazzie, R. & Zion, J.W. (1996). Navajo Restorative Justice: the law of equality and justice. In B. Galaway & J. Hudson (Eds.). *Restorative Justice: International Perspectives*. Monsey, NY: Criminal Justice Press

Zehr, H. (1998). Restoring Justice. In L.B. Lampman & M.D. Shattuck (Eds.). *God and the Victim: Theological Reflections on Evil, Victimization, Justice and Forgiveness*. Grand Rapids, MI: W.B. Eerdmans publishing

Zion, J.W. & Yazzie, R. (1997). Indigenous Law in North America: the Wake of Conquest. *Boston Law Review, 20*, 55-84

Zion, J.W. (1998). The Dynamics of Navajo Peacemaking. *Journal of Contemporary Criminal Justice, 14*, 58-74

Authors

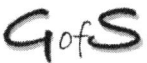

Governance of Security – http://www.gofs.ugent.be

SVA – Social Analysis of Security, Ghent University

Universiteitsstraat 4
B9000 Gent
Belgium
http://www.sva.ugent.be

Marc Cools
Wim Hardyns
Patrick Moreau (former researcher)
Lieven Pauwels
Paul Ponsaers
Evelien Van den Herrewegen
Maarten Van de Velde
Gerwinde Vynckier

Affiliates :
Caroline Mellgren, doctoral candidate at the Faculty of Health and Society, Malmö University, Sweden
Marie Torstensson Levander, Professor of Criminology at Malmö University, Sweden

IRCP – Institute for International Research on Criminal Policy, Ghent University
Universiteitsstraat 4
B9000 Gent
Belgium
http://www.ircp.org

Charlotte Colman
Brice De Ruyver
Neil Paterson
Nikolaos Stamatakis
Stijn Van Daele
Liesbeth Vandam
Tom Vander Beken
Freya Vander Laenen
Gert Vermeulen

**GAPS – Governing and Policing Security,
University College Ghent**
Voskenslaan 270
B9000 Gent
Belgium
http://www.gofs.ugent.be/

Arne Dormaels
Marleen Easton
Gudrun Vande Walle

Affiliate :
Dominique Van Ryckeghem, Policy Advisor, General
Directorate of
Administrative Police, Federal Police Belgium

Participating Reviewers

The editorial board acknowledges the participating reviewers:

Prof. dr. Alyson Bailes – Visiting Professor, University of Iceland

Prof. dr. Karoly Bard – Pro-Rector for Hungarian and EU Affairs, Chair of the Human Rights Program, Legal Studies Department, Central European University

Dr. Wim Bernasco – Senior researcher, Netherlands Institute for the Study of Crime and Law Enforcement

Prof. dr. Sven Biscop – Director Security & Global Governance Programme, EGMONT, Royal Institute for International Relations

Dr. Elias Kekong Bisong – Nigeria

Prof. dr. Serge Brochu – Titular Professor, School of Criminology, University of Montreal

Prof. dr. Willy Bruggeman – Chairman Governing Council For the Federal Police force, Head lecturer, Benelux university centre, Eindhoven

Prof. dr. David Canter – Professor, School of Psychology, University of Liverpool

Prof. dr. Mathieu Deflem – Associate Professor, University of South Carolina, Department of Sociology

Prof. dr. Nicholas Dorn – Faculty of Law, Erasmus University Rotterdam

Prof. dr. Stephen Farrall – School of Law, The University of Sheffield

Prof. dr. Bill Gilmore – Professor of International Criminal Law, School of Law, University of Edinburgh

Prof. dr. Sabine Gless – Faculty of Law, University of Basel

Prof. dr. Adam Gorski – Adjunct Professor and Chair of Criminal Procedure at Jagiellonian University, Poland

Prof. dr. Jacqueline Harvey – Professor, Accounting and Financial Management, Newcastle Business School

Dr. Badi Hasisi – The Institute of Criminology, Faculty of Law, The Hebrew University

Dr. Albert R. Hauber – Faculty of Social Sciences, Leiden University

Prof. dr. Fu Hualing – Professor, Department of Law, Faculty of Law, The University of Hong Kong

Prof. dr. Mike Hough – Director King's College London

Prof. dr. Wim Huisman – Professor of Criminology, Department of Criminal Law and Criminology, Faculty of Law, VU University Amsterdam

Dr. Michael Kilchling – Senior researcher, Max Planck Institute for Foreign and International Criminal Law

Prof. dr. Dirk Korf – Director Bonger Institute, University of Amsterdam

Prof. dr. Anna Leander – Professor, Department of Intercultural Communication and Management, Copenhagen Business School

Prof. dr. Michael Levi – Professor, Cardiff University School of Social Sciences

Prof. dr. Martin Lindström – Associate Professor, Department of Community Health, Malmö University

Prof. dr. Ed Lloyd-Cape – Professor of Law, Director of the Centre for Legal Research, the University of the West of England

Prof. dr. Mike Maguire – Professor of Criminology and Criminal Justice, Cardiff University School of Social Sciences

Prof. dr. Michael Merlingen – Associate Professor Department of International Relations and European Studies, Central European University

Dr. Nicola Padfield – Tutor and Director of Studies in Law of Fitzwilliam College, University of Cambridge

Prof. dr. Letizia Paoli – Professor, Faculty of Law, K.U.Leuven

Dr. Stefaan Pleysier – Faculty of Law, K.U.Leuven, Center of Expertise Societal Security (Katho)

Mgr. Veronika Polisenska – Institute of Psychology, Academy of Sciences of the Czech Republic

Dr. David Prior – Senior Research Fellow, Institute of Applied Social Studies, University of Birmingham

Prof. dr. George F. Rengert – Department of Criminal Justice, College of Liberal Arts, Temple University

Prof. dr. Robert Svensson – Associate Professor, Faculty of Health and Society, Malmö University

Dr. Bill Tupman – Senior Lecturer, Department of Politics, University of Exeter

Mr. Yves Van Den Berge – Prosecution Office Dendermonde, Belgium

Drs. Jasper J. van der Kemp – Assistant professor, Department of Criminal Law & Criminology, Faculty of Law, VU University Amsterdam

Dr. Gabry Vanderveen – Assistant professor of Criminology, Institute for Criminal Law and Criminology, Faculty of Law, Leiden University

Dr. Steven Van de Walle – Associate professor of public administration, Department of Public Administration, Erasmus University Rotterdam

Prof. dr. Harmen van der Wilt – Professor, Faculty of Law, University of Amsterdam

Dr. Judith van Erp – Senior researcher, Faculty of Law, Erasmus University Rotterdam

Dr. Bas van Stokkom – The Centre for Ethics, Radboud University Nijmegen

Prof. dr. Nicole Vettenburg – Department of Social welfare studies, Ghent University

Dr. Frank Weerman – Senior Researcher, Netherlands Institute for the Study of Crime and Law Enforcement